THE DARK POSTHUMAN

BEFORE YOU START TO READ THIS BOOK, take this moment to think about making a donation to punctum books, an independent non-profit press,

@ https://punctumbooks.com/support/

If you're reading the e-book, you can click on the image below to go directly to our donations site. Any amount, no matter the size, is appreciated and will help us to keep our ship of fools afloat. Contributions from dedicated readers will also help us to keep our commons open and to cultivate new work that can't find a welcoming port elsewhere. Our adventure is not possible without your support.

Vive la Open Access.

Fig. 1. Detail from Hieronymus Bosch, *Ship of Fools* (1490–1500)

First published in 2022 by punctum books, Earth, Milky Way.
https://punctumbooks.com

ISBN-13: 978-1-68571-070-5 (print)
ISBN-13: 978-1-68571-071-2 (ePDF)

DOI: 10.53288/0381.1.00

LCCN: 2022946161
Library of Congress Cataloging Data is available from the Library of Congress

Book design: Vincent W.J. van Gerven Oei

p. **punctum**books

spontaneous acts of scholarly combustion

HIC SVNT MONSTRA

Stephanie
Polsky
**The Dark
Posthuman**
Dehumanization,
Technology,
and the Atlantic World

p.

Contents

Acknowledgments

Thanks to Dr. Cecilia Wee for encouraging me to be experimental with the design of this project through the long night of Covid lockdowns and the uncertainty at times of how to balance its dimensions against events happening on the ground, in terms of social justice. Thank you for appreciating how the sweeping concerns regarding global history, personal identity, and futurist provocations played a role in shaping this offering; ironically made largely in isolation within the banal confines of my bedroom office.

Thanks to Dr. Deirdre Daly, for being the source of thought-provoking events recommendations to shape the course of this project, as well as wonderfully inventive conversations about its critical direction. I also appreciate your role of being my "title whisperer" for this project. Thank you most of all for never letting me forget our shared philosophical training at Goldsmiths as the foundational background to it all, and how this too, required its own forms of critical and contextual reappraisal. Knowledge should, indeed, be a source of liberation and congenially shared.

Thanks to Mohammad Mehdi Kimiagari for engaging me critically on some of the thorniest issues represented in this work and for making me aware of the need for it to engage with a much larger conversation around slow violence, American exceptionalism, and coloniality. Exploring various positions of alterity in regards to these conversations made for some of the most rewarding aspects of constructing this work. Happily, evidence of our lively exchanges of ideas can be found at various points in this volume.

Thanks to my amazing team at punctum books, Elieen A. Fradenburg Joy and Vincent W.J. van Gerven Oei who embraced this project with such enthusiasm from the beginning and stayed the course to make it into such a sophisticated offering. To start they provided me with an outstanding reviewer who immediately enhanced the structure of the book. Vincent and Lily Brewer, in particular, put in an enormous amount of thought and energy into its execution. Vincent's cover design was inspired and I am very proud for that to grace this book as it so brilliantly captures the spirit of the work itself. Lily's thoughtful and highly engaged copying editing made a tangible contribution to the ultimate shape of the text. It is rare for an academic press to center their efforts in this way and for that I am sincerely grateful. It has made this book truly what it is and made this process feel like a gift as well as an offering.

Thanks to Heidi Coker, Claire Lazarus, Loretta Walker, Erica Singleton, Libby Davis, and Brian Hogan who inspired, challenged, and fostered within me wildly improbable goals that were ultimately realized beyond my maddest dreams. As my fellow Wayfinders, they never let me play small as I was imagining the dimensions of what I could produce here. I am very grateful to have been on this journey with them.

Finally, and most especially, I would like to thank Eugene Polsky, my grandfather who is the North Star of all of my work. The deep bond shared between us as the Drs. Polsky remains alive through these works, much beyond the limits of his physical passing. I often think of how much he would have enjoyed engaging with a whole new series of ontological and epistemological thinkers in this volume, their darker takes on the subject of "American independence" and the costs posed to our understanding of being. In many ways, "The End of the Future" starts with "The Dark Posthuman," as also sincere companions in thought.

Black Posting and Human Pacing

Katherine McKittrick describes the work of Sylvia Wynter "on race and raciology" as an "analytics of invention."[1] In terms of the algorithmic estate, Blackness stands "as a variable in the problem solving equation before the question is asked," thus implying that the neoliberal system of governance "requires racism" to even begin "to work through and towards the problem."[2] The problem is that algorithms themselves appear in line with a reality that assumes from the offing that it will discover non-humanity within its midst and that it must discriminate for that variable as a matter of order where Blackness becomes synonymous perhaps not what with McKittrick calls "lifelessness," but perhaps something more obscure; a liveliness-in-death.[3] My evidence for this claim resides in the presumed biofacticity of Blackness itself. This over-coding of Blackness as both a biological and mathematical property, suggests that the original "Black Code," the *Code noir* of 1687, remains relevant to this contemporary discussion insofar as the code ordered slave masters on the Islands of French America to take care of the sick and old.

---

1   Katherine McKittrick, *Dear Science and Other Stories* (Durham: Duke University Press, 2020), 2.

2   Ibid., 111.

3   Ibid., 112.

While it is certainly true that the slaves themselves had virtually no rights according to Louis XIV's edict, this exception proves the rule and, in so doing, contradicts McKittrick's claim about "black life" being something that is "not relevant to the 'things people care about.'"[4] This situation is revealed in Article XXVII of the *Code noir*:

> [S]laves who are infirm due to age, sickness or other reason, whether the sickness is curable or not, shall be nourished and cared for by their masters. In the case that they be abandoned, said slaves shall be awarded to the hospital, to which their master shall be required to pay six sols per day for the care and feeding of each slave […].[5]

These exceptional conditions of care resonate with Gilles Deleuze's reading of immanence in Dickens as something that both confirms and denies the mathematics of transforming humanity into property. Such a life finds its potential everywhere in the biopolitics of contingent value and coercive demise. Deleuze declares in his essay, "Immanence: A Life," that "no one has described *a* life better than Charles Dickens."[6] The novel Deleuze chooses to explore to make his point is Dickens's latest completed work, *Our Mutual Friend*. His focus is on the lowly character Roger "Rogue" Riderhood whose dubious profession is to salvage missing corpses from the river Thames. Riderhood becomes the object of a near-fatal drowning toward the end of the novel. His life, however blighted in this moment of liminality, takes on a valence of transcendence as those around him contemplate his sudden change from a third-class to a first-class person worthy of costly medical attention. Riderhood's person, once just a "flabby lump of mortality," is now capable of pro-

---

4   Ibid., 114.
5   Jack R. Censer and Lynn Hunt, "The 'Code Noir' (The Black Code)," in *Liberty Equality, and Fraternity: Exploring the French Revolution* (University Park: Penn State University Press, 2001), CD-ROM.
6   Gilles Deleuze, *Pure Immanence: Essays on a Life,* trans. Anne Boyman (New York: Zone Books, 2001), 28.

foundly narrating how each one of his would be rescuers contemplates the mysterious soul housed within their own corporeal bodies, even as they consider how the fragility of life haunts all of their life paths.[7] Those on the scene are aghast, not by being in the presence of a dying Riderhood per se, but to a body as it is closest to the presence of death. It is only in that moment that it is conceivable to understand that *a* life that is equal to any other at an impersonal and even cosmic level of consciousness. In this moment of recognition, it is then possible to grasp the significance of a singular, yet non-arbitrary life. This pivotal scene from *Our Mutual Friend* allows Deleuze to emphasize a life acting as both a cypher and as part of a larger equation. A life that only assumes meaning if we take "the indefinite article as an index of the transcendental."[8]

The character of this life is expressly black(ened) when "a disreputable man, a rogue, held in contempt by everyone" is found as he lays dying.[9] Much like the slave in Louis XIV's edict, those around him nevertheless find themselves culturally, socially, and politically implored to attempt to save his life. Deleuze describes the scene as one where "those taking care of him manifest an eagerness, respect, even love for the slightest sign of life," only for that care to brutally recede as he comes back to life.[10] Once he has recovered, his condition of treatment will "become once again mean and crude."[11] What is critical for Deleuze is the period of liminality joining these two states of being where it becomes possible to witness "a moment where *a* life plays with death."[12] From within this state of play there emerges a temporary set of conditions where the singular and impersonal characteristics of *a* life mingle. Within this state it is possible, as Neel Ahuja recognizes, "to radically stretch the body beyond

---

7 Charles Dickens, *Our Mutual Friend* (London: Chapman and Hall, 1903), 31.

8 Deleuze, *Pure Immanence*, 28.

9 Ibid.

10 Ibid.

11 Ibid.

12 Ibid.

the organic lifetime, and into evolutionary, environmental and informational domains where life/death distinctions blur."[13] Deleuze asserts that "*a* life contains only virtuals."[14] This body's awkward temporal assertion causes those around it to question the very nature of what Ahuja refers to as "the imperial interest in the racialized matter we call bodies."[15] In its particularity it exists as a degraded life of no consequence. However, such bodies, by extension, are conceptually lacking in nothing if the parts of these bodies are actualized, and those which remain indefinite are temporally accounted for. This liminal situation brings into perspective the impossible task of accounting for the affect a life beyond its speculative capacity for extinction.

Tiffany Lethabo King refers to a state of "Black immanence that is worked out moment to moment."[16] The ontological position of Black fungibility relates to this at the level of representation "through the unfettered use of Black bodies for the self-actualization of the human."[17] Such fungibility does not act simply as space holder for "Black death, accumulation, dereliction, and limits," but also as "an opening for alterity and possibility" to make their way through so that category of the human is always a defensively anterior one.[18] For King, this is the motivation toward an ever-receding frontier for whiteness, so that it can dominate not only future territory, but also past meaning. This also is why posthumanism is indicted by King, as being at heart a project of white self-actualization that insists on space as its final conquerable frontier. King refers to this as "this logic and mode of conquistador thought undergirds the Deleuzian

13 Neel Ahuja, *Bioinsecurities: Disease Interventions, Empire, and the Government of Species* (Durham: Duke University Press, 2016), xi.
14 Deleuze, *Pure Immanence*, 31.
15 Ahuja, *Biosecurities*, xvi.
16 Tiffany Lethabo King, *The Black Shoals: Offshore Formations of Black and Native Studies* (Durham: Duke University Press, 2019), 223.
17 Ibid.
18 Ibid.

and Guattarian ethos of experimental and rhizomatic lines of flight."[19]

King argues that the philosophy of Deleuze and Guattari in *A Thousand Plateaus* is "colonialist on (at least) two accounts: in its need to render the Indian already and inevitably (ontologically) dead as 'it' has no ancestors or living community to whom one needs to be accountable; and in its invocation of the vanishing 'Indian,' which opens up the possibility of an 'ever-receding frontier' and inspiration for the metaphor of the rhizome."[20] If we go back to the passage in *A Thousand Plateaus* that King specifically interrogates in order to draw her conclusions, it is possible to construe her criticism in the context of directions, or rather *mis*directions, as they are interpreted in the New World as opposed to the Old one. Deleuze and Guattari acknowledge these as imaginary spaces that have the effect of materially shifting maps. They refer to America as "a special case" where "the literature," "the quest for national identity," for a "European ancestry," and "genealogy" eventually grows over everything that is deemed important.[21] In terms of their argument, what becomes chiefly important here is not the despotic administration of colonialism but rather its democratic reinvention through unfettered capitalism.

It is the flow of capitalism "that produces an immense channel, a quantification of power [...] that invents an eastern face and a western face, and reshapes them both — all for the worst."[22] In this drastic reshaping of reality into "quanta" or "internal exterminations and liquidations (not only of the Indians, but of farmers etc.) and successive ways of immigration from the outside" become the means through which everything comes together in America as lines of flight.[23] Deleuze and Guattari themselves acknowledge that, far from promoting freedom,

---

19 Ibid., 100.

20 Ibid.

21 Gilles Deleuze and Félix Guattari, *A Thousand Plateaus: Capitalism and Schizophrenia,* trans. Brian Massumi (London: Continuum, 2009), 21.

22 Ibid., 22.

23 Ibid., 23.

these same lines end in "variation, expansion, conquest, cap-ture" as a tyranny born of compulsive mapping.[24] These pas-sages open up a means through which to negotiate the ethos of Deleuze and Guattari's philosophy another way. While I agree with King that there certainly is a strong argument for their line of thinking following logic and mode of colonial "acquiescence" that makes "the epistemic revolutions internal to white Euro-pean humanity possible and seem natural as they dehumanize and kill Indigenous and Black people," it is possible to interpret their reading of America as veering somewhere significantly be-yond the British military ruse of supposed non-engagement, no contact and bodiless conquest.[25] Deleuze and Guattari's rhizom-atic reading of America is one that does not faithfully reproduce British colonialism but rather tentatively introduces another means through which to situation our understanding of the pathways of the corruption inherent within empire. Their asser-tion that the directions of perception of humanity are made dif-ferential in America, signals entry into understanding the ways that subtly compromise and unsettle the imaginary of European Man in his universalism. This is not to absolve British imperial-ism of its murderous impulse, but it opens up the possibility of allowing for other types of alterity to surface and persist in be-ing constantly within its midst. These energetic bodies are much more than simply nodes of transit. As King asserts, "Blackness remains a quivering live field, that enfleshed as 'fugitive fungi-bility' creates anxiety for the White master who tries to contain it."[26] This flesh becomes the site of radical "rearrangement" and "reorganization" that literally and figuratively changes the direc-tion of everything.[27] It changes fundamentally how America will see its Indigenous peoples, as well as its diasporic African peo-ples, not solely as bereft of ancestry, but temporal and spatial intermediaries in the birth of a New World, rife with possibility,

---

24 Ibid.
25 Tiffany Lethabo King, "Humans Involved: Lurking in the Lines of Posthu-manist Flight," *Critical Ethnic Studies* 3, no. 1 (Spring 2017): 173.
26 King, *The Black Shoals*, 110.
27 Ibid.

instability, and doubt as it shifts and exchanges the contours of the human with regard to is fundamental orientation.

As Deleuze and Guattari observe, "America reversed the directions: it put its Orient in the West, as if precisely in America that the earth came full circle; its West is the edge of the East."[28] This suggests that the first protean character of a uniquely American brand of liberalism might not be a white settler, but rather the mobile "human-like" figures of its conceptually malleable definition of what constituted its black(ened) bodies. As James Baldwin remarked in his 1962 essay for *The New Yorker* magazine, "Letter from a Region in My Mind," the "Negro" "is the key figure in his country, and the American future is precisely as bright or as dark as his."[29] The essay's subtitle bears this quote from Baldwin: "Whatever white people do not know about Negroes reveals, precisely and inexorably, what they do not know about themselves."[30] What Baldwin concludes is that "the spectacle of human history in general, and American Negro history in particular […] testifies to nothing less than the perpetual achievement of the impossible."[31] Such a situation might bear the potential to overlap with Wynter's scholarship and, in the words of McKittrick, works toward "the possibility of undoing and unsettling — not replacing or occupying — Western conceptions of what it means to be human."[32]

The Ambiguous Movement of Freedom

Bearing King's criticism of Deleuze and Guattari in mind, there may be another way to see lines of flight in their work as intersecting at another point with the space within the Atlantic

28 Deleuze and Guattari, *A Thousand Plateaus*, 21.
29 James Baldwin, "Letter from a Region of My Mind," *The New Yorker*, November 9, 1962, https://www.newyorker.com/magazine/1962/11/17/letter-from-a-region-in-my-mind.
30 Ibid.
31 Ibid.
32 Katherine McKittrick, *Sylvia Wynter: On Being Human as Praxis* (Durham: Duke University Press, 2015), 2.

consciousness that lends itself to a more fugitive type of meaning. This way of being emerges apart from the power relations
of slavery in the British Empire and the white settler world on
the other side of the Atlantic, from which the Black body is incrementally barred. It focuses on another way to interpret and
apply Deleuze and Guattari's line of flight to map how free people of African descent a used discursive language of intimacy
and kinship to construct and enact freedom in the Atlantic
world through the acquisition of property and social status in
Africa and the Americans over the eighteenth century. Jessica
Marie Johnson argues that is through their ability to lay claim
to their own labor that "free Africans and peoples of African
descent negotiated, challenged, and appropriated categories of
difference in the Atlantic world."[33] Freedom in this instance remains a contested category insofar as there remains an element
of coercion in the way these black(ened) bodies "were forced
to engage in intimate relations across gender and race, with individuals enslaved and free" and yet, despite the violence implied by those situations, these individuals were able to establish
"families beyond biological kin, and across race and status; in
ways that allowed them to accumulate property and distributed
legacies across generations."[34] Johnson makes the case that these
people manipulated the terms of intimacy and kinship in order
to rework the terms of their freedom. In so doing, they became
themselves key strategists in the remaking of their image to discursively determine how and what their freedom looked like on
a conventional basis. These individuals might come to represent
some of the first modern dark posthumans in that the relative
novelty of their being transcends the expectations of racialization, engenderment, and individuality and thus bring us into
another space of immanence.

---

33 Jessica Marie Johnson, "Introduction: The Women in the Water," in
    Wicked Flesh: Black Women, Intimacy, and Freedom in the Atlantic World
    (Philadelphia: University of Pennsylvania Press, 2020), 1.
34 Ibid.

Johnson makes clear that the legal, social, and political created the status of a "free person of African descent."[35] It could
be said that this situation both invites and refuses a series of
"Afro-fabulations" to use Tavia Nyong'o's term, as a form of
"counter-conduct" toward "the reality of present conditions
on the ground so that they radically introduce something else,
something other, something more" into the equation of subjective being.[36] Johnson raises an interesting distinction here to
complement this potential: "although freedom would emerge as
the quintessential struggle of the nineteenth century, free status in the eighteenth century remained a new and unfamiliar
state of being."[37] Their understanding of freedom required to use
Nyong'o's terms, a certain degree of "tactical fictionalizing of a
world, that is from the point of view of Black social life, already
false."[38] Freedom and free status in this sense are brought to the
fore as a performance, as something that is fashioned in tandem
with "slave owners, slave traders, imperial authorities, Africans
and peoples of African descent" engaging in a moment by moment monumental struggle to define the objective limits of both
chattel slavery and what is understood as humanity.[39] This work
goes far beyond the bounds of mere opposition and toward the
invention of a new category of personhood and of humanity
itself. This takes place largely through obscurity, through a series of obscured practices that these beings come to light and,
indeed, life. What I am referring to here is the myriad of "intimate acts mated with edits, codes, and imperial jurisprudence"
that produce a class of bodies that make the category of human
itself something that is "unresolvable," one that to this day, continues to haunt the project of the posthuman insofar as if offers the prospect that if human beings can be transformed into

---

35 Ibid., 2.
36 Tavia Nyong'o, *Afro-Fabulations: The Queer Drama of Black Life* (New
    York: New York University Press, 2019), 6.
37 Johnson, "Introduction," 2.
38 Nyong'o, *Afro-Fabulations*, 6.
39 Johnson, "Introduction," 2.

chattel property; so too can they be unmade from the troubling remainder of that project: "the unclear nature of free status."[40]

Here, we encounter a multiplier effect when it comes to the equation of need for black(ened) bodies to produce more than their fair share in terms of how we construct the limits of gender and sexuality because they are not made to congress simply with other human beings, but with the land itself through the equation of plantation economics. The hybridity of being between bodies and land demanded of this body that it produce something beyond conventional capacity. In this instance, the prospect of the dark posthuman emerges as a unique entity capacitated to achieve unnatural reproductive status. Here, both reproductive labor and productive labor are stripped of their boundaries. As such, these bodies become preoccupied with the law at one level and, at another, by the bodies of slaveowners, traders, and colonial officials at the level not of status but use, giving life itself a form of wholly unnatural extension through this categorical shift. As use, these bodies performed differently with regard to belief, emotion, and attachment, all of which were made artificially adjunct to it. These bodies become products of invention occupying lives that come to require companion fictions for the very fact that they are blatantly cast as beings "who arise out of the indeterminacy of living and dying, with life being perhaps the greatest fiction of all."[41] Nyong'o concludes that there is no way to recover such bodies using the materials of historical reconstruction, and this concurs with Johnson's argument that what we can recover from the experiences these bodies endured has the character of texture versus matrix. It is the remainder from a mathematics borne necessarily of predation. So where does that leave the object of freedom? It turns it into performance, into a practice that comprises humanity by willfully exceeding it.

Fantasies of plantation comingle everywhere with fantasies of implantation within the institutionalization of race which took

40 Ibid.
41 Nyong'o, *Afro-Fabulations,* 7.

place seemingly everywhere throughout the eighteenth century. When we attempt to account for this period, we must do so with the express ambition of revealing the aporia of colonially administrated societies to reveal their temporal porosity as well as the violent tendency toward manufacturing spatial diaspora. Along these axes, we will find a new critical genealogy for how we evaluate contemporary forms of humanity and posthumanity from the position of what shadows them epistemologically. Nyong'o refers to such work as operating "at a lower frequency."[42] I would offer that it operates as a virtual signaling that disrupts the reality we ascribe to intelligence, imagination, and reason. Their undersides and entrails are what interest me in tracing their broken up lineages and how those awkwardly scatter out like some shot in the dark. I don't class these as convivial acts, nor do I believe they come from a place of hope; rather I believe they come from a place of harassment, exploitation, and harm, made tolerable only to the degree that this stress can be made the stuff of a new type of immaterialism, a greater appreciation for the shadow that surrounds and defines the form.

For me, this is the problem of new materialism. It doesn't want to enter into relations with that which cannot be made into a classification, a phenotype, a quantity, and a measure. So here I am presenting something far different: life dancing on the edge of its extinguishment and rather than ceding to that destiny, making of the parameter something violently denatured, uncoupled, and disreputable. The raw, rogue state of being. Finally, to Nyong'o's point that "we do not know what a human outside of anti-black world would could be, do or look like," I would counter that these forms are embedded everywhere in the developmental time and linear timeline of history.[43]

If we take seriously Deleuze's proposition that "it is a pure virtuality on the plane of immanence that leads us into a life," within the singular impersonality of a life, dwells the potential

---

42 Ibid., 7.
43 Ibid., 26.

for violence to be incarnated.[44] However, for it to be actualized it must dwell within "a state of things or life."[45] That situation bears within itself the potential of transcendence at the most intimate levels of being and through the queerest of kinships. If, as Fred Moten has argued, freedom is a fundamentally criminal concept, it must also be understood that humanity too is a concept born out of delinquency. The existence of Blackness is part of a failed structure of social organization that demands of the object its imposition, that at once performs against the rigor of the experimental achievement that defines American as a transgressive property and, at the same time, preserves the tradition of oppression characteristic of an advanced, European civilization. "Blackness is, therefore, a special site and resources for a task of articulation where immanence is structured by an irreducibly improvisatory exteriority that can occasion something very much like sadness and something very much like devilish enjoyment."[46]

This is where we find the human in all its dereliction, in the gap between what is (biological) essentialism and what is (standard) performance, and it finds itself everywhere lacking containment. Here the human undulates through time and space "by way of the affirmative force of ruthless negation, the out and rooted critical lyricism of screams, prayers, curses, gestures, steps (to and away) — the long, frenzied tumult of a nonexclusionary" bias against its straightforward progress.[47] Through its obvious shortcomings, the human is nonetheless laboring worlds into being and in the shadow of that endeavor lies what Shane Vogel refers to as "the laboring into existence of queer worlds, and worlds of racial amelioration through performance."[48] Here we arrive at a more than one of an indefi-

---

44 Deleuze, *Pure Immanence,* 31.
45 Ibid.
46 Fred Moten, *In the Break: The Aesthetics of the Black Radical Tradition* (Minneapolis: University of Minnesota Press, 2003), 255.
47 Ibid.
48 Shane Vogel, *The Scene of Harlem Cabaret: Race, Sexuality, Performance* (Chicago: University of Chicago Press, 2009), 25.

nite article that is life. At the same time, they are being born, they are "decomposing in the preternatural world" of "racial and imperial histories" that are at least partially responsible for the parentage and, therefore, as the dark progenitors of their dilemma as precursory forms to the posthuman.[49] The posthuman is a species neither posthistorical nor artificial; rather, the posthuman is a product of the breakdown between actual and virtual being by repeated migration between the categories such that they are rendered meaningless. They are the product of dirty pain held up through bodies across centuries that recede as we attempt to confine them through traditional periodization that can no longer bear that weight against the event horizon of planetary abolition. Here again, the line of history breaks apart and something introjects itself into proceedings that we have failed to remember. In the quick line of succession between determining what is human and what is posthuman, sitting amidst the fragments of a confused accounting of world history, we must grapple with a legal, social, and political system that has failed to render judgment on a definable difference held between free status and freedom.

Heavy Weather

In order to understand this, it becomes crucial to understand how antebellum African Americans and a multiracial coalition of abolitionists in the United States, seized upon the publication of Dickens's novel *Bleak House* and "put it to work in a surprising number of ways," to confabulate, through a series of "allusions and appropriations" and to introject "a larger deployment of British literature in antebellum antislavery discourse."[50] Daniel Hack makes the case that "analysis of the uses to which *Bleak House* was put by antebellum African Americans and abolitionists — as well as their opponents — helps us to map the contours

---

49 Nyong'o, *Afro-Fabulations,* 140.
50 Daniel Hack, "Close Reading at a Distance: The African Americanization of 'Bleak House,'" *Critical Inquiry* 34, no. 4 (Summer 2008): 729.

of this underexplored transatlantic, interracial encounter."[51] If we take that prospect on discursively as well as materially we can begin to appreciate the contours of a dark posthumanism coming into recognition through the fugitive status of mapping itself, as well as the figurative status of mapping as a work of the afterlife of conquest. It also functions as something of a promissory note to make good on past theft. A similar thrust of appropriation is evident through *Bleak House*'s unauthorized African Americanization of space. This occurs through a triply estranging geographical momentum generating out of the text at least three interrelated bodies of interpellation that cannot be easily reconciled to one another. They are nonetheless the product of intimacy and kinship, making evident the fictional nature of claims to national identity on one level and, at another, how imagined of communities novelly produce their inclusion through the reading into being of texts of exclusion.

One clear instance of this is the *Frederick Douglass' Paper*'s editorial decision to reprint Dickens's mammoth *Bleak House* in its entirety during a period from April 1852 to December 1853. What is remarkable about Dickens's novel is that there are no apparent people of color present within its pages that explicitly dwell in London at the time of its setting; a situation that would have been historically incongruous on the ground. Black slavery in the novel is considered a distant problem of the colonies, whereas in Britain the chief concern should be one of white poverty. The abolition of slavery was a cosmopolitan preoccupation that, for Dickens, blinded the middle class from recognition of its metropolitan problem — the crisis of white destitution at home. This, for Dickens, constituted that class's moral and imaginary deficit that would cost it dearly should it be allowed to persist. That British liberal abolitionists chose to locate their attention elsewhere was the nature of the crisis that is represented everywhere in the novel.

It is not the institution of *slavery* per se that gains from Dickens's argument in favor of a domestic policy of white preser-

---

51 Ibid.

vation as much as it is the institution of *class* within English Victorian society. What is at stake is the direction of liberal philanthropy and the inward gaze. Throughout the novel, it is obvious that Dickens is not addressing the British Empire's policy toward America but its morally fraught relationship to Africa. This is an important distinction because it speaks to a fundamental aporia in the British cultural imagination that persists to this day; that chattel slavery was an exclusively American institution that bore no immediate connection to British colonialism. This perception arises as a consequence of the exceptional status of the United States as the one existing white settler colony that had liberated itself from British rule.

Douglass can envision himself as equally distanced from the novel narrative's control to the degree that he can make himself "the active participation in these processes of reproduction and appropriation of those who have been abjected from the text."[52] The character of Jo, the street sweeper, becomes the way into this local economy of abjection for Douglass, and his reader, because in his predicament as a nullified human being, he is the closest Dickens comes to acknowledging that these dark and distant others have some commonality with the situation of the white, English underclass. In many ways, Douglass cannot stand for them because he is made to stand against them, and never more so than at this moment when he understands that any numbers those white others might grow up to abstractly admire, if not materially participate, in the more violent aspects of Britain's colonial administration abroad; a bloody and blundering process made largely unimaginable to the ranks of the average subject.

Hack argues that Jo is made an exception to that when Dickens portrays him as "'refunctioning'" the doorstep of the Society for the Propagation of the Gospel in Foreign Parts 'as a place to sit and eat breakfast.'"[53] What this scene illustrates is a Jo whose improvised strategies for survival and conditional illiteracy make him unwitting kin to his Black enslaved counterpart. Jo

---

52 Ibid., 743.
53 Ibid.

is a brutalized specimen of humanity, however degraded, that is identified almost exclusively through his capacity for toil. He wanted nothing else. It is through this material condition that they share a type of ontological affinity that allows their narratives to intersect at the point where the rise of racial science and the consolidation of modern biological racism remain in their nascence.

Bleak House is set in the year 1827. Cristin Ellis described them as sharing a condition of "proto-posthumanism" that was developed in the 1850s in order to resist the hardening of these conditions of racialized, classificatory perception.[54] Douglass's 1850s writings, concomitant to Dickens's, "suggest, far from distinguishing the free (human beings) from the materially determined (nonhuman beings), this natural freedom percolates throughout the world, finding expression in snakes and birds no less than in transatlantic commerce and violent weather. And where this irrepressible urge breaks out — in stampedes, uprisings, and cyclones — it confronts those in its way with their own freedom: fight, fly, accommodate, or die."[55] According to Ellis, what Douglass is referring to is "an emergent systemic rather than strictly human force whose fugitive and deterritorializing freedom of circulation Douglass most directly identifies with 'agents' of matter itself."[56]

These agents, both human and nonhuman beings, will come to see themselves as "interrelated through biological, geophysical, ideological, and economic systems whose multiplied complexity no individual nor nation could hope to finally control."[57] The strange prescience of Douglass's remarks in his 1852 oration of his speech, "What to the Slave Is the Fourth of July," is not a work of antiracist prophecy so much as it is a forecast of convergence. It is posthumanist, but only to the degree that animality forms but one of the parts in his post-natural, imagined environ-

---

54 Cristin Ellis, *Antebellum Posthuman: Race and Materiality in the Mid-nineteenth Century* (New York: Fordham University Press, 2018), 135.
55 Ibid., 60.
56 Ibid.
57 Ibid.

ment. Technology, in the form of oceanic navigation, railroads, and telegraphs, as well as human and animal life, eventually will become possessed of sympathetic intelligence toward one another by forming connections across speciological divides and sentient differences.

It may be true that that beginning must wait for the globally wrought apocalypse by an unregenerate nation to pass, but it no less will have its day in the sun. Slavery fails because it compromises all earthly beings. The body to come will require a more capacious form. The prospect of a multiracial, national community is similarly too limited in this place of biological hierarchy, which insists on allegiance to its founder's constitution while ignoring the co-constitution of political and ecological processes that everywhere surround it and will ultimately overtake its ramparts. A certain appreciation of automaticity is at play here in Douglass's antislavery materialism and is required if we are going to imagine another type of nature through which agency can be applied to posthuman life. It is one that does not revert to the radical passivity of passive resistance but goes elsewhere to "invoke a physical cosmos" capable of acknowledging "dense interdependencies" that in this manner "preclude the erection of racial and speciological hierarchy."[58] What is critical to note here is that Douglass, as a free Black person, has no interest in moving "beyond the human" nor opposing it.[59] He does so not to mount a refusal against race and in particular Blackness but rather construe the human body as a position that cannot be escaped, but must rise to the challenge of pertaining to its animality, objecthood, and thingliness in such a way as to radically shift the terms of the existential predicament of modern, racial Blackness.

If posthumanist, object-oriented, and new materialist literature fail in their treatments of race it is precisely because they refuse to admit it into consciousness as remarkable *through its errancy,* nor to include *errancy* in their appeal to incorporate

---

58 Ibid., 132.
59 Ibid.

animals, machines, plants, and objects into the midst of what might constitute the realm of mutual perception. In their appeals to environmental optimization, the emphasis of knowing is always one of *errancy in service to correction.* The nature of actualization or the agency of matter is still tightly controlled by an exteriorizing force coded as universal. The posthuman and the inhuman cannot be understood as interchangeable terms precisely due to these counter imperatives and because Blackness constitutes a fundamental cleavage of the matter at hand. For me the darkness that surrounds the category of the posthuman, unlike the inhuman, is not a property nor is it a condition of "interstitiality, [or] unrepresentability beyond the positive registers of light and name and reason," nor is it "as a state of transformable negativity, as a groundless primordial resource," as Susan Stryker describes it, concerning queer materialization.[60]

I sense that the dark posthuman has, by contrast, very little to with something truly new and everything to what has come before that have dwelt in queer ecologies that were never strictly posthumous settings. That perspective on darkness is not so much about death in life, because finitude alone can never be world-forming. Rather than search for the cracks of light, our focus should be on the shadowed aspects that are often missed in the way we focus our attention on the habituation of what we have come to recognize as life. When left to their own devices, it is those obscurities that ultimately crack up systems. What results from that might not be so much celebratory as it is fabulous. That said, I don't agree with Jonathan Beller's assertion that the human, "once posthuman is always already cyborg" simply because I don't believe that any process runs that smoothly.[61] Similarly, there might be nothing progressive about certain formerly negated, racialized, and gendered human be-

60 José Esteban Muñoz et al., "Theorizing Queer Inhumanisms," *GLQ* 21, nos. 2–3 (2015): 228.

61 Jonathan Beller, *The World Computer: Derivative Conditions of Racial Capitalism* (Durham: Duke University Press, 2021), 131.

ings that are now derived and rendered into content provid-
ers, as Beller acknowledges. This daunting circumstance raises
the possibility of reversal and a need to potentially disrupt this
framework of social totality without proscribing radical disap-
pearance as its remedy. Here, I take King's point regarding the
refusal of that political meaning that "for some reason, White
critical theory cannot seem to fathom that self-annihilation is
something White people need to figure out by themselves. In
other words, 'they can have that.'"[62] I am less convinced that "the
selfless, subjectless, posthuman persists as the realm of life [due
to] the annihilation of Indigenous and Black life" because I am
not convinced that white people have, or will ever, arrive wholly
to this category.[63]

King asserts that "the crafting of the human is a process of
relations, specifically the relations of negation," and "moving
beyond the violence of the human is also a relational process."[64]
It would seem to follow, then, that their arrival into the category of the posthuman would mean that the human would have
"transcendence" insofar as this "is a relational process of accountability. So, it follows that posthumanism can't happen for
"White subjects' because they cannot transcend identity (e.g.,
whiteness, queerness), the subject (self-writing and autonomy),
or the human (self-actualization) without ending Native genocide and anti-Black racism."[65] The alternative to genocide, that
theories of posthumanism like Rosi Braidotti's propose, is that
Black and Indigenous people are fully incorporated into the category of the human. This is offered as the means to resolve the
problems of identities and subjects as they are currently configured so that the posthuman come finally come into formation
through a process of the universal negation of the human. The
failure of this process is inevitable, as King says, "if there is no
plan to enable Black and Indigenous life, then there is no tran-

---

62 King, "Humans Involved," 167.
63 Ibid., 177.
64 Ibid., 179.
65 Ibid.

scending the violence of the human" and therefore, ending it.[66] The part of this that becomes interesting for me is a "no plan to enable" as a measure of political aptitude in dark times. How can that lack of plan "trouble," *pace* Donna Haraway, the concept of capacity we have come to associate with emergency and parlay it into another type of thought regarding bios? This question prompts me to return to Dickens and to the structure of his novel *Bleak House.* Throughout the novel, Dickens compels his readers to consider what implicit connections can be made between the different strands of his narrative, situating them as if they were all information workers reading the book with their eyes, while their minds simultaneously decoded it and made meaning of it through the algorithmic problem-solving operations implied through serialisation. Everything here seemed to be running on slight delay including their expectations of gratification, and so it was that the format carried with it the echo of genealogy carried over from a natural order of things.

Jesse Oak Taylor cannily observes that "genealogy only answers one of the connections in the novel, however: the link between the supposedly orphaned Esther and Lady Dedlock, who is revealed to be her [biological] mother."[67] However, the question, *what connection can there be,* is asked in relation not to Esther but to Jo, the "outlaw with the broom," whose "genealogy remains unknown and irrelevant."[68] Here again we find ourselves at the edge of Frederick Douglass's territory and in a place where the ultimate forms of connection are found not in genealogy but in climate and its inherent capacity to carry with in it the seeds of demise. Both Jo and Lady Dedlock will succumb to it, as if it were an extended exposure to the elements. The meanings of their bodies are cast down as though they are there to remind the reader that, should London remain an ungenerate atmosphere, the fog will eventually reach them all to

66 Ibid.
67 Jesse Oak Taylor, "The Novel as Climate Model: Realism and the Greenhouse Effect in 'Bleak House,'" *Novel: A Forum on Fiction* 46, no. 1 (2013): 14.
68 Ibid.

dole out its long await justice, "drawing the city, its environs, and its inhabitants together within a shared atmosphere."[69] The arc of the moral universe is long, and it bends towards justice; just not in the ways we imagine but through "causes and effects" that "are distributed across time and space in ways that defy comprehension."[70] In the end, humanity will go the way of the other terrible beasts that have gone before it.

This scene of extinction "presents a slippage between the language of science and that of myth" and makes our recent scientific discoveries of humankind's terrible impact on the planet something that is paradoxically "new" to understanding.[71] This renders the modern metropolis a prelude to the prehistoric realm it will become to others as they witness the end of the world from the position of the species that will survive us. Taylor makes the case for the Victorians being the first to have "grappled with the yawing gulf of geologic time."[72] Similarly, "the 'mists' of deep time that had become a prevalent image in the scientific imagination" of that era are present everywhere in the novel.[73] Under a record of progress, there is a darker truth buried beneath the mud and the bone, which, when it rises, periodically affects a kind of detemporalization of what the modern world appears to be. Caught simultaneously between the unimaginable past and the unimaginable future, it appears exposed and disordered. Dickens depicts the city not in historical time but in climatic time. Within such a framework, he can see the city of London within a scope in "which glaciers flow and recede, species evolve and go extinct, and civilizations rise and fall."[74] It is "a timescale on which change can be interminably slow or shockingly rapid, but one in which individual, human

---

69 Ibid.
70 Ibid.
71 Ibid., 12.
72 Ibid.
73 Ibid.
74 Ibid.

agency appears laughably insignificant even while humanity's aggregated effluence has become a force of nature."[75]

As a term, humanity "itself remained in flux between its residual and emergent connotations (spiritual taint and environmental impurity) throughout the nineteenth century."[76] This says a great deal about the gulf that exists between freedom and free persons and why this concern occupies the whole of the nineteenth century with its numerous revolutions and uprisings. Similar tensions emerge through the novel, perhaps most memorably when Lady Dedlock, who is disguised as her maid and guided by Jo, the crossing sweeper to the desolate gravesite of the mysterious "Nemo," asks, "[i]s this place of abomination, consecrated ground?"[77] From this place and from numerous others, the dark posthuman will be born of the material and discursive economies of filth and disease that maintain the germ of novelty and transcendence. Posthumanity is a construct that imbues this body equally with unnatural sentience and agency, and in so doing highlights the paradoxical sense in which black(ened) bodies appear to live on — not by themselves but through the interpenetration of all types of being into one great network. Indeed, the principal effect of the climate of smog seems to be a breakdown in both the real and imaginary our perception of it, such that, as Dickens would write in *Our Mutual Friend,* "inanimate London was a sooty spectre, divided in purpose between being visible and invisible, and so being wholly neither."[78]

Building, a Movement

That is where Deleuze finds his affinity with Dickens and so too might Douglass find meaning in repurposing his local habitus so that it confirms, once again, a placeholder for inanimacy;

75 Ibid.

76 Ibid.

77 Charles Dickens, *Bleak House* (London: Bradbury and Evans, 1853), 160.

78 Dickens, *Our Mutual Friend,* 3.

where a security of animacy is seldom sought nor met. This all relates to what Mel Y. Chen refers to as "the brutal hierarchies of sentience in which only some privileged humans are granted the status of thinking subject"; through that privilege, they get to determine the borders of that privilege of consciousness and, indeed, whether a place where self-awareness is conducted is considered abomination or consecrated ground.[79] "Twenty-five years later, in a crowning extension and disruption — that is, disruptive extension — of this pattern, Frederick Douglass himself followed suit; jumping national, racial, and even ontological divides, he erected a small one-room structure behind his home in Washington, D.C., and dubbed it 'The Growlery.'"[80]

The Growlery is where Douglass wrote some of his most famous works. Yet to look at the structure he built is to witness that its architectural resemblance to slave quarters is incontestable. For him, to conduct his most celebrated literary work from this site constitutes a haptic gesture of radical independence. For Douglass, to design, fund, and erect such a structure behind his own conventional property speaks to "a moment where a life plays with death" and transcends it.[81] The enslaved child that was Frederick August Washington Bailey, who had previously been in physical and mental bondage, had transformed the contours of his very person so that this modest place of dwelling might conjure itself into being in his place as a space of intellectual and political liberation that consciously transcended the expectations of his frailty as well as his finitude.

Krin Gabbard refers to this movement as one of "noun to verb," where it becomes possible for there to be "an identification or alliance with fugitive spirit."[82] In so doing, for Black folks to "insist on openness, experimentation and formal innovation"

79 Mel Y. Chen, *Animacies: Biopolitics, Racial Mattering, and Queer Affect* (Durham: Duke University Press, 2012), 43.

80 Hack, "Close Reading," 740.

81 Deleuze, *Pure Immanence*, 28.

82 Krin Gabbard, *Jazz Among the Discourses* (Durham: Duke University Press, 1995), 81–82.

and thereby practice a "politics of neologism."[83] "One thinks about the role literacy played in Frederick Douglass's escape, of Harriet Jacob's denunciations of the Fugitive Slave Law, of the importance of slave narratives to the antislavery movement. W.E.B. Du Bois refers to the essays in the *Souls of Black Folk* as "fugitive pieces."[84] Their occurrence corresponds with escape. In another language, French, it is referred to as *marronage,* meaning "an act of invention exemplifying independence."[85] There is a movement of fugitive slaves from the Caribbean, Africa, and the Americas, all converging in the action of escape and the creation of societies thereafter based on the verb of owning reality however broken, bent, deformed, and reformed that position may be. This is the position occupied by the dark posthuman, as "it initiates a break while remaining overshadowed by the conditions it seeks to go beyond."[86] Power and subversion are ever channeling each other against a backdrop of legal, technological, and social developments as the posthuman undulates between the less-than-human and the more-than-human. As Gabbard observes, "orders of marginality contend with one another here" even as "divergence, tilt and flight become matters of countertradition."[87]

Douglass takes up that line. Born enslaved as Frederick Augustus Washington Bailey, he renamed himself Frederick Douglass upon his escape. Rather than choosing from the cacophony of names assigned to him, he decided to take the surname of one of the rebellious heroes of Sir Walter Scott's 1810 narrative poem, "The Lady of the Lake." With this choice, Douglass aimed to insert himself into the Romantic tradition of Scott's James Douglas, a character who is imprisoned as an enemy of the Scottish king but is eventually pardoned and celebrated. This was a gamble, considering the status of his own freedom that would be taken up again after the publication of his memoir, *Narrative*

83 Ibid., 82.
84 Ibid., 81.
85 Ibid., 82.
86 Ibid., 84.
87 Ibid., 95–97.

of the Life of Frederick Douglass, an American Slave. Written by Himself in 1845. Shortly thereafter, Douglass fled abroad to avoid fugitive slave hunters. Upon his arrival, he was embraced by British abolitionists and enjoyed giving lectures in major British cities. But he perhaps enjoyed, even more, the conversations and social occasions with the British antislavery crowd where he could showcase his knowledge of British literature, especially Shakespeare and Dickens.

During the 1850s, he serialized *Bleak House* in his newspaper with the idea of positioning himself inside the imagined communities spurred into being through great literature. Upon his return, his Growlery was similarly erected as a space for venting anger, posing inventive responses, or offering repairable solutions to often dispiriting developments beyond his control. Within this, Douglass becomes his own precious commodity, his slave narrative becoming the point of connection he has to his white audiences, even if what is "of supreme importance to the person who is represented in it, the road to freedom and the experience of freedom is exactly what cannot be communicated to a general audience of white people."[88] Jonathan Lamb draws a parallel between this insufficiency and the "it-narratives" of the late eighteenth century, "where things and humans change places," causing instrumentality and subjectivity to become intimately acquainted with one another.[89] The moral confusion of the two being comingled might well have led to the preponderance of biographies of rupees, banknotes, and gold to account for the sentimental burden of the "peculiar institution" in the object literature of the late eighteen century. François G. Richard argues that "by extension, we can view slave narratives as an expression of modern political subjectivity and the birth of the author," and therefore, "paradoxically, objects — the supposed

---

88 Jonathan Lamb, *The Things Things Say* (Princeton: Princeton University Press, 2021), 238.
89 Ibid.

antithesis of subjects — then would appear to be directly implicated in the emergence of critical consciousness."[90]

In the case of a figure like Douglass, it is helpful to question
the relationship between the objectification of humans as private property and the objectification of intellectuals as public
property. If we take seriously what David Graeber has argued,
that industrial capitalism is but a derivation of slavery, it becomes equally crucial to consider his argument that "the modern ideal of political liberty, in fact, has historically tended to
emerge from societies with extreme forms of chattel slavery";
and thus, the relationship between the two formations is similarly one of derivation.[91] This means we are dealing with a transformation. For Graeber, when we are speaking of slavery and
industrial capitalism, "we are dealing with the same terms, differently arranged, so that rather than one class of people being
able to imagine themselves as absolutely 'free' because others are
absolutely unfree, we have the same individuals moving back
and forth between these two positions over the course of the
week and working day. So, in effect, a transfer effected just once,
by sale, under a regime of slavery is transformed into one that is
repeated over and over again under capitalism."[92] For Graeber,
"what is accomplished once, and violently and catastrophically,
in one variant, is repeated with endless mind-numbing drudg-
ery in the other."[93]

Graeber is quick to add his interpretation of the relationship of these conditions is not historically based, meaning that
one form permutates into the other successively, but that they
coexist in time as capitalism comes into its maturity, at which
point the two fused in the figure not of the worker but the con-

90 François G. Richard, "Traveling Theory: Mark Leone, Slavery, and Archae-
    ology's Critical Imagination," *Journal of African Diaspora Archaeology and
    Heritage* 3, no. 1 (2014): 92.
91 David Graeber, "Turning Modes of Production Inside Out: Or, Why
    Capitalism Is a Transformation of Slavery," *Critique of Anthropology* 26, no.
    1 (2006): 79.
92 Ibid., 80.
93 Ibid.

sumer. Graeber contends that, unlike wage labor, slavery could not position itself "to create efficient markets for consumption: one cannot sell much of anything to slaves."[94] It was inefficient for owners to cultivate them as labor because human beings "are largely useless as laborers for the first ten or fifteen years of their existence."[95] The desire to recoup on that investment by bringing it literally in house on the plantation, rather than simply "kidnapping the products — and then, often as not, working them fairly rapidly to death," generated the idea that it was more "economically viable to breed slaves" than it was to buy or rent them.[96] This market logic emerged under extraordinary conditions when the US cotton boom was created by the British Industrial Revolution. That demand caused the Southern United States to become a zone of exception when it came to effectively stealing productive labor from other societies. This was not a sustainable method of conjuring labor because it comprised not only the separation of a domestic sphere and a workplace, but of an equally geographic separation, meaning that the producers and consumers of slavery no longer occupied different continents, and thus it became difficult to keep white populations and Black populations, free and enslaved, separate from the other. The plantation and the corporation also faced similar dilemmas related to their intimacy with one another.

It could be argued that with the development of the corporate form the era of idealization and exceptionalism applied to the human being had passed. Unlike any social body that had gone before, "capitalist enterprises were immortal persons free of the need to be born, marry or die."[97] Their ability to now dominate life meant that "the economic domain was effectively excised from the domain of transformation and the mutual shaping of human beings and came to be seen as something transcendent."[98] Given slavery's intimate relationship with the

---

94 Ibid., 81.
95 Ibid., 78.
96 Ibid.
97 Ibid., 81.
98 Ibid.

process of capitalism's maturation into an immortalized body, my analyses of the posthuman will be strategically pitched at a broad level of impersonality. They will engage with the features of life at the broad level of generality, spanning both time and space to help illuminate some of the specific historical connections between social practices, labor, and political economy showcased in this volume. Rather than clarifying their distinctions, the work of the dark posthuman is to disclose the dark legacies of global capitalism and slavery's relationship with an ever-more commodified politics of race, gender, class, and sexual identity in the present.

As I endeavor to contextualize particular expressions of the posthuman, I do so in relation to a broader conceptualization of the processes of labor, domination, and racialization that were to become inseparable from the development of a greater ideological project we now have come to appreciate as humanist modernity. Douglass describes his long "career" as a slave as "dark."[99] That "career" only truly ends with his imagination turning to the aspiration of becoming a Man. While it could be argued that that was a conceptual impossibility in the times that he occupied, something more becomes possible; he becomes entirely his own product and, in so doing, largely through fiction, he emerges in reality as something of a corporate merger between the truth and fiction of racialized identity taken to the ends of their political utility. Through literary recognition, Douglass squares the equation to allow him to experiment in the oceanic space held between truth and affect. He does so not be recreating the binary held between man and animal, nor seeking a way out of the darkness of slavery by taking refuge in Man; rather, he does so through adopting a definition of humanity that affirms what Nathan Snaza calls "a vital link — at once ontogenic, affective, and political — between humans and other animals, which has some attunement to the fact that both are constantly affected and being affected by a host of other agencies

---

99 Frederick Douglass, *Narrative of the Life of Frederick Douglass, An American Slave, Written by Himself* (New York: W.W. Norton, 1997), 6.

both vital and non-vital."[100] This assemblage of humankind challenges our understanding of what constitutes "dehumanization, objectification, thingliness and reification" through the technology of literacy.[101]

In terms of promoting mobilities that go against type, this is applied today to the category of digital literacy. Snaza asserts that "literacy links a human to fragile, diffuse, ever-shifting multispecies and multiobject networks."[102] Could this be an alternative dark web, where it becomes possible to appreciate what learning is possible and always has been possible outside of "human" contexts? Can such development encourage fugitive forms of corporeality to emerge and be supported in ways that slip the grasp of both states and corporations? It is worth remembering here that Douglass wrote three autobiographies throughout his life and that this force multiplier spoke to the constant need for him to perform and reperform his liberation for the benefit of demonstrating that condition of possibility for a multitude of audiences to witness in their own complexity and unevenness of status regarding that ideal category of freedom. For Wynter, "this then enables us to understand what had been the defining characteristics of our hybrid human origin: the fully completed co-evolution, with the human brain, of the faculties of language and of storytelling."[103]

What will become crucial to our further understanding, for Wynter, is an acknowledgment of the plurality inherent to "our genres of being human."[104] Only through redefining the contours of being human on terms of both myth and life, narrative and code, will humans be able to finally identify with themselves as beings conceived of hybridity and, thus, capable of engaging with a greater sense of being as a dynamic form from and within itself. The way to begin to appreciate the darker aspects of post-

---

100 Nathan Snaza, *Animate Literacies: Literature, Affect, and the Politics of Humanism* (Durham: Duke University Press, 2019), 46.

101 Ibid., 47.

102 Ibid., 64.

103 McKittrick, *Sylvia Wynter,* 72.

104 Ibid., 31.

human are "therefore in terms that draw attention to the relativity and original multiplicity of our genres of being human."[105] For Wynter this allows "discursive formations, aesthetic fields, and 'performative enactment' to overtake the previous significance of systems of knowledge in the elaboration of an understanding of humanity that is 'genre-specific (and/or culture-specific).'"[106] Through access to such resource, human beings will be able to motivate themselves "semantically-neurochemically, in positive/negative symbolic life/symbolic death terms," while at the same time, being able to appreciate how an "ensemble of individual and collective behaviors are needed to dynamically enact and stably replicate each such fictively made eusocial human order as an autopoietic, autonomously functioning, languaging, living system."[107]

This new ecosystem comes into actualization at the moment when what McKittrick refers as the "story-lie of ostensibly human development" falls away.[108] What also must collapse are "teleological underpinnings of that narrative" the articulate the "symbolic death of the denizens of the 'planet of slums'" in the same breath as they enunciate "the reality of climate change/instability, to which, inter alia, it gives rise."[109] To part with this story is to make way for a version of the "human [that] is not only a languaging being but also a storytelling species," what Wynter terms "the *homo narrans*."[110] This is a figure of singularity and of coevolution that is not one born of maps and territories but fashioned from the stuff of analogy and sociality. That being is based on a fiction of human origins that allow for synthetic, kin-recognizing subjects to emerge to challenge the rational principle of dominion and domination that currently stands at the center of a "neoliberal/neo-imperial, secularly biocentric, global

105 Ibid.
106 Ibid., 32.
107 Ibid.
108 Ibid., 19.
109 Ibid.
110 Ibid., 25.

order of words and of things."[111] At the edge of that world stands Deleuze's conception of a plane of immanence that already includes "life and death," as well as all "the deviations, redundancies, destructions, cruelties, or contingencies, as accidents that befall a life and remains at the surface of all of that movement."[112]

That unknowable condition applies to "the slave, the dog or generic 'worker' [who] are the more-than-human labouring figures of this potential [narrative] difference that defines control, it is the coloniser who provides the figure for the disruption of control's totalisation and, simultaneously, its total dissolution."[113] Turning our attention to this figure, in relation to the coloniality racialized body, can "be productive if it helps to elucidate how the deep histories of racial violence energise the embodied plasticity of racial form, its biopoliticisation and productive capacity."[114]

If our goal is to unsettle the progress of a "long-entrenched schema of phenotypic hierarchy" that has been instantiated by "a geographically dispersed violence," we have to view the project of the dark posthuman as one capable of mingling with both the deep time of slavery and coloniality, and the obscure worlds of scientific mutation, in order to pose ways to dissipate the fog that hangs over our contemporary understanding of the fragmentation and reconfiguration of the human form.[115] On one level, the posthuman becomes a new foray into controlling the terrain of a racialized and gendered matter, compromising of both time and space. On another level, it is about the modulation and affective management of planetary life at the species level and the control of invigoration itself. This perhaps is something that goes beyond the racial form as we have come to know

111 Ibid., 30.
112 Claire Colebrook, *Deleuze: A Guide for the Perplexed* (London: Continuum, 2006), 3.
113 Neel Ahuja, "Post-mortem on Race and Control," in *Control Culture: Foucault and Deleuze after Discipline,* ed. Frida Beckman (Edinburgh: Edinburgh University Press, 2018), 35.
114 Ibid., 40.
115 Ibid., 41–42.

it, and it poses significant concerns for across disparate sentient phenomena where racial formulation cannot act solely as our logos through which to anticipate what might yet come to define a posthuman relation to media, technical, and interspecies environments. In the contexts of posthumanism, we have few answers that concern what might be developed and assembled from the material remains of these interlocking planetary ecologies. Deleuzian and Guattarian "theories of control can play a part in developing such an understanding, as long as the relation of control and race can be rethought from the inside out" and from light to dark.[116] Throughout these pages you will find the posthuman acting as tool to appropriate darkness in all of its forms for the purposes of discernment.

---

116 Ibid., 42.

The Posthuman at Empire's Expiration

We live in a time when race, gender, climate, and technology find themselves in constant intersection with the financialization of life. Contagion and corruption threaten the future of the human as a source of categorical being. The construction of various types of fencing around this as a privileged category appears to be under constant strain as humans are increasingly construed as a class of resource that is both essential and expendable. This is taking place through codes of gradation that have their beginnings some five centuries ago with patterns of exploration and migration that would divide the world into old and new, white and Black, metropole and periphery. These acts of sorting enclosed the earth into the principle of a world and funneled the activities of world-making into periodized acts of devastation and cultivation, ruination and plantation. Being and dwelling emerged from these assaults as the products of markets and property, material the stuff of speculation, and technology the purview of mastery. Assignations of subjectivity, objectivity, sentiment and sentience, productivity and reproduction have over time become ways in which to variegate the quality of populations and peoples along lines of race, gender, and sex.

New colonial ventures involving the same coordinates of domain and destruction appear to arise daily in the midst of the largest global pandemic the world has witnessed in a century with the advent of COVID-19 coterminous with the contagion of social media conspiracies, currently overwhelming global democracy's capacity to effectively quarantine reality against the hazards of viral insinuation. It is by no means obvious what side will prevail between the forces of technological discorporation and world governance, nor what form of a settlement agreement might take amongst them when the time comes for that. For now, we exist in a world within a world in which beginnings coincide with a maturation of the neoliberal project in the late 1970s. The incorporation of the colonial periphery into the imperial metropole created a type of superstate where the market effectively became the world and the world assumed itself as a finite resource. That new version of the world subsumed all potential into the extension of life beyond its present understanding and territorial limit. The posthuman became its focus of social reproduction at a moment of historical conjecture that in many ways, lines up with similar moments in history including new world colonization, the enclosure movement, mercantilism, capitalism, the transatlantic slave trade, the two World Wars, the Cold War, decolonization, and globalization.

This book charts how the liberal ethos of racial differentiation and territorial ownership developed over the last five centuries starting with the Colombian Exchange are currently being grafted onto a new biopolitical category of the human-as-species. It argues that in many ways the Anthropocene culminates with the advent of the digital world and inaugurates a new era of the post-digital where the vast interior space of the human body is identified as its next terra incognita ripe for commercial domination. As a consequence of the rapid melting of the material world into the virtual world, a radical realignment of human rights and property rights is taking place within an atmosphere of radical divestment from the liberal sanctity of human life. This situation is taking place through the technical progress of the Internet of Things, wearables, implantables, artificial intelligence, data mining, and machine learning. Gross slippages are occurring between the eras of the past, present, and future that establish an ahistorical, "anywhen" potential for the universal mining of human behavior as a raw material that will later be refined into the product of congregate data.

The initial conceptualization of the human as its own enterprise in the postwar period prompted an alteration in the conditions for the production of subjectivity to include nature, artificial intelligence, animal life, and machine learning as a means through which to subsume competition, maximize profit, and universalize systems. The posthuman subject that ultimately is positioned to subsume it is fastened on to the framework of the life sciences and made to articulate itself through the language of enterprise so that he propels himself forward, not as a body in space but as a collection of cells, tissues, and organisms operating in simultaneity. Its reproduction is a matter of coordination, which is complemented by both diversification and demassification of its constituent parts that promote novel hybridity over common consistency. What is of concern now is not the existence of subjects, but how they behave.

At this juncture, the market functions as the ultimate morphological structure to which the human remains tethered at the level of ideal engagement. Within its thrall, the human subject both disengages from society and accedes to objectification. The human emerges from this scene of wreckage as the posthuman — meaning as a type of systematized being that has moved on from the boundaries of classification and into a new entanglement of conditional differentiation. In reducing human value down to this comparatively diminutive form, the market is essentially determined to break it down even further and sell its constituent parts for future servicing. The posthuman lives on past the human to the degree that its advancement can be packaged, sampled, mined, or banked. The screen becomes the limit of its domain, where once fencing stood. Its environment is seemingly continuous among the material and virtual making its substantive value abiding within a market without end. For its part, the posthuman remains trapped within the ever-present that demands that its prototypical value be both short-lived and soon-over. It must attend to itself amidst a technical division of hastened acceleration and premature demise. The market places a premium on its cognitive function over any other aspects of its capabilities for survival. Within this dynamic, race, gender, ability, and communicability persist as mechanisms to parse out exception from normativity and act as a means through which capital produces political economy by rendering certain individuals as surplus to value as a matter of both anticipatory profit and postindustrial conversion.

This book narrates the technological manipulation of the posthuman body through the conduits of race, gender, sexuality, and ability. It distinguishes the posthuman's place within both the liberal and neoliberal imaginary and reveals the ways in which its appearance first entrenched itself through the avarice of English settler colonialism and, subsequently, through the paranoia of American slavery. Posthuman played a crucial role in the functional adaptation of Cold War behavioral cybernetics and, thereafter, in the fetishization of technology within the era of global financialization. All of this allowed for the political reality of a dysfunctional planetary ecology to portend "the dystopian foreclosure of the future for anyone without access to wealth, resources and white privilege."[1] The posthuman is very much the product of world-building narratives that have their beginnings in commercial franchise and are fundamentally rooted in science, governance, and economics around the hegemonic appropriation of environments and commodification of bodies that initially fueled white settler lifeworlds and continue to be operational in the way we conceive of these worlds as continuous, ontological formations.

This book grounds itself in the humanities and is theoretically informed by some of the seminal texts of critical race theory. It focuses on issues related to how European networks of knowledge produced race over some five centuries. It explores how

---

1   tobias c. van Veen and Reynaldo Anderson, "Future Movements: Black Lives, Black Politics, Black Futures — An Introduction," *Topia: Canadian Journal of Cultural Studies* 39 (2018): 6.

slavery first enlivened raced and gendered global cartographies and how memory, race, class, and other aspects of social identity founded in these theories and practices made for the advent of the category of the posthuman through the dimensions of cultural, geographic, political, social, and scientific classification. Through the continuation of such practices, people of African descent have moved in the context of globalized economics and transnational spaces in the twenty-first century burdened with the legacy of those same confluence of logics.

The posthuman has always had to wear may selves at once on its person through responding to a multiplicity of labels, names, genders, and sexualities throughout their lifetimes. The posthuman had always has always been given to dwell within the ruins of a past and present, placed within the context of a future that is always fundamentally broken, where cause and effect rarely encounter one another in anything like a smooth line of history. The want has always been for ownership of any of these dimensions of being without regard to condition, to not remain stranded as the subsidiary of another's being, to another's claim to humanity, and finally to escape the suffocating confines of an ontological instrumental ontology that suggests a subcategory of humanity without rights onto itself. The posthuman struggles from its conception with a mandate not to simply perform at some higher level per se but on a more basic level to work. In many ways, the posthuman appropriates the tools of settler "discovery" onto itself as a way of making their dehumanization comfortable for others to witness.

For centuries, humanity has made sense of these technologies of race and gender according to how they serve them; thus they are made on one level to service the fantasy of servitude and, at another, to extend the dimensions of gendered and racialized figuration onto the prototype of a technologically nonhuman being who can assume sensory awareness but without the need for self-preservation. What is less obvious are the ways in which the desire to be something else altogether as a feature only given over to humans, and that dreaming of wanting to be somewhere altogether is tied to a similar profoundly limited purview. Mark Bould observes that, while

> Extropians, Transhumanists, and other rich white guys can reimagine white flight not in terms of suburbs, gated communities, or 'off-world colonies,' but of libertarian, pro-market, digital disembodiment, the overwhelming majority of the global population can only play in the ruins they leave behind.[2]

All of this is a matter of what Tricia Rose observes as the politic of taking up space, in a context of "social and psychological containment that makes it impossible" for Black men and women "to see themselves as major actors in a technological world."[3] The same situation pertained, "particularly in black folks whose access to technology is limited."[4] Her commentary relates to early 1980s, an era in which Black, inner city communities faced mass unemployment and were left with little opportunity for work beyond "becoming hidden workers for service industries or computer repair people."[5] These professions assume an affinity between invisible servitude and postindustrial reparation. Their corresponding, degraded opportunities related to "a place of understanding themselves *as already having been robots,*" first as slaves and servants and second as low wage carers and manual laborers.[6] Thus, adopting "the robot" into their ranks "represented a response to an existing condition: namely that they were labor for capitalism, that they had very little value as people in this society."[7] In many ways, robots pick up where

---

2 Mark Bould, "The Ships Landed Long Ago: Afrofuturism and Black sf," *Science Fiction Studies* 34, no. 2 (2007): 81.

3 Mark Dery, "Black to the Future: Interviews with Samuel R. Delany, Greg Tate, and Tricia Rose," in *Flame Wars: The Discourse of Cyberculture,* ed. Mark Dery (Durham: Duke University Press, 1994), 216.

4 Ibid., 213.

5 Ibid.

6 Ibid.

7 Ibid., 213–14.

their tortured labor histories left off as "the ultimate exploitable 'other,' a human-like being who does not need to be afforded the rights of humanity because it was created by human hands instead of human loins."[8] The next logical progression, the android, will do away with the need for humans in the act of reproduction, replacing the hands with other means of creation and other bodies with human equivalency, all amid a heated atmosphere of environmental denigration and new world-building. Welcome to Planet Rock, welcome to the dark posthuman.

Choreographic Demise

The advent of the novel coronavirus in 2019 presented corporations and governments with an exceptional opportunity to further develop and consecrate the value of information while cultivating new terrains of threat containment, biological determinism, human optimization, and neoliberal responsibilization protocols. The great unravelling of the Anthropocene begins with the neocolonial narration of COVID-19 as its great "un-differentiation" event, where novel, communicable agents will overtake the power of humans in the twenty-first century and, in doing so, reeducate them to conform to their superior operating systems. Within those, any differential aspect to the category of whiteness remains wholly unexplored. This position reifies a continuous association of humanity with whiteness and posthumanity with the ambiguous appearance of others as competing life forms.

COVID-19 becomes a cypher of democratic debility that suggests a decaying biopolitical lifeworld, which in turn incites corruption and invites terror and eventually leads to a cyclical escalation of both. The social world itself has been subsumed into a simulation able to predict pandemic, riot, and unrest, but not to calculate the means of preventing them. Data collection in this sense literally goes nowhere. Measuring without interfer-

---

8   Marisa Brandt, "Janelle Monae's Liberationist Posthuman Pop," *Difference Engines,* March 5, 2014, http://www.differenceengines.com/?m=201403.

ence, these systems can be relied upon to track the increase in coronavirus infections or the frequency of death threats against public figures on social media platforms, but not to meaningfully interrupt such trajectories. Intelligence and character become associated with probabilities for survival amid what the German visual artist and cultural critic Hito Steyerl refers to as "the infernal loop between empire and expire."[9] The same loop that begins to play itself out with Columbus's arrival to the New World now twists around arcs of planetary survival.

Steyerl's newest work "SocialSim" (2020) takes the form of a video installation made for the exhibition "I Will Survive" at the Düsseldorf Kunstsammlung (K21). The COVID lockdown situation in Germany in late 2020 forced her to transfer the central part of "SocialSim," the video installation entitled "Dancing Mania," from its gallery location into virtual space using augmented, virtual, and mixed-reality technologies. Her "Dancing Mania XR" version positioned the human being at the center of social action, currently paused at the level of surviving. In this way, it becomes a social simulation modelled on a social simulation where the traditional paradigms used by behavioral scientists are scaled up to incorporate the differential parameters of quarantined social space as well as to quantify social activity in ways that reconfigure the variables of energy, movement, infection, and death in order to correlate them with a greater prognosis concerning the viability of life on earth.

Suggestively, these elements are made accessible to participants solely as projections, as simulations of what might happen when contagion spreads over a wider swath of society and can no longer be regulated in terms of promoting optimal behaviors; rather, the pathogens represent themselves an illness, mania, overdose, or conspiracy that subverts the mathematics of a population who cannot stay one step ahead of them in terms of accurate accounting.

---

9   Hito Steyerl, "Dramaturgies of Resistance: Collectivity, Performance, Dialectics," performance at the Jackman Humanities Institute's Program for the Arts at the University of Toronto, Ontario, Canada, January 22, 2021.

Shortly after entering the immersive exhibition space, a compass is provided to the player in the right-hand corner of the screen that can be modulated for things like spread of contagion and speed of death as variables. The 360-degree video projection associated with this scenario is then rapidly populated by hyper-masculinist white police officers, who find they are descended upon by soldiers theatrically labelled as "people." Their seemingly posthuman-modelled, blue complexion escapes clear racial signification; their gender similarly ambiguous in their slight contouring. Simply through proximity to their mesmerizing, generic show of force, the conventional police officers assume the posture of being arrested — except there are no handcuffs. Instead, their posture becomes one of simulating the act of arrest, as opposed to being actually held under arrest. There is some degree of potential deniability seeded within this, as is their dancing in company to what appears at least, in theory, a totally homosocial theatre of engagement. Within such a turbulent arena, it is possible to associate dancing mania with other types of fevered progress and patriarchal rage.

A third player comes on the scene in the form of red lines that aggressively pull at the bodies of both the police officers and the soldiers. These red tracking lines appear to extend a farther grip as these respective bodies appear to dance out ceaselessly what we must assume are their final days and hours. These same lines progressively dissolve them until they are literally a shadow of their former selves, and then, nothing. This begs a few questions: is the virus ever an expression of singularity or is it rather always a product of hybridity and plurality of type? Where does the individual begin and the virus end? How does one separate out the (mis)steps of one to the other in terms of corrupted encoding?

Contagion makes itself known through the preponderance of red lines that literally swarm these bodies as a hypnotic, techno drumbeat loops around itself. The disco refrain that Steyerl references in the overarching title of her exhibition "I Will Survive" seems flagrantly outmoded in comparison to the always-on of post-digital connectedness that suggests a play that never ends, a game that never sleeps, and an audience that is habituated to forget the boundary between platform and setting, ignoring the existence of any technical "backstage" keeping it all going. The term "post-digital" describes the way in which material reality has been informed by digital technologies and networks and how the world has been redefined by its own image reflected back to it and recast by the digital to reveal different social textures and temporal mappings.

Steyerl's virtual installation of "Dancing Mania" in many ways remains foregrounded by the ecstasy-fueled escapism referenced in one of the biggest hits of the late 1980s, Frankie Knuckles's track, "Let the Music (Use You)." Techno would extend disco's life span by moving its influence beyond the dancefloor through the invention of the rave scene. The very concept of an "I" that "Will Survive" seems outmoded in comparison to the erotic sample looping techno that has come to replace its commercial dominance on the dancefloor. As opposed to disco, techno music requires a particular type of submission by the body. It requires a total environment through which to infect others with its tide and intensity. The rave materializes to become its preferred site of ecstatic and emphatic group submission, where it becomes possible for a collective body to become enthralled to a distinctively nonhuman or posthuman digital commander, in the posthuman figure of the typically male techno DJ.

From such a vantagepoint, "the digital prowess of the DJ becomes musical mastery of the natural, a cybernetic disciplining of uncontrollable biology by containing it in a MIDI channel. Each sexy computerized simulation of the black female diva brings the hetero-cyber male achingly closer to those essential truths of sex and race."[10] Even as the digital diva is granted precedence over the mic, it becomes evident that her voice is more a sign or sigh of resignation over the infinity of Black struggle to exist rather than an expression of satisfaction or assurance that

---

10 Susana Loza, "Sampling (Hetero)Sexuality: Diva-Ness and Discipline in Electronic Dance Music," *Popular Music* 20, no. 3 (2001): 353.

her liberal humanist declarations have been met. Rather, what that voice does is to codify the modern renderings of feminization and technologization while simultaneously sample testing metrosexuality and resurrecting differential racialization from the remains of a classical economics now situated within postcolonial space.

In many ways, rave culture was a product of 1980s Thatcherism, situated in abandoned warehouses and the open spaces surrounding them and made continuous within the urban and exurban landscape of deindustrialization and finance capitalism. Organizers, many of whom were young Conservatives, made commercial use of these newly disused spaces once occupied by heavy industry and situated at the edges of the innercity and at a crossroads of the divergent economic fortunes of Britain's postcolonial, postindustrial worlds, both north and south. These young entrepreneurs were taking advantage of other young people's disposable incomes to charge untaxable, cash-in-hand entrance fees, proceeds that would have normally gone to the traditional nightclub trade. They paired their operations with organized criminal gangs who furnished their illegal gatherings with fashionable, class-A drugs.

Rave and acid house sat at the very center of the seismic social changes that reshaped 1980s Britain. Both had their beginnings in the protest movements of the late 1970s before finding their way into abandoned warehouses that were themselves artifacts of the cooling-off period of postwar industrialization when acts of final refuse were spilling onto the dance floor through acts of chaotic release. The rave was not a space of political rebellion but of social withdrawal that signaled the rise of a postindustrial nomadism reflective of the dystopian elements of urban sprawl. It was a scene that promoted a stylish vision of a near future replete with aesthetic references to robots, virtual reality, video games, and electronic music. It was also a scene that promoted the riotous expression of subcultural identities including homosexuality and Blackness; albeit much of it contained within a predominantly white, heterosexual, socially mobile space. There was a darker side to this scene having to do with the enforce-

ment of public order through divided communities and suppression of certain bodies through socioeconomic hardship,
police brutality, and political disillusionment.

Sean Michael Feiner situates the rave's meaning within "the
oppressive totality that is late-capitalism and control society,"
where the rave acts as a placeholder for what is both "sacred"
and depraved about our times; here the dance floor figures as
"our last celebration of life, even as we know its end is just on
the horizon."[11] Feiner explains further that "within the chaos of
the ecstatic bodies of the rave-space, queer utopianisms and lost
futures come to join the ravers as specters of what might have
been, could be, and what was always already impossible. It is in
this ghostly waltz that we glimpse the simmer of a collectivity,
even if temporary, that imagines our bodies beyond their de-
lineations, divisions, and individualized containments."[12]

What Feiner appears to omit from his description is a prior,
historical sequencing of dancing and congregating in gay night-
clubs with the material risks of police apprehension, imprison-
ment, assault, or murder. At one level, he assumes that "a queer
body is always-already vulnerable, positioned in a liminal space
of visibility, opacity, and transitions in that the queer body is a
virtual subject," a becoming that may code-switch and translate
itself in order to survive "the real threats of violence of heterosexual patriarchal capital."[13] At another, he feels the need to contain this violence as though it could only be said "exist in the
late 1960s and 1970s."[14] For that time period specifically, Feiner
explains, "one had to often know their mere existence and expression as a human being meant legal and punitive threat and
violence and possible death."[15] In subsequent decades, such punishment has taken on the appearance of a "specter," wherein het-

---

11   Sean Michael Feiner, "A Rave at the End of World: The Politics of Queer
     Hauntology and Psychedelic Chronomancy" (PhD diss., State University
     of New York at Buffalo, 2019), 19.

12   Ibid., 20.

13   Ibid., 48–49.

14   Ibid.

15   Ibid.

erosexual violence now assumes a vaguer stance against queer bodies and queer life while at the same time remaining revenant to the present, as evidence of a contemporary fracturing of time itself, in which "the countless murders of trans women of color, and the increased hate crimes and activity among reactionary male affinity groups" remain ancillary to present conditions; yet the crimes ominously persist in order to "peer upon neoliberal trajectories of normative gay and lesbian politics," libidinally squaring the boundaries between engenderment and endangerment.[16]

Neoliberalism's economic experiment offered queer people access to a unique political assemblage where several elements of their identity could be addressed, but ultimately its influence on space remained limited in terms of its market philosophy. The designs it had on gay clubs and venues would engender material and technological conditions that structured the landscape in which an ordered vision of queerness emerged, which coalesced as a culture identity, celebrity, wealth, and exclusivity through several iterations of melding the margin into mainstream, the colony into the metropole, thereby exerting greater authority over it. The liberal, queer body may have well been a virtual subject, but its successor, the neoliberal gay, lesbian, and transgender body, now most definitely trades as the real. By the same token, the queer becoming that perpetuated itself through code-switch and self-translation now speaks materially and fluently to a new order of capital. Although this new order remains steeped in heterosexual and patriarchal expression, it nevertheless now appreciates that its survival depends upon capitalizing on inclusivity. Against the real threats of violence that persist within categorical difference, it pitches a superficial culture and defense of these communities.

Meanwhile, the vulnerability of the queer body has now become universalized through the positioning of the human-as-species with each of its subject operating within a liminal space of visibility, opacity, and transition that may have made

---

16 Ibid.

the queer body a beta tester for a virtual subjectivity that is now ideally at least assumed by all. In the millennial control society of the early 2000s, the queer congress between the genetic and the digital remained a formless possibility. Up until 2020, its potential still was construed as so innocuous and rare as to be of little concern to the global population. That former impression has been COVID'd out of existence with the arrival of a novel virus so invasive, that as humans, we are willing to offer our freedom of movement and bodily memories to biotechnology corporations and technocratic governments in exchange for the promise of mass inoculation and the privilege of proof of our right, thereafter, to circulate once again — as though the long held public policing strategies of track, trace, and isolate will somehow disappear along with the neutralization of the virus, as part of a "new normal." Coronavirus, in the meantime, has learned and adapted itself alongside the exploitation of personal data and biometric characteristics such as race, gender, age, and obesity, characteristics that, so far, seem to allow it to proliferate endlessly and mutate according to what it seizes from certain, compromised subpopulations.

There is an affinity between Steyerl's social choreography and the promiscuous behavior of the coronavirus. During her passing installation as well as theirs, the bodies contracted to one another fragment before our eyes; some hastily voided from the floor before the endless cycle of transmission starts again. The only way to stop it, presumably, is through changing the parameters associated with everyday life, implying that actual novelty delivers renewal, whereas corrupted innovation brings forth conspiracy. COVID-19 is very much the product of "resource depletion, climate change, and ecocatastrophe" in the way that its rapid global spread is narrated *as* progress and the ways in which it is adhering itself to capitalism's "insatiable desire to grow and subsume all."[17] Its willingness to sacrifice life itself to achieve greater motility not only conforms to the existing requirements of "the market and profit, but also its potential un-

---

17 Ibid., 11.

doing albeit in a nonhuman or post-human way."[18] Here, Feiner lands on something of great significance as he points the way towards the nonhuman and the posthuman as the elevated players of a capitalistic future, capable of surviving beyond humanity as newly productive monstrosities.

Steyerl's installation graphically maps the "mutual constitution" of humans and nonhumans through an artificial "field of ecological configuration and continual reconfiguration."[19] She refuses to compose these spaces of exhibition as sites of human agency. Here the nonhuman no longer can be relied upon "to be safely oriented in relation *to us,*" but it instead orients itself in relation to something alien, the virus whose presence to this territory assumes a kind of dangerous intimacy.[20] As Rob Coley observes, if ours is now "a world of vital materialism, this vitality does not result in a more enchanted world but a world utterly indifferent to our fate, a world that is newly terrifying."[21] Within such a landscape, no threat is ever truly over and as such it is entirely possible humans would recreate their same brutal history in outer space as intensive relationality and affective attunement give way to productive pathologies, the likes of which we can scarcely imagine. Steyerl's work proves even our imaginations can be fundamentally antisocial, and the trauma of exploitation an adhesive substance in its own right. The only mode of appropriation in our era may well be crisis, and therefore, malfunction is attendant to all contemporary being; a fact that we are recently attuning ourselves to as we give way to the performance of our algorithmic decoding and recording as vaguely posthumous, generic, posthuman flows. The rave scene of 1989 might have looked like another world. Nowadays it is one whose reactionary moments and oppressive totality in some ways appear grimly familiar.

---

18 Ibid.

19 Rob Coley, "'A World Where Nothing is Solved': Investigating the Anthropocene in 'True Detective,'" *The Journal of Popular Television* 5, no. 2 (2017): 137.

20 Ibid.

21 Ibid., 138.

The Posthuman in Elizabeth I's Empire

In order to appreciate this contemporary situation, we must turn towards the origins of the Atlantic world. In March of 1584, Queen Elizabeth I granted her favorite privateer, Sir Walter Raleigh, a charter to seize lands in the New World on behalf of England. From the beginning, this was to be a land bequest. The charter was in effect a licensing agreement for his heirs to forever lay claim to these territories as the products of their "free libertie" and for them to assume that would be the case, "at all times for ever hereafter."[22] America would be theirs "to discover, search, finde out, and view" under the proviso that "such remote, heathen and barbarous lands, countries, and territories" were "rightfully theirs for the taking."[23] The legal caveat here was that land seizure was permitted so long as it was "not actually possessed of any Christian Prince, nor inhabited by Christian People."[24] Beyond that, it was up to Raleigh and his company alone to determine if those lands that "shall seeme good" enough "to haue, horde, occupie and enjoy."[25] If they seemed that way to them, such property came assuredly "with all prerogatives, commodities, jurisdictions, royalties, privileges, franchises, and preheminences" as could be reasonably guaranteed by the Crown.[26]

Before setting foot in this new territory, Elizabeth already figured on a plurality of bodies; political and corporate joined in their task to make license of England's white settlers, their heirs, and successors, who would on "to inhabite or remaine, there to build and fortifie" what constituted the greater "Realme of England."[27] Her concerns centered on the establishment of a permanent English settlement in the New World and the forma-

22 Elizabeth I, "Charter to Sir Walter Raleigh: 1584," *Yale Law School Lillian Goldman Law Library, The Avalon Project,* https://avalon.law.yale.edu/16th_century/raleigh.asp.
23 Ibid.
24 Ibid.
25 Ibid.
26 Ibid.
27 Ibid.

tion of a base from which English privateers could continue to plunder Spanish ships. Elizabeth wasn't interested in obtaining land rights, per se, but a guaranteed return on her investment according to the agreed upon terms of procurement. Elizabeth was to receive one fifth of gold and silver secured during the seven-year period of Raleigh's initial foray into the New World in order to fatten her royal coffers. Hence, the rationale behind Raleigh's enterprise was "primarily the exploitation of gold deposits believed to be present in Virginia."[28]

Within its passages, Elizabeth's Charter displays two imagined directives for Raleigh's course; the first being the political desire to forge "full-fledged colonial plantations inhabited by English men, women and children under English law to form an agrarian society" and the second, a mercantile desire to "establish smaller commercial centers through which to trade and exploit commodities."[29] This duality would set up the essential conflict within the discourse of English imperialism for centuries to come. The initial impetus for a North American empire comes to England as the result of external pressures to compete with Spain. Spain in the 1570s and 1580s was quickly emerging as the New World hegemon, establishing colonial dominion over Southern America, Mexico, Central and South America, and the West Indies. Through them, Spain was able to plunder an incalculable amount of gold that fueled their ascent as the premier power with Europe. Theirs was a Catholic empire that threatened to dominate culture, finance, and defense through a strategic alliance with the Papacy in Rome. Had Elizabeth failed to act, Protestant England would have faced existential ruin devastating the fortunes of "a small group of English politicians, courtiers, merchants, clergymen and entrepreneurs."[30]

On the other side of the class spectrum, was a burgeoning population comprised mostly of peasantry who required a

---

28 Michael G. Moran, *Inventing Virginia: Sir Walter Raleigh and the Rhetoric of Colonization, 1584–1590* (Bern: Peter Lang, 2007), 447.

29 Ibid., 3.

30 Ibid., 4.

level of subsistence beyond what England's current landholdings could produce. Poverty ensued and extended as a consequence of the land-enclosure movement, which stripped the peasantry of its right to cultivate its nourishment from common land. Through the enclosure movement, such acts of cultivation now constituted trespass. Landlords had effectively privatized the land solely for their own use and chartered benefit. The plantation can be said to have its true beginnings in this type of insertion of commerce into property and land into hire as the immediate consequence of this movement. Secondary to it was the migration of great numbers of rural poor to England's cities in hopes of finding alternative means through which to support themselves. The cities simply could not absorb them in their sheer numbers, putting great pressure on Elizabeth to find a means through which to address their unwanted presence. Eventually, New World colonization was proposed as a way in which to remove the poor from view by exporting them abroad as expendable labor.

In the years previous to his American colonial venture, Raleigh occupied himself through soldiering in Ireland. Milan Rai contends that "the invasion of North America began with Ireland."[31] Such an enterprise begins with the "plantation of people," which, in the case of Ireland, meant the "plantation of Southern Ireland with English and Scottish Protestants."[32] The establishment of the colony of Virginia was built upon this same approach requiring the displacement of indigenous peoples through the implantation of hostile competing populations. The logic followed that indigenous populations would have to be besieged in order to ensure England's progress as the overarching occupier of these lands. These methods, once "derived and perfected in the conquest of Ireland, were then transported to the American colonies" along with "the ideology

---

31   Milan Rai, "Columbus in Ireland," *Race & Class* 34, no. 4 (1993): 25.
32   Ibid.

to justify confiscation and genocide."[33] The belief was that as Ireland fell to English conquest, so too would the Indigenous inhabitants of the New World.

One of the key tactics used to decimate North American Native nations was economic warfare. Rai compares "the deliberate destruction of the buffalo as the social and economic basis of many American nations" to a process that had taken place in Ireland.[34] In Ireland, the objective was to destroy value of cattle herds impacting "a large part of the population, living in "creaghts" or roving communities of herders, [where] cattle were the basis of wealth and society" and thus they were collectively starved of both revenue and status.[35] In America, it would not be enough to destroy indigenous buffalo and cattle stores, cultivated fields and orchards, homes, and towns; rather, the English would have to go so far as to annihilate whole communities in order to impose a complete order of destitution upon Native nations.

Irish men, women, and children who were able to survive the vast hardships imposed by England's colonial occupation faced further jeopardy when they were subsequently transported to the Americas as a class of livestock. This took place through the development of "thriving trade in Irish indentured servants, sold in Barbados and elsewhere in the Americas" to work on English plantations.[36] These Irish bodies would have crossed paths with African bodies on their way to the New World. What separated one from the other as a class of commodity was the feature of their indenture versus their enslavement. Rai indicates that the African slave in this economy actually laid claim to better valuation than his Irish counterpart. This was the case "because the planters would have to pay for them" and thus, "they would have an interest in preserving their lives, which was wanting in the

---

33 Bill Rolston, "The Training Ground: Ireland, Conquest and Decolonisation," *Race & Class* 34, no. 4 (1993): 17.

34 Rai, "Columbus," 30.

35 Ibid.

36 Ibid., 31.

case of bond servants."[37] By whatever means, the accumulation of bodies on these plantations was designed to enact a transference of resources to fortify an elite English planter class and to sustain them in their claims to wealth and power. Such status was conferred primarily through the acquisition of control over an overwhelming portion of the settlement population. A majority of would-be colonists drawn to both Ireland and America were comprised of "the seagoing gentry of southwest England," who simply moved their brand of enterprise from one "on to the other."[38] The direction of travel was far from singular. Indeed, "some settlers who had gone to the American colonies were attracted back to take advantage of the colonisation of Ireland."[39] The same personnel from one conflagration were directly transferred into this new frontier of colonial warfare in order to enact much of the same methods of conquest on America's Indigenous nations.

From the time of Sir Walter Raleigh's first colonial voyages into the Americas, the Native American was viewed as an object to be inventoried by way of declaring ownership over it. The approach mimicked what was previously being done by the Spanish in the New World. Plants, animals, minerals, and peoples were to be documented and classified to produce a visual record of what the English crown now owned. It was from the works of these painters, cartographers, and collectors that the educated, white, European male was able to picture the New World in the sixteenth century, capturing his imagination for what constituted its alien forms of life. Michael Moran maintains that this took place at two levels. The first was through "panoramas of Virginia that conveyed a sense of mastery over the unknown territory," while at the same time implicitly claiming "control of ownership over all that was pictured."[40] The second was through the objectification of the Native American body, capturing the "the

---

37 Ibid.
38 Ibid., 26.
39 Ibid.
40 Moran, *Inventing Virginia,* 140.

intimate actings of daily life including eating, dancing and conversing" in the form of illustrations that would make their way back to the collections of the English merchants, who were keen to inspect what was effectively part the merchandise.[41] These surveyors of lands and places approached them as a continuous entity whose existence down to the individual was something "rightly belonging to the civilised English."[42] It was they who would go on to exploit this relationship to what was thought of as "the natural world" through a combination of technology and military might. Things did not quite go to plan some thirty years later in what eventually became the site of the Virginia colony, Jamestown.

In 1606, what would come to be known as Jamestown was little more than a marshy landscape that swarmed with mosquitoes and flies promoting disease and provided no potable drinking water. The territory was chosen because of its defensive capabilities. It was "surrounded by water on three sides and no Indigenous tribes were settled there" because it was for all intents and purposes practically uninhabitable.[43] So, arrived the "100 colonists that had been indentured for the task" of settling this place, "by the Virginia Company of London, a merchant venture for New World colonies under royal charter, which would later reconstitute itself as a joint-stock holding."[44] As much as they were wed to theories about how to conduct their business, "the Jamestown settlers were utterly disconnected from their environment, and unlike the Native Americans possessed no socio-ecological memory of how to survive there."[45] Neil Price refers to the fact that "an imbalanced proportion of the settlers were

41 Ibid.

42 Ibid., 141.

43 Martin Kelly, "History and Founding of Virginia Colony," *ThoughtCo*, December 4, 2020, https://www.thoughtco.com/virginia-colony-103882.

44 Neil Price, "'James His Towne' and Village Nations: Cognitive Urbanism in Early Colonial America," in *The Urban Mind: Cultural and Environmental Dynamics*, eds. Paul J. Sinclair et al. (Uppsala: Uppsala University Press, 2010), 478.

45 Ibid., 481.

'gentleman investors' who preferred to physically accompany their money to see its proper profitable use in the founding of the colony."[46] Manual labor of any kind had been wholly foreign to their lives in England, and as a consequence, these men were utterly unskilled when it came to actually constructing a colony. To that end, they had made no provision to bring along "carpenters, farmers, journeymen and generally anyone who actually knew how to make a living from the land or to physically construct a suitable place to dwell."[47] Instead, they assumed that Native people would support them in a similar fashion to what the laboring classes had always done for them in England.

Often these men resorted to "antagonising them through excessive demands or unnecessary force," with many of them literally "expecting to be fed" by Native people according to the earliest reports of interactions in the colony sent back to England.[48] For their part, the colonists spent their days not planting to support their needs, but rather "passed their time bowling in the streets and drinking."[49] The only crop that emerged of any real interest to them was tobacco for which it was understood there was a growing market back home in England. The colonist still required corn for their subsistence and, thus, resorted to the threat of arms to coerce trade with "tribes" people. Basic foodstuff was procured in exchange for "beads, copper and other 'Trucking Stuffe'" only through show of force.[50] This situation formed the basis of the colonial myth that Indigenous people were too ignorant to know they were the obvious losers in trade. Nonetheless, Native resistance had its own value insofar as it was used to justify land expropriation, Native enslavement, and gradual clearance of them through armed combat. This white settler campaign of hostilities would span more than a century

---

46 Ibid.
47 Ibid.
48 Moran, *Inventing Virginia,* 17.
49 Anthony S. Parent, *Foul Means: The Formation of a Slave Society in Virginia, 1660–1740* (Chapel Hill: University of North Carolina Press, 2003), 17.
50 Ibid.

in the run up to the war for independence from English crown. In the eyes of what would be Americans, England treaty directives were far too protective of Indigenous rights. To be American was to find oneself on the side of free land and unfettered trade. The posthuman is essentially born within this conflict over the constitution of labor that would insure both.

By the mid-seventeenth century it was becoming obvious there was a labor shortage of poor English and Irish to service Virginia's tobacco plantations. Charles II promoted the African slave trade as a means with which to fill this requirement. The plantations requirements for labor would be met by the newly formed Royal Africa Company, a joint-stock company open to investment by all of England's subjects. The year 1663 marks the beginning of the transatlantic slave trade within the English Empire. Enslaved African labor provided the basis for the emergence of a landed gentry within colonial Virginia that had come to depend on the Black body not just as a source of wealth, but as a contrasting symbol of their free status and their elevation within civil society as a new rentier class. They became a Southern aristocracy able to economically and socially dominate colonial America from the late seventeenth century onward precisely through transatlantic agricultural markets made wholly possible through African slavery.

England's Atlantic slave trade had provided planters with essential labor to produce tobacco, cotton, indigo, coffee, tea, cocoa, sugar cane, sisal, oil seeds, oil palms, hemp, rubber trees, and fruits, much of which travelled back to England to be refined into commercial goods. African slaves were an exceptional part of that inventory insofar as their bodies were proffered by the Royal Africa Company as both a valuable commodity and as a means to produce other eventual commodities. When these bodies started to be produced in the colonies, they became something altogether different in terms of their classification. This led to a radically changed concept of labor as something that, once transported over to Virginia, could thereafter essentially be cultivated at home and husbanded in a similar way to other plantation lifeforms. Against this diminution of African humanity raised a New World classification of whiteness. Thomas Jefferson, one of America's constitutional framers, stood very much at the forefront of this 2.0 version of white humanity capable of accommodating within its bounds both democracy and aristocracy, even as its subtle appearance was shadowed by the rigid enslavement of others. These planters designed ostentatious mansions in order to literally elevate them above their Black slaves and the surrounding white communities of workers. These new dwellings were built to stand as both imitations of English manor houses and as admonitions against those who would dare trespass their New World authority. Often, both of these preoccupations led down the path of their owner's bankruptcy in attempting to shore up their status, as well as, their security.

Expenditure didn't matter as much as maintaining an appearance of civility. This involved a sustained process of "self-colonization" enacted over a lifetime.[51] A crucial part of that process in the American South in particular involved securing dominion over a profoundly racialized and gendered society. Such an undertaking had immediate financial implication insofar as the control of slaves was in and of itself considered an asset. Its manipulation demonstrated a form of mastery believed to be far superior to that found in England in regard to the control of its servant class. The control of women operated as a separate but related asset. It was imperative that through control of its wives and mistresses, the household assumed itself as a small-scale model of domestic finance. Within that arrangement, the terms of master, slaveowner, and husband all become equivalent to one another in acting towards the goal of forming a more perfect union, in this case, between feudal patriarchy and market paternalism. The veil of benevolence that Southern planters like Thomas Jefferson assume in dictating all of its rules

---

51   Kenneth A. Lockridge, *On the Sources of Patriarchal Rage: The Commonplace Books of William Byrd and Thomas Jefferson and the Gendering of Power in the Eighteenth Century* (New York: New York University Press, 1992), 97.

is punctured by his resentment of constantly measuring his domestic order against that of England in order to lend credence
to his own internal rebellion against its perceived limitations;
he did so while remaining wholly aware that he was denying
that right to America's women, slaves, and Indigenous peoples
through the character of his very existence. Jefferson's democratic project absorbs into itself the qualities of social paranoia
and anarchic rage even as it attempts to steadily acquire identity
and society through the fashioning of a new sovereign state. Jefferson's literal and codified knowledge of the state of Virginia
does precious little to mitigate its appearance as an entity that
remains very much peripheral to London's metropole.

Julie Flavell's book *When London Was Capital of America*
recounts how in the mid-eighteenth century, the wealthiest of
Virginia's planters chose to live as absentee masters in London.[52]
The relay of Jamestown settlers back and forth to London to obtain desired supplies dated back to 1607. In the intervening century and a half, those that could not indulge themselves in travel
did so through replications of London in their choice of food,
books, furniture, and accessories that they estimated would appear "in the home of any English landed gentleman."[53] Once
in London, "white American colonists [...] were allowed to be
Englishmen" to the degree that most wealthy Americans still
"looked and sounded like Englishmen" throughout the eighteenth century.[54] This was by no means an accident but existed
by design. Most wealthy colonial youths were expressly sent to
England not only to acquire their formal education but, equally,
so they sounded like English gentlemen. "It was said that in the
eighteenth century that Virginia's first families sounded like
educated Londoners."[55] Their attempts to pass themselves off as
English gentlemen extended to the treatment of their African
American slaves while dwelling abroad. The severity of their

---

52 Julie Flavell, *When London Was Capital of America* (New Haven: Yale
University Press, 2010), 10.

53 Ibid.

54 Ibid.

55 Ibid., 71.

treatment had to be tempered to support the conceit that within the confines of their plantations, "slaves lived under better conditions than the [white] English poor."[56]

Great Britain's involvement in the slave trade "had given London its own population of blacks London-born, African and American."[57] That situation complicated transatlantic racial hierarchies as enslaved and free Black bodies conducted their own types of urban commerce with one another as well as with a multiclass, multi-ethnic white population. Georgian Londoners were far less fascinated with the appearance of Black bodies, as compared to Native American ones, because Black and Asian communities had existed in London dating back to the early modern era. The appearance of Native Americans in London, by contrast, is something that was highly coveted, alongside the acquisition of New World wildlife and plants. Almost from the beginning of the Jamestown settlement, Native Americans became popular "objects of import" in the city to the delight of its inhabitants.

One example of this is when Pocahontas "made a brief sojourn to England between 1616 and 1617," achieving lasting fame thereafter.[58] She arrived in London christened as Rebecca Rolfe in 1616, and her trip to London was made expressly to raise funds for Britain's struggling American colonies. Her enduring fame within the city perhaps was due in part to the fact that "Pocahontas" never exactly made it to the Old World, and by extension "Rebecca Rolfe" definitely never made her way back to the New one. In 1617, she contracted a communicable disease and ultimately died in Gravesend, aged 6 months, or 21 years, depending on how she is perceived against the category of human. To complicate matters further, at the time of her death, the term "American" conjured images of "Indians" in the average Londoners mind, whereas white people or Black people were

---

56 Ibid., 41.
57 Ibid., 3.
58 Ibid., 181.

classified as European, or African, respectively; all of them assigned varying degrees of human likeness.

Jefferson's vain attempts to figure Paris as a substitute capital for the American South as a way to disavow Britannia as America's motherland did little to erase the fact that "Virginia" was named for the express pleasure of an English Queen, whose advantage was taken by and for the fact that she "was already both fruitful and husbanded."[59] From the first instance, what was desired was that Virginia be founded upon land with no evidence of inhabitation by Christian people. The privileged term here was not a presumption of Christianity but of personhood. The settlers' inability to see the Indigenous people among them in Jamestown as people meant that their God-given humanity could readily be denied. In this gesture, they became America's first posthumans, a prototypical class reserved for cultivation and clearance, as beings both continuous with property and broken from land. That pattern of expropriation would continue for the successive centuries that would lead to a late-capitalist ordering of species according to the Elizabeth's colonial model of upward and outward expansion, conditional individuation, corporate domestication, and categorical segregation. Elizabeth's mercantile quest contributes meaningfully to birth of another of era of English expansionism and market confidence. Margaret Thatcher's foray into globalization takes a similar tack by surrounding herself with a court of flatterers and buccaneers, all ready to do her bidding, all dependent on her favor to gain license for the material exploitation of colonial Others.

The Posthuman in Thatcher's Empire

The dissolution of Britain's formal empire profoundly informed Thatcher's strategy to transition its international reputation into one synonymous with finance and service sectors. This approach radically exacerbated gaps in wealth between Britain's classes, concentrating its least vulnerable members within the

---

59 Parent, *Foul Means,* 9.

finance sector and its most within the service sector. Thatcher's economic policies were indeed groundbreaking, providing "the first example of a global neoliberal economic movement that re-shaped the world economy in the 1980s" and presented a "model for other neoliberal governments to implement market reforms such as Ronald Reagan in the United States and Deng Xiaoping in China."[60] Thatcher's greatest contribution was to move the economics of the City of London onward from a colonial model concentrated on banking and insurance to a globalist model driven by products and markets. This shift which led to a transformation of the financial sector was superficially fueled by a combination of economic liberalization and financial market deregulation. At a much deeper level, the financialization of the British economy was an ideological project that retooled tech-niques of colonial governance to bear on metropolitan trade.

As the 1980s wore on, it became clear that Britain's outsized contribution to a rapidly increasing the globalization, financial-ization, and deregulation of the world economy could only be sustained through a paired service economy reliant on an influx of immigrants, refugees, and casual workers. While the rich were incentivized to reposition themselves around the UK's burgeon-ing "financial centres, glistening high-tech enclaves, and quirky high-culture districts," the poor were hindered in their progress by their coerced displacement from public housing that lead to "the rapid increase in slums and homelessness."[61] Real estate be-came not a place for individuals to dwell within but an asset class for interests to exploit. The division between these two concerns fell along highly racialized lines that abstractly coded wealth as white and poverty as Black, making of the urban landscape a free-market zone of self-reinforcing segregation. Within this environment, it is capital that "produces race as a socio-political

---

60 Bradley W. Bateman, "There Are Many Alternatives: Margaret Thatcher in the History of Economic Thought," *Journal of the History of Economic Thought* 24, no. 3 (2002): 309.

61 Ida Danewid, "The Fire This Time: Grenfell, Racial Capitalism and the Urbanisation of Empire," *European Journal of International Relations* 26, no. 1 (March 2020): 294.

category of distinction and discrimination."[62] As a consequence, the free-market principle that supported Britain's former colonial empire was transferred wholesale into neoliberal ideological modes of urban governance allowing for "practices of urban planning, slum administration, and law-and-order policing" to become prime elements within its initial ventures into the New World of global capitalism.[63]

From the sixteenth century onward, market logics defined racial and colonial logistics on the ground in the New World. Thatcher's neoliberal project must be construed as something continuous with those structures of colonial governance that link not only the metropole to the periphery, but also differential value to human life. The potential loss of the last vestiges of formal empire prompted Thatcher to invent new tools of state intervention and methods of social ordering to fortify and complement Britain's position as a longstanding, global hegemon. Her ambition to create a network of tax havens for capital investment out of what remained of the British empire's colonial holdings was in many ways reflected in her ambition to create a network of such havens for investment out of what remained of Britain's urban real estate. Thatcher's was a project of aligning the interests of islands within islands, a sort of investor shell game to revitalize Britain's economy and stature. Britain's colonial economy was refined in the process of these two interlocking movements that make new types of planetary settlement possible through the financial seizure of peripheral jurisdictions and metropolitan neighborhoods. This allows for imperial states to reclaim ownership and remap the world to offer shelter and furnish wealth, once again, to an elite class of settler.

Britain's tax havens track their beginnings back to the 1950s and 1960s, an era where it and many other European countries began an effort to design financial instruments that would preserve the spoils of their centuries old plunder of Asia and Africa. The proliferation of offshore tax havens provided these parties

---

62 Ibid.
63 Ibid., 297.

with a means with which to shelter their personal wealth and ensure no disruption to "their business activity, investments and capital movements."[64] The new tax havens frustrate the postcolonial world's new leadership in their ambitions to "expropriate and nationalise" wealth related to "assets previously extracted from the land and labour of their colonised subjects."[65] White Britons and other Europeans maintained the conceit that this was their property from which such value was ultimately derived. Rather than risk its seizure, they elected to liquidate their assets by funneling the proceeds abroad. As a result, "an enormous outflow of money from the late colonial world began — much of it fuelling the expansion of tax-haven business."[66]

At this time the British empire was morphing into what would become "the modern offshore system" we recognize today.[67] Its third empire emerged from a network of jurisdictions that remained either under British colonial rule or recently became independent colonies or held the status of British Overseas Territories and Crown Dependencies. Places such as Hong Kong, Singapore, the Bahamas, Cyprus, Bahrain, Dubai, the Cayman Islands, Bermuda, the British Virgin Islands, Turks and Caicos, Gibraltar, Jersey, Guernsey, and the Isle of Man all became tax havens for an affluent white settler class now in the process of formally retreating from empire. The City of London emerged as the core of this new empire with these financial-colonial outposts functioning as its colonies.

Left behind was the special status afforded to Britain's imperial metropoles of London, Manchester, Liverpool, and Sheffield as national tax rates rose to objectionable levels, targeting, in particular, wealthy earners whose revenues were used to finance

---

64 Vanessa Ogle, "The End of Empire and the Rise of Tax Havens," *The New Statesman*, December 18, 2020, https://www.newstatesman.com/international/2020/12/end-empire-and-rise-tax-havens.

65 Ibid.

66 Ibid.

67 Nicholas Shaxson, *Treasure Islands: Uncovering the Damage of Offshore Banking and Tax Havens* (New York: St. Martin's Publishing Group, 2011), 8.

Britain's welfare state and to effect wealth redistribution across the nation in the immediate decades following World War II. Those metropolitan centers had increasingly come to resemble Britain's former colonies through a principle of racialized, urban segregation that created a new version of the colonial world, complete with a national program to facilitate white tax avoidance to deprive these populations of state revenues. At the same time, British elites had effectively evolved the empire so that it remained a place where the sun never sets and where they could continue to grow rich from the former empire's remnant properties in the Caribbean, Asia, and the Pacific. These small island nations would now be used to help launder money before it got back to the City of London and distributed to other large finance centers. "For the City, it was a beautiful self-reinforcing dynamic: The more that countries opened their financial systems, the more business would float around internationally, ready to be caught in the nearby nodes of the British offshore spiderweb and then sent up to be serviced by the City and its allies on Wall Street.[68]

When Thatcher calls contemporary Britain a "living tapestry" rather than a society, she is referring not to an interweaving of the lives of its common men and women throughout time but the role of its noble members in composing England's illustrious history. It is their mythical lives that featured in the type of lavish embroideries that were displayed whenever foreign ambassadors came to see Elizabeth I during England's first golden age as a colonial power. They were there to evoke a network of associations into the political fabric of the court amid the nascent political economy of global imperialism. At the time, colonialism itself was something of a cutting-edge technology that called for a complex network of associations in order to eventually become wholly operational. For Thatcher, England as a society, "can only be recognized as such in the past tense, after the fact, and never without myriad mediators — instruments,

---

68 Ibid., 85.

institutions, statements, and techniques."[69] Similarly, Thatcher's rejection of classical laissez-faire liberalism led to the advent of a new imperial age for England that centered on the processing of complex financial information. This situation allowed for the City of London to configure itself as the premier site of global competition and exchange and for history to record Thatcher "as an unyielding free-marketeer single-mindedly committed to limiting the size and scope of the state."[70] Not one of these events could be depicted, or indeed predicted, without accounting for their foregrounding in a postwar scientific landscape of cybernetics, systems theory, evolutionary biology, and cognitive psychology, which lead to a consensus belief that British society required a radical transmutation of its imperial identity in order to exert dominion over international credit and financial markets, as opposed to material production and tangible spending. As a consequence, England's third iteration of empire would have to instigate and operate according to a series of abstract rules that "individuals need not even know they are following," meaning that their governance was basically evolving toward becoming algorithmic in nature.[71]

This situation allowed for social forms and institutions to emerge as self-organizing entities capable of absorbing previous types of social formations into their operations. Herein, "human agents become part and parcel of the dynamic and metastable biological, physical, and technological systems with which they constantly interact."[72] England's third empire would be an economic one fully integrated with its previous imperial biosphere and would function as a logical outgrowth of its previous methods for exploitation of colonial resources. This time, however, the focus would be on the creativity of their refinement through ever-expanding webs of market diversification and the introduction of complex systems into their management. It was

---

69 Ricky D'Andrea Crano, "Posthuman Capital: Neoliberalism, Telematics, and the Project of Self-Control" (PhD diss., Ohio State University, 2014), 4.

70 Ibid.

71 Ibid., 5.

72 Ibid.

understood that economics and biology would converge at the point of their shared concern for optimization even in times of uncertainty. The global financial market "provided the perfect atmosphere for this new type of synthetic social order" to develop and support itself.[73]

John Hinkson reminds us that "by the 1980s high technology had reshaped the market into an institution of a significantly new kind: the global market, which reconstructs the world we have taken for granted through a renovation of all our institutions, including those that ground our humanity."[74] That period of radical renovation profoundly shifted the ground that formerly staged humanity. It did so as "a function of an institutional revolution" that first became "evident in the emergent capacities of the high-tech sciences arriving out of the Second World War."[75] In the 1980s, the capacities of the global market were joined to that movement to uproot the logics subtending humanity. By the time of the global financial crisis of 2008, the grounds for humanity's engagement with the world tilted further still into a position of near total upheaval. What Hinkson refers to a "novel phase of politics," "which propelled Thatcher and Reagan into power," gains its rites of innovation through a series of events that altered the DNA of humanity such that it remained essentially open to crises thereafter.[76] As the market depended on seismic expansion, so too did it necessitate humanity's categorical rupture in order to accommodate that volatile condition.[77] The posthuman has been there throughout, positioned in its shadows, waiting for its part to be taken up in and among the contemporary artifacts of technological change — the computer, mobile phone, Internet, social media, and so on, which now aim to "personify" other types of appearance through these new sets of conditions. Therefore, this book

---

73 Ibid., 4.
74 John Hinkson, "Trump's Trajectory?" *Arena Magazine* 145 (December 2016), https://arena.org.au/trumps-trajectory-by-john-hinkson/.
75 Ibid.
76 Ibid.
77 Ibid.

is as much about evolution as it is revolution when it comes to examining these competing forms.

Chapter 1, "The Dark Posthuman: Social Reproduction, Social Justice, and Artificial Ecology" speculates on W.E.B. Du Bois as the first dark posthuman of the twentieth century, as a self-described "race man". My historical plotting of Du Bois cuts across several geographies including of the African American South, Wilhelmine Germany, the African Global South, National Socialist Germany, and the Algorithmic South to shape and support his intellectual contributions to the categories of the "New Negro" and "Pan-Africanism." Utilizing the work of Sylvia Wynter, Philip Butler, Rosi Braidotti, Christina Sharpe, Donna Haraway, and Katherine McKittrick it situates a reading of Du Bois that allows for the positing of race within the greater category of the posthuman as something that is not only a biological category, but equally the productive formation of a geography, economy, and ecology that coincide with the advent of sociology itself as an imperial disciplinary form. Du Bois acts as a locus for these developments through his early work onwards leading him to become the America's leading authority on race in the early part of the twentieth century, during a time when Victorian eugenic principles characteristic of race prevailed and in many ways, corrupted his perception regarding "the uplift of the race" through the application of social scientific principles.

This chapter revisits the German South, West Africa, and Liberia as locations that acted as a prototypical space of fascism and genocide that are hard to breakdown and almost never talked about when discussing the legacies of slavery in the United States. While Du Bois is often lauded as a beloved figure for Afrofuturism, his life-long romance with Germany is seldom critically addressed, an omission that is rather glaring considering his lifetime pseudo-appearance as a (white) German aristocrat. Du Bois's unwavering affection for Germany was based on a long-standing impression that German immigrants in the United States, were more accepting of Blacks than white Americans. This which is an idea that Du Bois believed and promoted for much of his life. Du Bois's perception of Germans went hand in hand with a belief that Jews were slave owners disproportionate to their population numbers in the American South. This belief led to a long-standing presence of anti-Semitism in the African American community which exists to this day. There is some truth to this insofar Judah Benjamin, a prominent Jewish planter was the Confederate's Secretary of State and was a legislator of slavery for many years prior to that. Benjamin incidentally was also gay. During the Civil War, the British were on the side of the Confederacy and so when the North won the war, Benjamin was granted exile and when on to establish a lucrative law practice law in the North of England staying, as it were, amongst his fellows.

American theorists of race tend to comment very little about German colonialism, whereas a great deal of critical race theory focuses its attention on Spanish, Portuguese, French, British, and Belgian sources. Very little is written about German and Dutch involvement and, even more evidently, very little in the twentieth century. My argument is that has something to do with the founding of the country as a white, agrarian estate by German and Dutch settlers in particular. The founding of Liberia by these same constituent parties raises the prospect of another historical omission — a situation of Indigenous genocide at the hands of a group of biracial African Americans in cahoots with conservative, white industrialist interests.

These histories are far from obvious and seldom topics of discussion with the content of the painful subject of how reproduction and extermination go hand and hand in the management of the population. The American case for producing rather than buying new slaves when they are "used up" in terms of the energy they can output in their lifetimes is somewhat of an anomaly. The two practices went on side by side. Moreover, there was the third economy in slave rental which later makes its way into the model of the "chain gang," which bears on the establishment of white settler time and the property ambiguity of rental.

German colonial genocide wasn't about making labor free, as it was in the American South in its relationship to chattel slavery, but it was more about making labor subordinate and invisible by physically distancing it from the category of whiteness prior to the model of clearance of these territories through genocide. Germans like to think of themselves as much more enlightened in their colonialism than other European powers involved at the time of their initial investment in southwest Africa because it was based at least initially on a nominal payment system. It is the native population's resistance that they fail to anticipate leading to the genocide much in the same way as it went in other parts of Africa. Germany's involvement in slavery goes back to the seventeenth century, which is opposed to the understanding that it was a latecomer to this practice of colonialism.

Du Bois's failure to appreciate that this historical situation parallels the colonial founding of America bears on his positioning of the condition of Blackness as somehow "maladaptive." This sets him on a course to imagine Blackness as something that can be made superior by nature, and through biological enhancements. His openness to embracing and elevating the "New Negro" as a posthuman figure, has it figured in relation to its simultaneous framing as both fungible and plastic. Du Bois's prototypical figure takes on the valance of an engineered product, which backs onto the question of human agency. As a consequence, Du Bois is haunted by the prospect of some forms of Blackness remaining willfully maladaptive, hence the "New Negro" as overwhelmingly structured towards the design goal of approximating whiteness. Similarly, the queerness of the "New Negro" scene must be internally reconciled, through the promotion of heterosexual, interracial coupling between white and Black "mulattoes" in particular, which is interesting given Du Bois's branding of himself through a certain type of familial eccentricity and his confined iteration of himself through public narrative and image.

Chapter 2, "Aspirational Theft: Technocapitalism, Cognitive Augmentation Devices, and Algorithmic Coloniality" explores the new market in wearable technologies and the effect of these devices on sovereign forms of thought and communication. It brings this discussion in line with the work of Shoshana Zuboff and the phenomenon of surveillance capitalism and its implication for the ways in which wearable devices will play their part in the harvesting of data through human connection. Such labor output will likely remain completely unremunerative. At the same time, its effects will be profoundly limiting to the imagination of its human consumers who will become increasingly reliant on these technologies to correct their behavioral and mnemonic failings.

As bodies become further tethered to these devices, something reminiscent of the "slave collar" of yesteryear emerges suggesting there is something profoundly territorial happening here. While these wearable devices omit some of the more fearful ostentation of their slave-collar predecessors, they nevertheless continue to make operational some of their original purpose, which was to impede illicit movement and constraint their labor within the confines of artificially manufactured environment. Similarly, their contemporary versions are designed to prevent errant wearers from moving beyond a certain parameter of admissible behavior, through what we currently refer to a location-aware function.

From the earlies days of the new American republic there was a growing tension between its agrarian and industrial models of economy. The question of which would prevail haunted Southern planters like Thomas Jefferson, who feared the encroachment of northern industrialization further westward into the territory of the new nation, portended freedom for African American slaves. Jefferson was convinced that should the Southern plantation system ever be disassembled, the presence of freed African American slaves within the country would pose an existential threat to their economic preeminence. Jefferson's agrarian solution to this impending disaster involved "interbreeding" between Black and white individuals in order to create what could have been called an intellectually superior hybrid product that could then be expropriated from its owner and brought to bear on transatlantic trade in a novel way. These bodies would be shipped back to Africa through a concerted campaign to establish them as the synthetic leaders of a newly established series of American colonies. Jefferson's vision set the stage for a coloniality of cognition we continue to reckon with today through the use of technologies to that purport to better augment human capacities. The human, in question, always manages somehow to remain an affluent male body whose social awkwardness can be forgiven due to his potential to success in the marrying the ambitions of master and servant to one another through the invention of one continuous body of influence. Through this type of intervention, it is possible to form a feedback loop between colony and metropole through technology that at once acts as a means of establishing intimacy and distantiation.

This chapter examines two proposed wearable technologies, both called "AlterEgo," that assume certain powers from their retooling of older forms of colonial exchange and cognitive dominion. The chapter goes on to explore how autism has been brought into the new imperial equation and how it relates to an understanding of the Anthropocene. Both the autistic subject and the planetary subject are cast here as sources of native intelligence that might be deployed to effectively save the rest of us from ourselves through the sophisticated refinement of cognitive capitalism. This will require an intersubjective relationship to emerge between humanity-as-species and machine learning. The technology of race does not disappear from this configuration but is instead brought forward to adapt the category of Blackness to confirm to the rankings of numerous objects not assigned sentience. This is occurring at the same time that the meaning of life is plummeting in value, and seemingly overtaken by concerns for elevated response. The world to come is one that privileges animistic awareness and compromises the limitations of humanoid language and human intelligence in favor of a version of existence teeming with a multiplicity of informational outputs. The datafication of reality will perpetuate racialization beyond the material standards of Blackness or whiteness by colonizing the very fabric of thought and meaning. This scenario will take place through posthuman framings of property and propriety.

Chapter 3, "The Virocene: The Ill-Informed Nature of Non-human Contagion" looks at the ways in which metaphors of infection, extinction, environmental apocalypse, and global pandemic cluster around race and gender as categorically vulnerable types of situated within the greater category of humanity as a species. This chapter charts the progress of resource expropriation, colonial dispossession, and environmental ruination on a planetary scale, which harbors as its consequence the burdening of life with commercial valuation. Its trajectory shifted focus onto a new normal of comprised living wherein the mass acts as multiplatform collateral to shore up a market-based society projected against the twinned formations of slavery and colonization installed, as it were, in order to shadow a liberal ideology bent to the task of defining itself in freedom through the bondage of others. That is to say, it is only through the acquisition of others as property that whiteness can be made equivalent to universal sovereignty.

The financialized logic of conserving life and letting die, that subtended the management of American slavery in many ways, endures through the Anthropocenean categorization of humanity as an unevenly endangered species. The advent of the New World as the climatic initiation event we now recognize as the Anthropocene may be thought of as the great differentiation event. It is now bookended by the advent of COVID-19 as the great un-differentiation event, wherein the concept of humanity is postscripted to span the breach between the technical and biological being, leaving open the question of why the terms of innovation themselves remain unreformed in their colonial definition of planetary threat. If life is programmable in these sorts of scenarios, then so too is death in the form of managed extinctions of life forms deemed expendable.

The desire to remove Black communities from the United States dates back to the founding of the American republic. It is concurrent with the movement towards colonial expropriation. It also intersects with the scientific fascination with eradication of racial difference, as though it itself were a looming source of contagion circulating within the body politic. The chapter examines the early republican writings of Thomas Jefferson, Hector St. John Crevecoeur, and Benjamin Rush in order to picture how a transition from natural philosophy to biological racism occurred in the early decades of the nineteenth century to afford an understanding of the abolition of slavery from the standpoint of racialized eradication and expropriation. From an economic perspective, black(ened)ness was construed as perhaps the most radical form of capital dispossession, and thus, the continuous project of its invisiblization through the subsequent century of American history was a matter of balancing loss against acquisition within the United States's greater biopolitical economy.

This chapter concludes with the postwar transition from liberalism to neoliberalism that took place within an atmosphere of accelerated biological research in the areas of molecular biology, cell biology, and microbiology that allowed for the manufacture of genetic, microbial, and cellular level life to form the basis of a new global economy under which these microorganisms were mobilized as their own asset class, unfettered by conventional understandings of what constitutes racialized productivity. The origins of the HeLa cell line act as a placeholder at mid-twentieth century for the culmination of a new appreciation of life as information and its persistence as data, which together act as a new class of resource open to be extracted by the forces of neoliberal entrepreneurialism.

Here, life resides somewhere within the boundaries of the genetic, in the affective domain that now reproduces itself through infection rather than impregnation. This allows for its lineages to break with the legal and social histories that have previously defined race in terms of blood and ancestry and connect with new categories of lively classification with their basis in association. It suggests that racism and colonial nostalgia continue to color biological innovation that credits the advancement of technology for its capacity to inoculate "humanity" from the worst aspects of those projects through the introduction of a new universalism through object orientated ontologies. With regard to contagious diseases such as COVID-19, nature is not the locus of their situation but forms its ground for deployment as a device of accumulation by dispossession. If the pandemic it has wrought is "solved" by microbiologists, it will do little to address racial disparities in health outcomes; instead, it is likely to move us forward to a future where biotechnology is naturalized along previous lines of universal assumption and implicit bias.

The move towards the enslavement, not of bodies per se but of microorganisms and molecules, figures the terms of the next frontier of biocapitalism. Even more subtle perhaps than at the level of organs, or tissue, this market readily bypasses the concern for exploitation at this level of being. It's ripple effects with the human, animal, and plant world will be profound as dependencies on immunotherapies that can last decades, and endocrinological molecules of testosterone, estrogen, and so forth all figure into a market that is all about capacities in its predictive sense. Biotechnology will become a futures market in capacities that must, by definition, grow beyond the limit of the human alone. The human becomes the consumer of other life in this model, but it also becomes, essentially, its raw form.

Chapter 4, "Trans-Substantiations: The Artificially Engineered Presentation of Others" starts with rereading Alan Turing's queer self-presentation alongside a rereading of his contributions to the scientific field of artificial intelligence. Turing was highly engaged with the intersubjectivity and embodiment of his thinking machine. Thus, from the very beginnings of its artificial inception, he was aware of the risks involved should it be fundamentally denied its claim to proper intelligence and appropriate response. As such, he always modeled the human mind with room for its artificial counterpart to intellectually and emotionally accompany it in ways that mimicked the homosocial norms of his privileged youth as a product of both England's public schooling and imperial administration. Essentially, this means that Turing was all over the place with his loyalties and affections, and as a consequence he believed it wholly possible to share his liminal desires among genders, humans, and machines.

At the same time, Turing understood that sentimental capacity for love, rather than the biological act of sexual intercourse, was the feature that had previously been held up to deny full claims to humanity to those who were differentially raced, abled, and gendered within society. Affinity among these categories comes through an assumption at one level that they share in a type of social disability, while at another one they share in a kind of potential that accedes to the possibility of emergent instantiations of being. The computer analytics that Turing introduces into the world fundamentally disrupt and alter the location of gender, so that it assumes the position of the posthuman. In many senses, this becomes the source of what could effectively become a postgender reality resulting from conjoining artificial intelligence and mathematical biology.

This chapter progresses with rereading McKenzie Wark's trans memoir *Reverse Cowgirl,* which uses fragments of biography and auto-fiction mixed in with emails and Facebook posts to engineer a tell-all exposé of her transition. Wark categorizes her undertaking as "less an act of self-disclosure than one of self-dispersal."[78] What differentiates Turing's era from Wark's is the rapid progress of electronic communication and the near-universal expansion of the digital environment. Prior to *Reverse Cowgirl* in 2020, Wark publicized another sort of transition narrative with reference to a much earlier electronic phase in the mid-nineties. This previous work involved a sexually charged email correspondence with the writer Kathy Acker. Wark only chose to curate their letters into an academic publication in 2015. The book entitle*d I'm Very into You* refers to a quote in an email from Acker to Wark. Acker in many ways exoticizes Wark, as a pseudo-foreign body from Australia, a sister to the United States in its history as one of Britain's white settler colonies, which Acker somehow manages to class as delayed in its civilizational progress as compared to the United States.

Wark resists Acker's projection of Australia as a country that remains somehow a frontier culture. Rather, she suggests it is just the desert, a space of cultural voidance and demise where

---

78 "Extract: Reverse Cowgirl," *Tank Magazine,* n.d., https://tankmagazine. com/tank/2020/02/reverse-cowgirl/.

there is no "there" there when you get to the heart of it. Presumably due to the centuries-old progress of British colonization, everything that once was erased, evacuated, blacked out from within itself, including its racialized history that somehow became unreadable at least in any contemporary sense. The black hole of the real that Wark insists upon belies the fuller bodily expression of a white hole that previously allowed figures like her, Acker, and other settler-colonial type "to enter, as and when" territories of racial contestation within a deeper understanding of unequal access. What prevails, over time, in their correspondence is an inability to account for the profound emptiness of sexual, political, and national identities once they are unleashed into the free for all of the virtual world. There they become, unreadable, imperceptible, and unknowable to the degree that they fail to acknowledge the difference to between epistolary fictions, bodily risk, material desire, and clinical morbidity.

For this to happen something must be rendered from meaning so that Wark still controls the sentence, making gender and desire correspond with a formal reality as opposed to a sublimated fantasy. Wark's unreserved enthusiasm for hormone therapy to optimize human being echoes the rhetoric used to endorse endocrinology in the United States in the 1930s, when hormonal science was first being widely conducted. The concept of the "glandular self" was used to promote endocrinology as a new field of medicine. Endocrinology prided itself on its ability to provide a biological explanation for a number of facets of individual identity, from race and gender to sexuality and class. Through the decades leading up to and including World War II, hormones became interventionist technologies, acting as a means through which to register and modify one's self. What is significant about endocrinology is that it laid the groundwork for individuals to figure themselves as amenable to psychological and biomedical intervention. In the postwar period the emphasis shifted to a discipline of humanity's glandular systems such that it might then be freed from the negative burdens of biological heredity.

The emerging postwar science of cybernetics posed an entirely new, non-determinist model of biological life as plastic and alterable. This approach introduced a new standard for understanding embodiment, which redefined all bodies in terms of their capacities. This new standard of judgment applied to racially normative and gender normative cisgender bodies, now mediated through their relationships to hormonal optimization. By contrast, it forced both racial and transgender bodies to rely on conceptual protocols of embodiment based on their historical relationship to animacy and contemporary engagement with hormonal technicity. This use of hormonal technology to mollify the body can be construed as a means of devaluing it by judging it only through a lens of innate dysfunctionality. As such, endocrinology produced a world that retains meaning in our current iterations of categorical designations of racial hierarchies, class stratifications, gender binaries, mind–body splits all based on an original failure of white settler, colonial societies to be healed from their founding racist trauma. The imperial West's natural history begins with the taxonomy of the globe and ends with the biological differentiation of humanity following that same exteriorized grouping of descending others, made wholly serviceable to science for the very fact of their difference.

Chapter 5, "Capital Gains: Object Ontologies, Settler-Colonialism, and Financialized Futures" confronts the malleability of the posthuman body and its relationship to neoliberal mandates regarding productive, capacitated bodies such as biodiversity and neurodiversity. It argues that historical differentiation has always served as a prerequisite for capitalist advancement. What is novel about the neoliberal era is its tendency to disaggregate bodies in order to draw value from them and that its focus has shifted from a concern for capitalizing on materiality to biomateriality with the consequence being that all life down to the molecular level must now perform the duty to transform themselves into commodities. The emergence biotechnologies in the latter half of the twentieth century brought into being through genetic engineering, assisted reproductive technologies, human genome sequencing, and phenotypical variation forced the categories of race and gender to cut differently into the path of enterprise than they had done previously. One consequence of this was that the categories of race, gender, and species can now be compelled to multiply and, in so doing, express themselves as unique in and through their pluralization. Their job, as it were, is now to actualization vitality at any cost. The initial othering of these bodies was something posterior to the development of empire. In this position, as the back side of empire, is also possible to identify them as the first iteration of a posthumanism, wherein nature becomes the obverse of culture, and animal the obverse of human.

Within a neoliberal speculative economy that takes as its starting point an individual agency fundamentally detached from patriarchal and colonial histories, it becomes increasingly difficult to conceptualize what, if any, form of voluntary submission would be tolerable to the objectified bodies poised to act once again as collateral for exchanges which circumnavigate around various well-ordered sites of mutual production developed through straits of settler colonialism and financialized capital accumulation. These have recently been adapted, mediated, and intensified demands on the human as a species amongst others in order to meet the specific needs of a neoliberal present. This one conversant with a new ontology that insists upon a new world where new locations of empirical social order have now emerged. Its coupling of novelty with property leads us back precisely to the urge to make objects comportments of colonialism and settler colonialism.

The molecular turn in ontology signaled a kind of deregulation of life itself, making it possible, in Margaret Thatcher's words, for money to act as "the great driving engine, the driving force of life."[79] The money as a molecular object operates much like any other chattel class within the market, that is, as an abstraction, and as such, it comes to function as fungible,

---

79 Douglas Keay, "Margaret Thatcher: Interview for 'Woman's Own' ('No Such Thing as Society')," *Margaret Thatcher Foundation,* September 23, 1987, https://www.margaretthatcher.org/document/106689.

across different disciplinary terrains. What is forcibly omitted from the record of such transactions are the historical relationships that make possible the "abstraction of the molecular as a contemporary iteration of settler logic."[80] This assertion suggests that global monetization might fulfil the property that racialization once occupied. Rather than privilege the molecular as a code that marks a departure from a previous historical impetus, the molecular reinforces a connection between life and value that refers back to a previous tendency to invest only certain materialities with livelihood while deadening others. It is possible to appreciate neoliberalism as an offshoot of the project of racialization in order to recode governmentality in order to correspond with a need to protect the global market from these types of external pressures, at least for a time.

The process of converting land into money and bodies into legal tender compromised the definition of life itself under conditions of capitalist domination, allowing it only to flourish within the parameters of investment, accumulation, speculation, renewal, and habitation. Its converse is that which is emptied of populations and devoid of naming and are treated as surplus to requirement. The digital is ancillary to the process of valuation, which locates its trajectory not in analogy but further back towards the stuff of materialization. Neoliberalism becomes in its own way a kind of species, enlivened through information that synthetically adapts to fill the institutional spaces it has driven into extinction while seeding an ideological climate whose patterns of growth and migration may be said to have reconstructed race for the purpose of allowing new markets in nature to thrive under certain conditions of lending value having to do with the acknowledgement of the end of empire and the emergence of decolonization movements in the late 1980s.

The work of Jason W. Moore and Kathryn Yusoff illustrates how changing social ecologies in Britain made for the anxious admittance of the colony into the metropole of its capital cities

---

80 Jordana Rosenberg, "The Molecularization of Sexuality: On Some Primitives of the Present," *Theory and Event* 17, no. 2 (2014): n.p.

and how this related to larger concerns about how to conserve stature for a Britain without empire. For Thatcher, the answer rests on Britain's ability to acquire a new overseas empire. It ultimately does so, through servicing the wealth plundered from its former colonies and parleying those revenues into the city of London, making it once more the financial center of the world's global markets. Through such wealth management, Thatcher was able to transpose the language of economic decline into a new ethos of market ascendency in terms of Britain's relationship to globalization. The posthuman conceptualization of Britain reaches its apotheosis in Brexit. Britain's exit from the European Union thus acts as a means through to eclipse the Britain's troubled past of colonialism by making of possession something apart from property. Possession here figures as the source of affective links that have their origin not in the precise defining national borders but in the rough netting of capital flows from its former imperial territories.

If we appreciate the understanding Blackness as a transatlantic phenomenon, it must be positioned as a working order that spans the two consciously white worlds. The visible and invisible, black(ened) bodies that exist today in public space remain very much a product of the English empire, even as they were being in many ways fully annexed into Britain in a neoliberal context for which Thatcherism at a particular time of urban apartheid was posed as the solution. This also dovetails perfectly with the advent of what we are now calling casually, as though it were value-free, "gentrification."

At the start, from Enoch Powell to Margaret Thatcher and most recently to Theresa May, was a passionate demand for attention to be paid to the white body as the national body that white Britons felt responsible to defend against the presence of Others. This required an ongoing exercise in colonial amnesia which then metastasized into Brexit. Sovereignty isn't for everyone after all. And so, it goes into the prizing of Extinction Rebellion among that same satisfied Establishment of white middle-class bodies, who believe that climate crisis is everyone's responsibility when it is convenient to do so. This, without admitting that the nation relies so heavily on the care provided by its Others to maintain its precious and privileged environment. Hence Boris Johnson's life-changing visit to the National Health Service (NHS) hospital narrates as a near-death experience, where he finally wakes up temporarily to that fact, before lapsing back into his colonial amnesia. Care remains thus the province of those who have means as well as needs, whereas the "foreigners" with unpronounceable names making up the frontline simply must care.

This is precisely the intimacy of violence that comes into being through the absence of the right to exercise full independence from a system that entraps racialized people by its definition of nature into alienation. Where you stand determines what you can perceive. Contamination becomes the result of what is witnessed being overcome by a spectrum of toxic storytelling; some of which appears so innocuous that it is barely realizable until the minority are made ill from it, whereas the majority are merely badly affected.

Only those that dwell within a black(ened) world appreciate what it is like to live with a condition of what Christina Sharpe terms "total" weather. It is they that must grapple with the constant graeyness that is racialization, where mood assumes itself as matter. The matter being made not of the fact that was not being adequately represented anywhere but that the sadness of that situation was deeper still because it could not reach the surface of recognition. It exists instead as a state of fatality, as something coming from the obscure outer edges of being in a situation which reflects and enlarges the colonial periphery and its relationship to the metropole.

Against such polling, one has the choice only of allowing or withholding consent or enduring black(ened) subjectivity as a form of super violence or distancing oneself from it, as a life-saving mechanism. I question the terms of this is agreement of black(ened) being to the degree that "life-saving" has long been the preserve of a biopolitical discourse that counts the death of others as part of its efficacy and at the same time refuses to acknowledge others as having any sort of life to begin with — Black being something to see, to witness, but not to be. Can something that is not allowed full being be allowed full dying either? Has it to exist at some other level of survival? My sense is that this bare life, this mere survival as a category needs to be interrogated further going back to the origins of Darwinism as the racial science of the white West and natural philosophy before that.

This is present as a politic within the world certainly as a particular version of that world of representation and narrative development. Whiteness's inability to assimilate the emotional and psychological state of Blackness into its center in any meaningful way apart from a superficial signification of inclusion, has meant that the category of the posthuman comes up against a rather past tense, static form of understanding — that the refusal to admit others into the category of the human was, indeed, the very basis for humanity for so many centuries — up to and including now.

What is the meaning of an endless looping of the deaths of black(ened) bodies through brutal policing in the United States, and then amplified throughout the Western imagination, if it does not continue to satisfy some reassurance that others can be killed but not the I that claims itself as fully human? This extends a certain logic that that "I" can decide whether or not it cares if its actions kill others. That that "I" can kill others as its pleasure to do so, as it is its right. The same can be said for the handling of COVID-19, under the Trump administration the racial hierarchy of death from the virus became clear and this ended any sort of campaign of public health mitigation. The fact that disproportionate numbers of Black and Brown bodies were succumbing to the virus, as compared to white bodies, meant that the "China virus" was doing its job in reinforcing the belief among Trump and his base whiteness carried within itself a universal code of immunity.

Those who continued to attend his rallies and appear in public unmasked resonated with Trump's comment as a presidential candidate, that he could shoot someone on Fifth Avenue and getting away with it. The historical resonance of that attitude. For Americans, it was about white men killing with impunity as the mark of their distinction within the greater society, applying to minorities of any distinction. Being that it was Trump, it was a dog whistle that no one in the mainstream press would dare to narrate explicitly, but of course, it was there and remains so. No major white protest work will be made of that that is any more potent than the Trump blimp because that is another type of refusal that ends in the same place of threat and despair that is seldom articulates. Whiteness needs to keep reproducing that savagery as the other side of its coinage. Whiteness needs to manufacture consent. Whiteness needs to make a property of those black(ened) bodies, to make them desirable, available for pleasure to do with what it will in this world. This is a sexual issue as much as it is anything else.

Racialized people as the new matter of society relates to the new emphasis on object orientated ontologies, which operate as a new form of colonialism. It does so to the extent that is obsessed with the fraught problem of bringing liveliness to matter, a concern that goes back to Du Bois, to Douglass, to Jefferson, all of whom struggle with the categories of humanity and animality, all of whom struggle with the question of where the line of availability of people crosses into abolition, crosses into annihilation. Here it is important to recall that abolition was a product not of moral judgment but one of technical and economic reasoning. Social media on the internet continues in this tradition. Its need to make sexualization something that is a matter related to but separate from cognition, for whom it is possible to say form a system, a feedback loop that later becomes the basis for cybernetics and so on. Cognitive capitalism, biocapitalism, all require some artificial market in reproduction; it is always a question of resources.

Nobody "works" anymore they are just passively contributing their accentuated version of life force; that is, "living their best lives" at whatever level that is, playing their roles largely for others benefit, which is where Blackness is now being experienced by the white precariat who are now drawn to questioning if this all is really just a game. Making game of people of color used to be their sport, and now perhaps, it isn't going to be that way forever. Not for all of them anyway. Is this the end of capitalism? What of America's shadow families, the product of house slave versus field slave sexual hierarchies that now are suddenly being rediscovered through *Ancestry.com* as though they hadn't been hiding in plain sight for generations? What about the extension of the color line in the fetishization of mixed-race people as the faces of the fashionable Black bourgeoisie, as though that too were a new and laudatory phenomenon of a more inclusive twenty-first century?

All this feeds that libidinalization — its violence is considered so banal as to not be worthy of much polite attention. It is an era of unchecked narcissism, of zooming forward based on you curating your story as much as your look, that again is what sells. It is painful, even devastating to consider the other option, of not consenting to these standards of judgment. The standards are catching, and we don't want to be judged as bad at our jobs of continuing to appear relevant. Especially if we aren't exactly sure of the sustainability of our positioning. These times which are both ahistorical and atemporal as well as being apolitical and apathetic, make it that much harder to refuse. That said, civil society must experiment with holding space elsewhere even for a little while for the potential of something getting through and some alternative types of work being managed at a time where universalism means everyone has to be available as well as felicitous in order to survive. It is the new service after all.

1

The Dark Posthuman

Social Reproduction, Social Justice, and Artificial Ecology

For some five centuries, humanity has positioned itself at the fulcrum of sociopolitical order. The term "posthuman" suggests a capability of existing apart from that order, something situated beyond its former punitive borders of reason and sentience. Nonetheless, the posthuman, like its predecessor the human, has to take its place amongst the respective role allocation, social hierarchies, and divisions of labor that remain attendant to social being. Thusly a coloniality of power very much persists within this redefinition of mankind and race as a technology that remains both indispensable and irreplaceable to the reproduction of what qualifies as posthuman. The posthuman implies a humanity that has achieved an extracognitive function distancing itself further still from those who have failed to realize themselves as fully human. The new paradigm of the posthuman elaborates the domination formerly expressed in terms of humanity by extending a logic that racialized others can only be assimilated into the world wide web of life and internet of things as beings granted temporary status and defined against a criterion of developmental potential that begins and ends within the contours of white masculinity. Within this context, race must be critically reassessed to plot its iteration within the post-liberal concept of the posthuman and localize its portent within the critical ecological narratives of both the Plantationocene and the Anthropocene.

In order to posit race within the category of the posthuman, this chapter critically intersects groundworks laid by the theorists Sylvia Wynter, Philip Butler, Rosi Braidotti, Christina Sharpe, Donna Haraway, Anna Lowenhaupt Tsing, and Katherine McKittrick. It brings together the geographies, economies, and ecologies of the African American South, Wilhelmine Germany, the African Global South, National Socialist Germany, and the postwar Algorithmic North and South as well as the imperial and aggregate principles that animate them. It covers conflicts both interracial and intraracial that bear upon the the- orization of modernity, biology, and racial identity within the founding of the discipline of sociology. W.E.B. Du Bois acts as locus for these developments through W.E.B. DuBois's early work on formulating a concept of the "New Negro" and "Pan-Africanism." It reviews his ambition to globally position himself as America's leading racial authority and the ways in which this authority becomes corrupted through his interiorizing of Victorian eugenic principles characteristic of his formative age. Du Bois may well have cast himself as the first dark posthuman, and this situation allows us to read his data visualizations as evidence of how he plotted the uplift of his race through techniques of sociological reification, aesthetic rarefication, and controlled development. For this reason, he is also a historical precursor relevant to discussions of the Plantationocene and Anthropocene as they figure into a longer narrative of advancement and debility.

And in the Beginning, There Was Man

The celebrated decolonial theorist Sylvia Wynter outlines a historical transition from a "theocentric" description of man dating from the sixteenth century towards a "biocentric" one in the nineteenth century.[1] This description eventually branches into the maturation of the biological sciences, the social sciences, and the humanities throughout that same century. These disciplines promoted a specific set of "truths" and as their name suggests their schools of thought "commanded obedience."[2] Their forms and formulations of what constituted the domain of mankind "necessitated" the constant monitoring of "individual and collective behaviors" and in so doing established how such orders were maintained.[3] Through these various modes of inquiry, the very being of humanity was "brought into existence, produced, and stably reproduced."[4] This is consequential as the breakdown of these disciplines in the early twenty-first century appear to indicate a path through which the posthuman might emerge and allows us to conjecture the modes of discipline that might in future inscribe them as an "informacentric" being.[5]

Humanity throughout the modern era has balanced itself between the twinned formations of mastery and slavery, nondependence and dependence. These were terms cast from a universal term, human, based on rationality rather than irrationality. These terms suggest the existence of a posthuman capable of hyperrealization, a being who exists not as a product of biotechnology — the category of racism or the effect of racism in total — but a being that is something apart from that space of punitive distinction. The posthuman, nonetheless like his predecessor the human, has to take his place amongst the respective role allocation, social hierarchies, and divisions of labor which remain attendant to social being in our current age. Here a coloniality of power very much persists, and race, as a technology, remains both indispensable and irreplaceable to the re-

---

1 Sylvia Wynter, "Unsettling the Coloniality of Being/Power/Truth/Freedom: Toward the Human, after Man, its Overrepresentation — An Argument," *CR: The New Centennial Review* 3, no. 3 (2003): 269.

2 Ibid., 271.

3 Ibid.

4 Ibid.

5 Ibid.

production of what qualifies as a recognition of the posthuman. Equally, what does not qualify remains of value as a proximate function insofar as it demonstrates the existence of some innate *incapacity* within those who have failed to adapt and, therefore, merit being designated as an *anti-type.*

The posthuman implies a humanity that has achieved an extracognitive function distancing themselves further still from those who have failed to realize themselves as fully human. Furthermore, Wynter argues, Western man in this century has made it his mission to dominate "specific orders of consciousness or modes of mind."[6] In the advent of universalizing domains of control, such as the internet and artificial intelligence, we have increasingly "come to experience ourselves as this or that genre/mode of being human."[7] For the most part, these new processes of imperial governance have progressed "hitherto outside our conscious awareness, and thereby [are] leading us to be governed by the 'imagined ends' or postulates of being, truth, freedom that we likely put and keep in place, without realizing that it is we ourselves, and not extrahuman entities [at their helm], who prescribe them."[8] The consequence of this is "the subordination and impoverishment of the vast majority of the worlds to which they/we belong."[9]

Wynter refers here to the status of Japanese and other lighter-skinned peoples as "honorary humans" within this new algorithmic estate, whereas Black and Latinx people remain "forcibly proscribed" from attaining this status through the continued institutionalization of their poverty and detention.[10] The category of posthuman relies upon the being of mankind itself, in all of its "multiple self-inscripting, auto-instituting modalities."[11] Therefore, posthumanity presumes itself as the measure of the post-digital world in much the same way as its

---

6   Ibid., 328.
7   Ibid., 329.
8   Ibid.
9   Ibid.
10 Ibid.
11 Ibid., 330.

predecessor humanism. While posthumanism claims to have decentered systems of knowledge, in reality these continue to concentrate authority by carrying out their orders of autopoiesis. AI and cybernetics subtly exclude the factors of race and gender and deny the emergence of multiple sites of enunciation. Posthumanism's reliance on Western exceptionalism, technological fetishism, and ableism belies its commitment to a false narrative of inclusion inherent in the category of the human; on one level, this commitment promotes a superficial concept of inclusion through the lauding of the value of diversity, while on another level, it promotes thought processes that fail to question the knowledges, ideologies, and privileges that subtend the very "being" of its sociogenic invention.

This overreliance on universal proclamation with the very term posthuman confounds both the temporal and organic sense of what we might classify today as life. Indeed, "critical posthumanism claims to value a radical inclusivity (of nature, ecovitality, and nonhuman animal life), hyper-relativity, and complexity."[12] However, in so doing, it assumes equality of interest and participation within that inclusion as the basic level of existence. It also assumes that the experience of subjectivity is one open to enhancement of a technological nature, without taking responsibility for its previous, historic role in stipulating what constitutes "progress" towards a shared condition of humanity or, more precisely, recognition as part of humankind by these other entities. Here we find ourselves back in the territory of the Other as somehow maladaptive or requiring of corrective to fully participate in being within its territory, that is, within the scope of its relativity to humankind. This establishes a proprietary relationship from the very beginning of this project to extend "humanity" to others. That gesture of contemporary inclusion must necessarily imply a prior historical expulsion. It is this feature that haunts the category of the posthuman bringing

---

12 Philip Butler, "Making Enhancement Equitable: A Racial Analysis of the Term 'Human Animal' and the Inclusion of Black Bodies in Human Enhancement," *Journal of Posthuman Studies* 2, no. 1 (2018): 107.

it into proximity with an afterlife of anger, grief, and dispossession that is not readily absolved from memory. For this gesture to be achieved one has to willfully unmoor oneself from one's historical situation and one's cultural claims to exist and to not be, in Rosi Braidotti's words, "bound negatively by shared vulnerability, the guilt of ancestral communal violence, or the melancholia of unpayable ontological debts."[13]

Braidotti's argument centers on "interdependence" as a means through which to surmount the agony of a former dynamic of non-dependence and dependence "shared" amongst these various parties. Furthermore, she asks for compassion to become the affective currency through which the debts of the past may in the present be forgiven.[14] In many ways, this request implies that sympathy and empathy take the place of granting what has long been immaterially owed. This is a particularly grievous bargain, one that Philip Butler reads "as a power move" that effectively rewards "the executors of ancestral communal violence" by requiring both their assent to peaceably "move beyond it" and the complimentary discharge of their aggressors' "ontological debt."[15] It remains unclear what these aggrieved parties might get in return as there is no mention of any deal to challenge "the capitalist socioeconomic structure" that was built off their backs and was the source of their misery and exclusion dating back centuries.[16] Nor is there any acknowledgement that their subjugation might very well be required again in future to maintain the standards of what a certain mankind has always referred to as "civilization." It is very difficult to imagine that this arrangement would offer much in the way of refinement to that system of hierarchical needs, a system that might indeed benefit those who are seldom, if ever, afforded the system's comforts and securities. Nevertheless, Braidotti remains keen to place further demands on those who have only recently been honored

13 Ibid., 112.
14 Ibid.
15 Ibid.
16 Ibid.

with any sort of place at the table of humanity, to make better of their existence in line with its white hegemonic expectations.

Butler observes Braidotti's profound failure to recognize that white communities and Black communities have occupied "separate historicocultural temporalities" for some five hundred years and that, indeed, it has only been in the last fifty years or so that they have become socially recognized within the category of the human.[17] As such, within her economy of inclusion lingers the specter of expulsion and dismissal. It is that glimmer of violence that continues to subtly terrorize those same groups it so recently admitted to the property of the human. Coloniality remains firmly rooted in that same place of address that promises an unwanted return of the very elements of oppression that foreshadow the withering of "progress." The bitter irony is that figures like Braidotti, who are so quick to admit "strings, quarks, photons, bacteria, viruses, and all living organisms" into the family of man, recognize so little of the coevolution of natural sciences and Eurocentric racism in their work.[18] Their call for the recognition of universal sentience similarly excludes mention of any number of scientific discourses given over to proving that Black, African, and African Americans and animals had a fundamentally lesser capacity for feeling. The greater concern for the affordance of sentience to these entities is how their newly acquired status might absorb them into transnational flows of labor, technology, and capital and, as a consequence, how they might become institutionalized into positions of deployment that could potentially further dehumanize people of color. Instrumentalization seems to be dogging this posthuman discourse of conscious interrelativity. The empiricism of Braidotti's new materialism is also inseparable from the projects of colonialism and racism for the very reason that it seeks to make use of nature and matter as sources to reify persistent notions of what constitutes knowledge, experience, and value.

---

17 Ibid., 113.
18 Ibid.

Shadow Productivity and the Feeling for Life

Within Braidotti's onto-epistemological framework lurks a fundamental opacity; it's inability to account for the absence of Black feeling within the structure of contemporary civil society and the purview of representation. This inability to assign positive sentience to Blackness persists as a modality of violence that continuously "produces blackness as a locus of incapacities," Black affective responses are only legible as signs of pathology, further reifying Blackness-as-subhumanity on an "epistemic, material, metaphysical, ontological" level.[19] Tyrone S. Palmer situates "this inability to conceive of Black emotion, to imagine the Black as a sentient being with interiority," within "the history of racial chattel slavery," wherein white humanity was defined by its negation within the contours of the slave body.[20] Palmer argues that

> while not all "sentient beings" are endowed with subjectivity, sentience is itself a precondition for subjectivity within the modern field of representation — the Subject has consciousness; is self-knowing, self-reflecting, and feeling. Denying or contesting the Black's sentience, then, has a dual function: to write Black people outside of the Human and position them as immutably affectable, unfeeling repositories for brute force.[21]

The consequence of this in many ways is to socially deaden them. This situation is especially significant at a time when posthumanism socially and politically advocates for the agency of nonhuman beings seemingly without regard to a world where Black bodies have yet to achieve full lively personhood. Christina Sharpe asserts that these bodies must be conceived of

---

19 Tyrone S. Palmer, "'What Feels More Than Feeling?': Theorizing the Unthinkability of Black Affect," *Critical Ethnic Studies* 3, no. 2 (Fall 2017): 32.
20 Ibid., 46.
21 Ibid., 51.

through a prism of "containment, regulation, punishment, capture and captivity" as constitutive of a "total climate" through which the Black body must travel.[22] The contemporary middle passage of these bodies navigates between "ungrievable death" and "lives meant to be unliveable" and proceeds in such a way that it comes into intimate contact with the immigrant and the refugee, that is to say, stateless people who become the subject of forced movement and the object of surveillance.[23] This is indeed a form of consciousness but one pointedly with no innate virtue. Rather, its value must be externally assigned. This is not to say that there is no place where the Black body might yet thrive amid this harsh societal atmosphere. The trick is to keep breathing to literally maintain aspiration within a social ecology that constricts the basic atmosphere of life. At the same time, the state concocts new modes of violence that conform with the values of a neoliberal state seeking at every point to contain resistance and promote submission, continually registering bodies within a system where nonparticipation equals death.

The ghostly remnants of colonialism, racism, capitalism, and militarism haunt these systems of apprehension. Participants are therefore kept in a constant state of presentism to maintain the illusion that this system of reality is infinitely stable. Within such an atmosphere it is advantageous to feel otherwise and to generate innovative relationalities when dealing with the feelings these institutions require of the Black bodies they apprehend as objects of state dependency. Within this steady state model of control, crises have a specific purpose. That purpose, according to Kristen Simmons, is to rupture "history and install an event-based logic, rather than inviting structural interrogation."[24] The discourse of crisis itself performs a particular function insofar as it "seeks to stabilize an institution, practice, or reality rather than interrogate the historical conditions of possibility for that

22 Christina Sharpe, *In the Wake: On Blackness and Being* (Durham: Duke University Press, 2016), 22.

23 Ibid.

24 Kristen Simmons, "Settler Atmospherics," *Fieldsights,* November 20, 2017, https://culanth.org/fieldsights/settler-atmospherics.

endangerment to occur."[25] The implication from Simmons's argument is that crisis itself has become an institutional requirement to justify the ongoing policing and repression of various communities of color. This logic facilitates the construction and perpetuation of massive police-state coalitions that in turn produce ever more nuanced techniques of civic control. These technologies profit off race and generate new markets that are built on the historical appropriation and extraction of value from the Black body.

Environment figures here in the formulation of the "Plantationocene," as an alternative reading of the Anthropocene, that stresses the centrality of slave agriculture to the formation of human-made climate change. What is important to stress here is that the "plantation isn't just a material institution that has led to the planetary catastrophes of the Plantationocene; it's also a set of ideas, archives, ideologies" that have, on one hand, "become the foundation for Western capitalist endeavors at large" and on the other, have cultivated "what Vandana Shiva has called 'monocultures of the mind'" where "ideologies of the plantation [...] fundamentally shape how human beings relate to each other and to the natural world."[26] Anna Lowenhaupt Tsing defines the plantation as "those ecological simplifications in which living things are transformed into resources, future assets by removing them from their life worlds."[27] Tsing defines plantations as "machines of replication," and thereafter as "ecologies devoted to purification and the production of the same."[28] The pattern in which they progress relies on an overreliance on proliferation, so much so that it becomes concentrated, carceral, and corrosive in its stance towards the terrains of life where it

25 Ibid.
26 Natalie Aikens et al., "South to The Plantationocene," *ASAP Journal*, October 17, 2019, http://asapjournal.com/south-to-the-plantationocene-natalie-aikens-amy-clukey-amy-k-king-and-isadora-wagner/.
27 BCRW Videos, "Anna Lowenhaupt Tsing — A Feminist Approach to the Anthropocene: Earth Stalked by Man," *Vimeo*, December 18 2015, https://vimeo.com/149475243.
28 Ibid.

fundamentally requires labor. This pattern becomes performative and takes on a sort of subjectivity through the energy it subverts from this labor. Ultimately it transforms that energy into something akin to a hostile takeover. "Scholars have long understood that the slave plantation system was the model and motor for the carbon-greedy machine-based factory system that is often cited as an inflection point for the Anthropocene."[29] In modernity, plantations have become continuous and synonymous with the prospect both of propagation and of devastation. Along similar lines, Shiva asserts that plantation "monocultures spread not because they produce more, but because they control more."[30] It is within that remit of control that proliferation merges with pathology and mutually assured pathways towards environmental subjugation.

In the course of the early modern period, emerging economies instigated a convergence of "migratory forced labor […] intensive land usage, globalized commerce, and colonial regimes" which together "gave rise to the emergence of plantation" and its attendant forms of indenture and slavery.[31] The axis of this new commercial configuration was balanced between violence enacted against a newly racialized humanity and violation enacted against a newly dispossessed environment. Ancillary to this arrangement were slave gardens, or small plots of furtively cultivated land typically located near slave cabins, or the most remote uncleared boundaries of the plantation. These modest entities stood adjacent to the new world of capitalist accumulation and trade profiteering, yet they remained fundamentally enmeshed within a local atmosphere of consumption reliant upon the endless yielding of Black bodies and virgin earth. Concepts of coercive labor and subjective freedom existed as co-innovations of imperial liberal governmentality.

29 Donna Haraway, "Anthropocene, Capitalocene, Plantationocene, Chthulucene: Making Kin," *Environmental Humanities* 6, no. 1 (2015): 163n5.
30 Vandana Shiva, *Monocultures of the Mind: Perspectives on Biodiversity and Biotechnology* (London: Zed Books, 1993), 7.
31 Sophie Sapp Moore et al., "Plantation Legacies," *Edge Effects,* October 12, 2019, https://edgeeffects.net/plantation-legacies-plantationocene/.

As such, slave gardens were made contiguous with principles of growth and flourishing, as well as survival and resistance. The emergence of the slave garden points to the need to reframe contemporary discussions of the plantation and early forms of capitalism as outgrowths of liberal systems of rule where "slave labor and other forms of exploited, alienated, and usually spatially transported labor" were routinely made productive to include cultivated forms of counter-aggregation.[32] A new class of being comes into the equation to nurture them "in even the harshest circumstances" producing not only "crucial human food," but also refuge for biodiverse plants, animals, fungi, and soils to self-perpetuate and self-motivate well and apart from the socio-ecological crises happening all around them.[33] As such these slave gardens might offer a means with which to challenge

> ethical visions that minimize or obscure unequal relations of difference, and it might help conceptualize multispecies assemblages […] that [are] not just envisioned but lived and that simultaneously tend to the needs of social reproduction, social justice, and ecological care in ways that have for the most part remain under-recognised as formations in and of this world.[34]

Generativity, as we have come to accept it, remains very much cloaked in the rhetoric of a critique of the Plantationocene that focuses its attention on "the ever-greater ferocity in globalized factory meat production, monocrop agribusiness, and immense substitutions of crops like palm oil for multispecies forests and their products that sustain human and nonhuman" life.[35] The

---

32 Haraway, "Anthropocene, Capitalocene, Plantationocene, Chthulucene," 162n5.

33 Ibid.

34 Janae Davis et al., "Anthropocene, Capitalocene, Plantationocene? A Manifesto for Ecological Justice in an Age of Global Crises," *Geography Compass* 13, no. 5 (2019): e12438.

35 Haraway, "Anthropocene, Capitalocene, Plantationocene, Chthulucene," 162n5.

hidden force in the Anthropocene's hidden ecology, by contrast, may well be found in how they malfunction allowing for counter-intentional forms of life to reassert control by other means through "germ plasm, genomes, cuttings, and all other names and forms of part organisms" that were never granted fundamental recognition.[36] These Others — plants, animals, microbes — were all displaced and made operational within this founding economy of exploitation. What is largely omitted from this understanding is the centrality of the slave to the slave garden and how they perpetuate discreet hierarchies of life that remain informed by historical imperialisms that make these same slaves involuntarily responsible for the planting and caring of the plantation's gardens, as well as and their own. In this sense, the slave of the garden becomes somewhat reduced in understanding of their productive capacity. They were involuntarily supplanted into this environment, as were the very products of their labor making of them a sort of semblance, as peoples that came into being solely through productive relations to the formations and products which they alone produced.

Wynter writes, "'the Caribbean area is the classic plantation area since many of its units were 'planted' with people, not to form societies, but to carry on plantations whose aim was to produce single crops for the market. That is to say, the plantation-societies of the Caribbean came into being as adjuncts to the market system; their peoples came into being as adjuncts to the product.'"[37] This situation creates an environment of information where it becomes necessary for this peculiar dynamic of sociality between peoples and plants to itself become the raw material appropriated for the use of whiteness as capital. Moreover, as Paul Gilroy observes, "the plantation system made that way of dominating nature part of the slave's experience of unfreedom."[38] The Black body within such an equation exists as

---

36 Ibid.
37 Elizabeth Maddock Dillon, "Zombie Biopolitics," *American Quarterly* 71, no. 3 (2019): 634.
38 Paul Gilroy, *Against Race: Imagining Political Culture beyond the Color Line* (Cambridge: Belknap Press of Harvard University Press, 2000), 200.

"the afterlife of property" and, in that same calculation, emerges as both figuratively dead and unduly alive.[39] Slavery is a condition that was intended to transcend the imaginable when it came to its value because it represented the theft of personhood itself and, therefore, the profound loss of control over one's own body.

The terrifying specter of this condition extending beyond the realm of the living figures into the creation of the zombie as an individual whose life, even in death, remains extorted. As a figure, Kaiama Glover asserts that "the zombie exists only in the present of its exploitation. It represents the lowest being on the social scale: a thingified no-person reduced to its productive capacity."[40] Amy Wilentz defines the zombie as "the inanimate animated, the robot of industrial dystopias."[41] As such she concludes, the zombie is "great for fascism" because, as a discredited subject, the zombie persists "devoid of consciousness and therefore is unable to critique the system that has entrapped him. He's labor without grievance. He works free and never goes on strike. You don't have to feed him much."[42] Fascism organizes itself as an adjunct of capitalism insofar as it can fundamentally alienate humanity from society and reorient its persistence towards resourceful extremity and fundamental acts of endurance. In this way, the plantation and the work camp become relational to one another in the span of modern capitalist development. Some 500,000 individuals became slave laborers in National Socialist work camps during the twelve years they assumed total control of Germany. Their involuntary ranks included not only Jews, but also "Soviet POWs, political opponents, recalcitrant forced laborers, and common criminals," roughly half of whom "were

---

39 Sharpe, *In the Wake,* 22.
40 Dillon, "Zombie Biopolitics," 625.
41 Amy Wilentz, "A Zombie Is a Slave Forever," *The New York Times,* October 30, 2012, https://www.nytimes.com/2012/10/31/opinion/a-zombie-is-a-slave-forever.html.
42 Ibid.

hired out to private industry."[43] These forced laborers worked extensively for prominent "German companies such as BMW (aircraft/cars), Daimler-Benz (cars), I.G. Farben (chemicals), Krupp (steel), Messerschmitt (aircraft), Porsche/Volkswagen (cars), and Siemens (electronics)."[44] These companies were determined that through their use of slave labor they would be able to retain and even expand their profits and corporate assets during and beyond wartime.

The prisoners of Germany's concentration camps represent another form of labor practice contiguous with slavery. Their origin dates to 1884 when German colonial settlers in southwest Africa began "to establish lucrative plantations by exploiting the labor of local Herero and Nama (also known as Hottentot) indigenous peoples."[45] Those that resisted were subject to large scale campaigns of genocidal war and racial annihilation. Eventually, it became obvious to the Germans that they were essentially killing off their labor force through these campaigns, and they decided to shift their practices to the design of labor camps where a program of "'extermination by labor'" could be enacted up what remained of the Nama and Herero civilian populations. These included "women and children, [who] were knowingly and methodically worked to death."[46] This same model of operation came into the interior of Germany itself with the construction of the Dachau concentration camp.

The Plantation as Camp

In 1938, under the orders of the Reichsführer of the Schutzstaffel (SS), Heinrich Himmler, concentration camp prisoners at

---

43 Catherine A. Epstein, *Nazi Germany: Confronting the Myths* (Hoboken: John Wiley & Sons, 2014), 190.

44 Ibid.

45 Edwin Black, "In Germany's Extermination Program for Black Africans, a Template for the Holocaust," *The Times of Israel,* May 5, 2016, https://www.timesofisrael.com/in-germanys-extermination-program-for-black-africans-a-template-for-the-holocaust/.

46 Ibid.

Dachau were forced to build a plantation adjacent to the camp. The camp itself was constructed on land "formerly blanketed with coniferous forest" and the plantation to be built upon it was conceived explicitly as "an economic program."[47] The transformation of the land was done by camp slave laborers who were forced to work "without machinery" to convert "wetlands into arable soil, by excavating trenches, cutting peat to fill ponds, and spreading soil"; all essentially with their bare hands.[48] Those who survived this backbreaking work would later be charged with "the cultivation of local herbs was [so that] Germany should have no need to import foreign medicines and herbs. The economic importance of the work done by the prisoners in the herb garden increased as the war progressed."[49] The plantation was constructed to be the largest garden of medicinal herbs in the whole of Europe, whose bounty would service the medicinal needs of both Germany's military and civilian populations. The camp slaves who were charged with its cultivation were also the subjects of the camp's medical experiments. It was their bodies who were subject to systematic testing protocols involving roughly a thousand varieties of plants, in order to arrive at inexpensive and abundant treatments that could then be administered throughout the expanding territory of the Third Reich.

Opened in 1933, Dachau was the first Nazi concentration camp. It preceded the official start of World War II by some six years. The herbal plantation associated with it would last another five years beyond the official end of the war 1945, until 1949. This suggests that the herbal plantation performed an exceptional function within the German economy that ideologically transcended fascism by linking race, soil, and climate together. The Dachau plantation was not merely a means with which to

---

47 Judith Sumner, *Plants Go to War: A Botanical History of World War II* (Jefferson: McFarland and Co., 2019), 193.

48 Ibid.

49 "ss Experimental Agricultural Facility 'Herb Garden'/'Plantation,'" *kz-Gedenkstaette-Dachau,* n.d., https://www.kz-gedenkstaette-dachau.de/en/historical-site/virtual-tour-of-the-surrounding-area/ss-experimental-facility-herb-garden-plantation/.

exhaust the biological potential of camp prisoners through its cultivation and application, but also to develop techniques "to train 'settlers' for the Eastern territories, part of ss plans to use biodynamic cultivation in the environmental and ethnic re-ordering of the East."[50] Camp slaves were experimented on to determine if herbal and homoeopathic remedies could sterilize human beings, "a project that was deemed of great importance for controlling the predicted population growth in the East of the expanding Reich"[51] The East was eventually to be colonized by Germans as an expansion of their living space. It was under-stood that "ethnic German farmers in an agrarian empire under Nazi rule would eventually clear away the 'Slavic populations' currently residing there presumably using similar patterns of racialized expropriation used previously on West African and Jewish populations."[52]

These plantations were part of a biodynamic agricultural movement in Germany that promoted organic farming tech-niques couched in racist, nationalistic rhetoric that had flour-ished for decades before the rise of Nazism. Peter Staudenmaier contends that "there were substantial points of convergence between biodynamic philosophy and the tenets of blood and soil, some of them stemming from common roots in pre-Nazi culture" dedicated to the "conjoining of racial and rural discourse."[53] These lines of thought were used interchangeably as a justification for sustaining the German population through a program of mass migration, settler colonization, and racial genocide. "Heinrich Himmler's Generalplanost, or General Plan East, aimed to exploit Eastern Europe for raw materials, energy, food, and labor even if the process meant destroying lo-cal economies, uprooting communities, instituting slavery, and

---

50 Peter Staudenmaier, "Organic Farming in Nazi Germany: The Politics of Biodynamic Agriculture, 1933–1945," *Environmental History* 18, no. 2 (2013): 396.

51 Ibid.

52 Ibid., 394.

53 Ibid., 398.

murdering millions."[54] Those Slavs who remained alive would do as slaves to their new Aryan masters.

In a speech to ss generals on October 4, 1943, Himmler explicitly spoke of Slavs as slaves, emphasizing "that their right to exist depended upon service to the Volk: 'Whether nations live in prosperity or starve to death interests me only in so far as we need them as slaves for our Kultur.'"[55] The goal was to produce a decimated, enslaved population of 30 million through the systematic "killing or removal of some 45 million Slavs" within the Eastern Soviet territory.[56] Those that survived would have their labor extracted from them in ways similar to their counterparts at Shark Island and Dachau. The mass fatality of these populations only served to underscore the Nazi's ideological principle of their inherent racial inferiority. Himmler's plan was largely a failure because, even though millions of German men and women were keen to actively participate "in the racially motivated conquering expedition of large parts of Eastern Europe, which included the seizure of property, theft and murder," far fewer of these "soldiers, policemen, officials, teachers, nurses and auxiliaries" wished to make far reaching personal sacrifices to materially populate the National Socialist vision of a post-war 'German East.'"[57] As a consequence, this settlement campaign was indefinitely deferred.

Germans willing to pledge their labor during the destructive phase of this enterprise were far less keen to engage with the Reich's plan for a post-genocidal settlement involving efforts to possess and manage these expanded territories. Once cleared of its native peoples, there proved to be few Germans who desired to tend to the remaining land, trees, animals, birds, fish, and so on required to reestablish and sustain these new sovereign do-

---

54 Benjamin Madley, "From Africa to Auschwitz: How German South West Africa Incubated Ideas and Methods Adopted and Developed by the Nazis in Eastern Europe," *European History Quarterly* 35, no. 3 (2005): 435.

55 Ibid., 441.

56 Ibid., 443.

57 Daniel Siemens, "'Sword and Plough': Settling Nazi Stormtroopers in Eastern Europe, 1936–43," *Journal of Genocide Research* 19, no. 2 (2017): 209.

mains. German colonialism was about forming attachments to environments. For this to be achieved those humans that inhabited them had to be portrayed as indistinct and indistinguishable from their environment. German settlement in the east, therefore, required the effacement and denial of both cultures and identities in favor of a narrative of geographical settlement and management. Coloniality was enacted through the reinforcement of such indistinction.

This same approach prevailed during the German settlement of southwest Africa and was evident in its efforts to establish a regime of forced labor within that regional space. Soon after their arrival, German settlers worked to establish a system of plantations where land would be parceled out according to racial hierarchy. The worst soil was "naturally" to be given over to the native African population to subsist upon, whereas German settlers were to be granted the best. The greater aim of this settlement plan was for southwest Africa to become "one great German slave plantation operated on scientific lines and thriving on low-cost production of tropic staples for export."[58] African slave laborers on the plantation were forced to carry a "passport," which would "include the person's name, the name of his father, his approximate age, place and date of issuance, a thumb print, information on his state of health, etc." and "to also bare upon his person, a metal dog tag imprinted with the current number of his labour passport."[59]

This datafication of the enslaved individual would reemerge as a feature of surveillance when the first concentration camps were set up for Jews in Eastern Europe. It might have even been plausible for these two highly surveyed populations to have been combined during the subsequent era of German National Socialism in Africa. Indeed, "certain high-ranking members of the ss dreamed of making concentration-camp prisoners from Europe work in the mines of Southwest Africa and on road con-

---

58 Karsten Linne, "The 'New Labour Policy' in Nazi Colonial Planning for Africa," *International Review of Social History* 49, no. 2 (2004): 212.
59 Ibid., 214.

struction projects, while African workers were to be forced to labour on government plantations" perfecting a protocol of systematized racism across their colonial territories.[60] This would have the effect of clearing Jews from Europe to make way for German expansion while at the same time filling in labor gaps. The fact that this idea never materialized made keeping the Jewish laborers alive in Europe surplus to requirement. Such were the vagaries of racial capitalism that required the seizure of land and genocide of native peoples as a means of guaranteeing access to raw materials, markets, and investment areas in furtherance of an imperial project.

Colonialization of the Mind

At the end of the nineteenth century, this project entwined itself with the founding of the Verein für Socialpolitik (Social Policy Association), which granted German sociology its first institutional base.[61] Its express purpose, as an organization of academic economists, was to offer "advice to the German state on, amongst other topics, methods of colonizing the German Empire, that is, on settling those territories claimed by the German Imperial State with populations identified as German. This included expelling Poles from, or subordinating them in, the eastern parts of Prussia but, just as important, preventing German workers from leaving the region."[62] Prominent sociologists that remain giants of the discipline such as Georg Simmel, Max Weber, and Jürgen Habermas founded their careers on works that explicitly promoted the merits of the internal colonization in the German east and the overseas colonization in southwest Africa. "These national economists not only founded the discipline of sociology in Germany but also shaped the discipline in

---

60 Ibid., 212.
61 Andrew Zimmerman, "German Sociology and Empire: From Internal Colonization to Overseas Colonization and Back Again," in *Sociology and Empire: The Imperial Entanglements of a Discipline,* ed. George Steinmetz (Durham: Duke University Press, 2013), 168.
62 Ibid.

the United States through the ground-breaking work of Robert E. Park and the Chicago School."[63] Colonialism stood at the very foundation of sociology as an academic discipline, persuading a generation of scholars such as Booker T. Washington and W.E.B Du Bois to contemplate settler colonialism as one of the core tenants of both Pan-Africanism and Black nationalistic thought. Intellectually and institutionally these men were compelled to believe that colonialism was continuous with their projects to civilize Black Americans by having them act as the superiors of their African brethren, and in so doing, making the Black American race economically productive.

In the case of Liberia, Du Bois was keen to promote authoritarian models of development, wherein a "mulatto" American elite would come to near total hierarchical dominion over the native African population. Such a condition would be enforced through a combination of economic dispossession and ethnic violence. The founding of Liberia was originally a white supremacist project to send freed African Americans in the northern US back to West Africa, thereby setting it up as an American colony specifically designated for Black American emigrants to inhabit. The territory of Liberia itself was a former slave-trading base. From the time of its founding in 1847, the country duplicated many of the economic and social structures of the American South. "In Africa, they endeavored to recreate the only social and political order they knew, that of the antebellum South — with themselves as the master class. They erected buildings in the style of plantation mansions and dressed in formal nineteenth-century clothes despite the equatorial climate."[64] The richest amongst them established their own plantations, carving them out of the surrounding jungle, and forcing the native population into servitude "tilling their cotton, tobacco and coffee crops."[65] It was not just the land that they zealously culti-

---

63   Ibid., 168.
64   Ciment James, *Another America: The Story of Liberia and the Former Slaves Who Ruled It* (New York: Farrar, Straus and Giroux, 2013), xvii.
65   Ibid.

vated, but their own biological stock, through intermarriage of families that made up the founding "mulatto" elite. Within a few generations, a ruling class was established within the country who would over time come to be responsible for the management of a nationwide plantation system replete with slavery, forced labor, and concubinage. During these founding decades, they worked hard to establish and maintained diplomatic and trade relations with the United States, France, Germany, and the United Kingdom to increase their wealth and status.

The settlers, known as Americo-Liberians, had a complex relationship of dominion over Indigenous Africans who lived in the region. However, in the end, it all came down to how their bodies could be utilized in pursuit of profitable trade with the West. The most famous of these financial arrangements was the granting of the American Firestone corporation with a ninety-nine-year lease of one million acres of land in order to establish a rubber plantation within the country and, as well as to exploit any gold, diamonds, and other minerals found within that same territory. Du Bois crucially aided Firestone's entrance into Liberia. He was in a unique position to do so as he enjoyed felicitous relations with both the Liberian ruling class and the white American political elite, who keenly sought to establish such an enterprise to promote Liberia's modernization. "Du Bois's activities on behalf of 'Liberia,'" or more accurately on behalf of the ruling Americo-Liberians, took the forms of "diplomacy, propaganda-publicism and financial brokering."[66] Du Bois's sympathies lain very much with the master class in its responsibilities to govern its wayward natives. The proposed plantation workforce would be made up almost entirely of Indigenous labor pressed into service. The way forward would require "experimentation" by great men such as Harvey Firestone, who would be willing to invest in specific "reforms" within "partially developed countries" like

---

66 Cedric J. Robinson, "W.E.B. du Bois and Black Sovereignty," in *Imagining Home: Class, Culture, and Nationalism in the African Diaspora,* eds. Sidney Lemelle and Robin. D.G. Kelley (London: Verso, 1994), 149.

Liberia.[67] Firestone alone, to Du Bois's way of thinking, would be a position to colonially "invade it, reform it, and uplift it by incorporating the native born into the imported industry and thus make the industry part of the country."[68] The exact opposite effect was achieved, meaning that the country was made part of the American rubber industry and Firestone its chief feudal lord and modern master.

In a deal signed in 1926, the Firestone plantation came to operate as a state within a state, controlling a majority share of the country's economy. Liberia as an overall state would sign away its sovereignty in "a last-minute clause" which "dictated that the Liberian government take a $5 million dollar loan from Firestone, effectively putting it under control of the US government."[69] As a newly acquired American territory, Liberia naturally "would receive military protection against hostile neighbors" and offer itself as a strategic American "naval post off the coast of West Africa."[70] Through the terms of the deal, American capitalism and American neo-colonialism had effectively conspired with an exploitative African American ruling class to rapidly reconfigure Liberia as both a plantation economy and military base. Du Bois's collaboration with Firestone fatally stalled the project of African sovereignty in favor of expanding his project of African American colonialism in an attempt to rival that of its white European counterpart. In so doing, he allowed American industry to reap untold profit from the exploitation of Indigenous people and their native environments, which would lay the groundwork for forced labor, war, and disease to congregate in this fundamentally compromised biological, political, and social landscape that would continue for decades to come.

Throughout West Africa, Du Bois and Washington wanted to establish capital intensive agricultural projects that would rely

67 Ibid.

68 Ibid.

69 Garrett Febler, "Black Zionism, Reparations, and the 'Palestine Problem,'" *Black Perspectives*, August 28, 2016, https://www.aaihs.org/black-zionism-reparations-and-the-palestine-problem/.

70 Ibid.

upon the labor of its Indigenous populations as well as other foreign, imported laborers to produce Black American national wealth. In 1901, Washington's Tuskegee Institute sent an expedition to the German colony of Togo in West Africa to transform the region into a cotton economy of similar magnitude to that of the post-Reconstruction American South. Washington believed that under German rule the "Negro" in Africa could be taught to apply his "natural" ability there. He pronounced in the German agricultural trade journal *Tropenflanzer,* "[t]he Negro is a natural cotton grower," who, "if carefully lead, encouraged, supported, and protected," could become an important source of economic productivity.[71] What Washington was effectively doing through these remarks was transplanting his conceptualization of racial uplift into the German colonial context. This was done as a means through which to socially and economically discipline Africans in an ostensibly post-emancipation world through the installation of a superior, external authority. Whether it was a "mulatto" elite, or a white laborer was inconsequential. Rather, what was important was that the Black body was made to assume a position of subservience to a code of labor and morality thus far taught through the institution of slavery. Indigenous Africans who may yet have been unfamiliar with the value of hard labor would have been carefully managed in order to accede to the natural order of things.

Of course, that order was far from natural. It required that Africans be introduced "simultaneously to biological, social, and political mean[s] of control capable of retooling domestic slavery into colonial domination and keeping Blacks apart from those avenues of mobility that might result in autonomous being."[72] As Germany became a colonial power in the late nineteenth century in Africa, the training and discipline of the African body under American slavery served as the foundation of the German administrative model. The concern for the economic efficiency of Africans was much the same, as was a social

---

71 Zimmerman, "German Sociology and Empire," 182.
72 Ibid.

conservatism around issues of disciplining racialized populations. While there was no explicitly biological model of racism
evident in Du Bois's and Washington's key works, their thought
remained rooted in social and economic assumptions that German sociology would continue to promote through the colonial
and postcolonial eras within Africa, justifying global inequalities and domination along the lines of racial exceptionalism.

There was always a color line to adhere to when it came to
making racial and agricultural assumptions about slavery in
the American South in the nineteenth century. As a new century dawned, those assumptions were made to conform to new
standards of European colonization in order to create the conditions for a racialized global South to emerge. Washington
"praised what he identified as a tendency of Negros, not merely
to imitate, but to imitate the best in white people, while reject-
ing poorer models for imitation, such as the Chinese."[73] Washington, like Du Bois, and the ruling "mulattoes" in Liberia were
in an exceptional position to influence various castes of Black
laborers to either apply themselves to manual labor or aspire
to education as a means of patterning their existence. In the
words of Washington's institutional partner James N. Calloway,
all this was recommended so that they could ultimately "seek to
imitate their masters and become English gentlemen."[74] These
remarks implied that educated Africans were superficial imita-
tors, performing a version of imitation inferior in depth to that
of the African American. This principle implies that somewhere
within the African American community was presumably a
"Negro" who was capable of matching likeness with the political and economic interests of white European elites, such that
he could pass inspection as one of their own. Du Bois was that
man. The great sociologist Weber endorsed just such a belief
when he averred that "Du Bois exemplified not African Ameri-

---

73 George Zimmerman, *Alabama in Africa: Booker T. Washington, the Ger-
    man Empire, and the Globalization of the New South* (Princeton: Princeton
    University Press, 2010), 138.

74 Ibid.

cans […] but rather one of the half-Negros, quarter-Negros, and one-hundred part Negros whom no non-American can distinguish from white."[75]

While studying beside Weber in Berlin, Du Bois "adopted the gloves, cane, silk ties, and hats of the elites, which led Germans to respond to him very differently than [presumably other Black] Americans."[76] Based on his impeccable comportment, Du Bois was identified by Germans as a Black American of exceptional quality. For his part, Du Bois returned the favor. Du Bois remarked that it was only in Germany that "I began to realize white people were human."[77] During his time in Berlin, Du Bois consciously traded on his appearance of affluence. The only racist incident that he recalls while inhabiting the city occurred when he was overcharged for a cab ride to the university. Du Bois, even in that incident could not distinguish his treatment between the categories of race and class. Therefore, recalling the incident helped Du Bois introduce the problem of Blackness as one of differentiation. As such, it could be empirically resolved by the apparent erasure of economic difference.

On a similar level, Du Bois could find no affinity in Germany between the racial discrimination that American Blacks and that experienced by German Jews. Du Bois concluded that due to advanced education in Jewish communities, political activity, business ownership, and per capita affluence as well as their legal ability to intermarry with Germans, Jews simply did not qualify as subject to categorical racial discrimination. While hostility to Jews was prevalent everywhere within imperial Germany, it was for that very reason — their apparent class ascendency and assimilation — that Du Bois could remain willfully blind to it and its implication within a greater burgeoning project of German nationalism. It was this nationalism that was of particular interest to Du Bois. There was a strong aversion to assimilation

75 Zimmerman, "German Sociology and Empire," 186.
76 Kenneth Barkin, "W.E.B. Du Bois' Love Affair with Imperial Germany," *German Studies Review* 28, no. 2, (2005): 286.
77 Ibid.

within Germany of its others, and therefore, despite Du Bois counting it amongst the advantages Jews enjoyed in Germany at this time, their ability and willingness to do so was at the very heart of arguments requiring their permanent exclusion and eventual annihilation. This was the case insofar as Jews posed a visceral threat to the purity of the German Reich as an imperial domain of singular ethnicity. What Du Bois failed to recognize within his seemingly positive reception into German society was that it was premised on the fundamental perception of him as a foreigner granted temporary status within the nation-state. It was that category that formed the basis of his perception within Germany as a subject of both academic and economic exceptionalism. Du Bois was in essence perceived as much as a cultural curiosity as he was a racial one and perhaps more so, given German understanding of the role ethnicity played in defining an individual: Du Bois was not asked about his race but rather his "family."[78]

Du Bois connected with the elites of other countries in Europe during his academic sojourn in Berlin. This provided an opportunity for him to glimpse the differences between these powers when it came to codes of racial hierarchy and discrimination. Through friendship with Galician nobleman on his course, he was able to visit Galicia and witness first-hand how "farm laborers there were more impoverished than southern blacks in the US," shedding light on the conditions of neighboring Poles brutalized as a consequence of Austro-German racism against this population.[79] Du Bois had no more direct relation to the plight of these people than he did those farm laborers in the United States, and therefore it was perhaps easier for him to view them from a similarly detached perspective. Du Bois's failure to recognize a similar pattern of coercion against Jews in Berlin and Africans in southwest Africa during his first year of formal studies may well have been influenced by his overwhelming admiration for German ethnic pride. This admiration

---

78 Ibid., 288.
79 Ibid., 289.

allowed him to remain blind to how the newly formed nation's desire for expansion was already being coupled with a logic of physical removal of peoples deemed ethnic others from its living space. His apparent ability to be respected and even accepted by his German academic peers and superiors encouraged him to view their enthusiasm for colonial expansionist policy as necessary for the survival of the national economy.

Du Bois was made to understand that such principles of expansion were unquestioningly "in the public interest of the German Reich" as well as being sociologically sound.[80] These same individuals, when confronted with the brutality of the colonies, responded with either an "outright denial of any guilt on the grounds that such measures were either necessary or deviations of law that were punished; the strongest justification relied on the similar or even worse treatment of the natives by other European colonial Powers."[81] A young Du Bois arriving in Germany in 1892 would have encountered a country some seven years into its colonial project in southwest Africa. A "policy of total control over the natives was at the very heart of German colonial rule."[82] Germany's main preoccupation from the start concerned the prevention of racial mixing in its African colonies that would adulterate the superior quality of Aryan bloodlines. The very same German civilization that Du Bois so ardently admired was the same one that founded its authority on the preservation of its ethnic whiteness. From 1894 onwards, Germany pursued the physical separation of white people and native Africans in every conceivable sense as it formally established its settlement land holdings in southwest Africa. That philosophy of racial segregation eventually required that the native populations be forcibly restrained, removed, or murdered in order to clear the way for near-total German occupation of their former

80 Mohammad Shahabuddin, "The Colonial 'Other' in the Nineteenth Century German Colonisation of Africa, and International Law," *African Yearbook of International Law Online / Annuaire Africain de droit international online* 18, no. 1 (2010): 35.
81 Ibid.
82 Ibid., 31.

territories. Du Bois could be accepted, conceptually, in German academia by "distancing" his appearance from both 'Negroes" and native Africans and by limiting his intellectual contribution to sociology solely focused on the analysis of African Americans. Du Bois used the scientific methods he learned in Germany to promote the notion of a "talented tenth" to represent the progress of his race. That tenth was presumably at least, in part, blood-based which echoed very much in Weber's perception of him as an exceptional "Negro."

Birth of a Race Man

Like many Americans abroad Du Bois spent a great deal of effort to pass not racially but culturally and economically into German society. Kenneth Barkin observes that "Du Bois sought to avoid all contacts with other Americans in Europe."[83] Presumably, that also meant other African Americans. During that same period, he strove to assimilate to an appearance of European white normativity. From 1894 onward, Du Bois wore a Kaiser Wilhelm moustache. He eventually came to adopt in total the Victorian uniform of the German middle class: a woolen suit, vest, tie, gloves, cane, and moustache. These items never lost their allure for him. As though preserved in amber, Du Bois modeled his appearance on the German middle classes that he observed during his two years studying abroad in Germany for the remaining fifty-four years of his life. Du Bois's penchant for drawing a likeness between himself and the prevailing white European aristocracy of his youth did not end with it. Du Bois persisted in wearing this uniform throughout the whole of the Nazi era, a time when few African Americans sought to emulate German culture. His enduring sartorial affect suggests a man out of time and, indeed, unable to cope with the moral disparities of a changing world.

Du Bois was born just a quarter-century after the abolition of slavery in the United States. He would have come of age as

83 Barkin, "W.E.B. Du Bois' Love Affair with Imperial Germany," 294.

the first generation of young adults to make their way into the larger world following on from that wounding legacy. Du Bois rehabilitated himself beyond that formative injury by adopting the appearance not solely of a Berlin university student, but also perhaps more significantly of a vaguely aristocratic (white) Berliner. This was an instance of double consciousness, but not in the usual way it is portrayed as Du Bois's intellectual legacy, which identified with both a Black and white lens of perception simultaneously. Rather, it seems plausible that for Du Bois in his person, with his relatively fair-skinned complexion, had convinced himself that he could pass as a member of the German nobility, which comprised the top tenth of Wilhelmine society, hence his later notion of an African American "talented tenth." Those with advanced university education occupied an even higher status in society at this time and this combined with Du Bois's "patrician" German accent, would have furthered a mystification of his racial origins amongst at least the local proletarian population.[84]

What is perhaps most remarkable is that the basic features of this transformation were acquired so rapidly. After just a few months of a homestay in the small town of Eisenach with a German family, Du Bois was able to master a German accent. His three semesters studying at the Friedrich Wilhelm University of Berlin following that had allowed him to classically perfect it. Du Bois focused his studies during this period on a comparative analysis of agricultural smallholdings in the United States versus Germany. He was working towards a PhD with hopes of completing it there, and thus he was gaining a German qualification as opposed to an American one; perhaps towards settling in Germany permanently. The circumstances that lead to du Bois's "departure without achievement of that goal, as well as the loss of thesis itself, lend mystery to that formative two-year period of Du Bois's life."[85] In the end, it came down to a combi-

---

84 Ibid., 297.
85 Sieglinde Lemke, "Berlin and Boundaries: *sollen* versus *geschehen*," *boundary 2* 27, no. 3 (2000): 46.

nation of nationalism and bureaucracy on the part of both his German and American universities and the philanthropic institution funding him that denied him this opportunity. "The Slater Fund officials denied his request" for more time in Germany and recommended that Du Bois "return to Harvard and offer [himself] as a [PhD] candidate there [and] devote [his talent] to the good of the colored race."[86] This suggestion infuriated Du Bois, who would later revise the telling of this scenario. He would go on to draw profitably on accounts of both his personal and academic relationships to Berlin, allowing him to ascend to the pantheon of Black intellectual thought in the first half of the twentieth century.

Du Bois often credited Germany with conferring upon him his humanity, as opposed to race, and "this was primarily the result not so much of my study, as of my human companionship, unveiled by the accident of color."[87] This idea that his race was an accident, a misfortune visited upon him rather than something innate to his being, is something further borne out by his assertion that his relocation to Germany provided him with an existential do-over as it were. Du Bois was utterly convinced that Germany was where he could comport himself widely within the space of an environment free of racial precarity. Sieglinde Lemke remarks with incredulity that Du Bois could frequent turn-of-the-century Berlin "at a time when 'imported' Africans were exhibited at so-called *Menschenschauen* at the Berlin Zoo" and still maintain the opinion of it as a society free of categorical racism.[88] At another level of spectatorship, of course, it was conceivable to do so. Sat amongst exclusively white German academics, it is entirely possible that Du Bois could look upon this same institutional scene and feel reified in his humanity by present company. He was indeed not an "animal" at all, because, by their recognition of him as a human, such an equivalence was

---

86 Ibid.

87 W.E.B. Du Bois, "My Evolving Program for Negro Freedom," in *What the Negro Wants,* ed. Rayford W. Logan (Chapel Hill: The University of North Carolina Press, 1944), 35.

88 Ibid., 50.

made impossible. He was no African, but he was certainly an American. It was this nationalistic virtue that set him apart and made his appearance viable in Germany. The title of his thesis, "The Large and Small-Scale System of Agriculture in the Southern United States," does not explicitly propose race as its central tenant, and yet his funding for his sojourn to Germany was wholly dependent on it. He asserted in his funding application he was there to "scientifically study the Negro question past and present with a view of its best solution [...] and to see how far Negro students are capable of further independent study [and] research in the best scientific work of the day."[89] His American institution was led to believe that this work was explicitly scientific, while his German one was led to believe that his study was explicitly economic. The situation of keeping up these dual appearances concerning his research proved to be ultimately unsustainable.

When Du Bois returned to the United States somewhat predictably, he was forced to continue to limit his work to address the "Negro question" and make a career based on the science of doing so. Du Bois advocated vociferously for historical study, statistical investigation, anthropological measurement, and sociological interpretation to reveal the empirical truth of his race and solutions-based thinking on how to manipulate its progress towards optimization. He required massive data sets to move towards this goal and the institutional recognition to proceed. In 1897, Du Bois became a professor of economics and history at Atlanta University, a private Black university in Atlanta, Georgia. Shortly thereafter, he began his field research on the Black population of Philadelphia's Seventh Ward, which included interviewing roughly 2,500 households over the course of 835 hours."[90] If there were any doubt that Du Bois would be concretized as an exclusively Black sociologist by his professional peers, Weber's commissioning of him for an essay on the economic and political situation of Black Americans, entitled "Die

89 Ibid., 52.
90 Ibid., 56.

Negerfrage in den Vereinigten Staaten" ("The Negro Question in the United States"), inserted the question of race firmly at the equation of his further academic recognition. Like his brethren, he would be forced to demonstrate that he was a respectable and civilized scholar capable of complying with the prevailing thought of his day and knowing his place — in Germany. DuBois's concept of double consciousness comes through here as conscious complicity with an ethnic multiplicity inscribed into the experience of African diasporic people. Therefore, the "Negro problem" is always already an issue having to do with a plurality misconstrued as a singularity, wherein the construction of Blackness has been forced to reveal itself over a distributed network of privation. In this sense, its appearance works to both deflect and buttress structures of power including racism, nationalism, and imperialism. Du Bois's genius therein relies upon an understanding of how Black selves are differently situated both within and outside of this network, through an acceptance that it is always answerable to a logic that presumes it can codify Blackness: past, present, and future.

Black liberation from Du Bois onwards remains about capturing Black subjectivity through cultural encoding and redeploying it through the upward recoding of the color line. Nonetheless, it remains a privileged subjectivity that is actualized through violence and the decimation of the Black population, the other side of the scale from the talented tenth, linking its progress not to the promise of bright new futures but to biological abominations, genocidal campaigns, and environmental catastrophe. Blackness within this bottom range remains a product identified as a "social problem" and sociology once again, offers that product its remedy. Du Bois's early work must be understood as taking shape during a period where political and civil rights for Black people had for all intents and purposes bottomed out in the post-Reconstruction era with the advent of Jim Crow legislation, prompting a period of intense African American struggle to counteract a substantially worsening atmosphere of racial hatred.

During his years in Berlin, Du Bois christens himself a "race man." What is significant about this definition is not the predicate of race but rather of manhood. For Du Bois to ascend to this category, others have to descend, namely an uncivilized class of Black people who refuse to toe the line of respectability and therefore, are unworthy of partaking in the spoils of Black ethnic citizenship. Foremost, such recognition is achieved through promotion into the category of human being. This rests firstly with the privileged Black bourgeoisie, and secondarily with the white middle class. It was the acquisition of class that afforded human recognition. Du Bois was convinced that Black life was organized according to a closed system. In opening that racial environment to include white elites, it would become possible to mitigate the climate in which racial prejudice flourished. By establishing an affinity between white elites and their Black counterparts it would be possible to then unite against the presumably shared moral failings of their respective lower classes. He lamented that "the best of the whites and the best of the Negroes almost never live in anything like close proximity."[91] Du Bois was convinced that if those "who by nature and training are the aristocracy and leader of the blacks" were able to mingle with "the best element of whites [...] a natural affinity in terms of leadership would develop amongst them."[92] Du Bois's concern here was for the sociological and ecological, as reality's determinative elements.

Du Bois based these assumptions on his experience of being a Black, foreign, postgraduate university student ostensibly accepted by German elites during his short stay there. What he failed to recognize was that Germany's policies of both ethnic nationalism and colonial segregation were fundamentally at odds with his model of elite cooperation amongst races. Du Bois's concept of the "talented tenth" of the "Negro" race directly intersects with a French and English logic of colonialism.

---

91 W.E.B. Du Bois, *The Souls of Black Folk* (Mineola: Dover Publications, 2012), 101.
92 Ibid.

Both nations believed that their colonial interests could best be served if local ethnic elites were recognized in their status. Du Bois concurred with their position, that these figures ought to be "made leaders of thought and missionaries of culture amongst their people" to encourage the ascent of their populations to be governed externally by these superior world powers.[93] Colonization was to be undertaken based on the governing principles of liberal universalism, which promoted individualism, egalitarianism, and the belief that the world can be made better by human effort. The catch in all this is that you had to be considered fully human to benefit from the full extent of these principles. German colonialism radically departed from this model of imperial governance, choosing instead to promote ethnic nationalism above these principles of liberal universalism. Hence, they positioned the preservation of their authentic culture as the central feature of their plans for colonial expansion.

These differences are reflected in their respective national attitudes towards assimilation versus containment of ethnic outsiders dwelling within the enlarged confines of what constituted their nation-states. All were concerned about limiting the influence of these minority ethnic cultures for fear that they would weaken and adulterate the force of the prevailing majority ethnic culture. Those cultures that stood at the forefront of colonization were portrayed as racially superior and thus innately better at organizing societies than the Indigenous members of the communities they would eventually come to subjugate. It was simply a matter of evolution that some races came to dominate others. Germany's imperial stance was unique, insofar as it was determined to exclude the subjugate race rather than incorporate it into the national body, for fear that direct contact with it would weaken their civilization.

In the years that Du Bois inhabited Berlin, Germany advanced a policy of total control over native Africans, which fell in line with its efforts to create an ethnically homogeneous community-dwelling within its colonial territories. This dogma,

93   Ibid., 104.

Jurgen Zimmerer notes, "was at the very heart of German colonial rule from the very first consideration of how working relationships should be regulated as early as 1894."[94] Germany was a state fundamentally rooted in race rather than in any concept of universal civilization. Therefore, it could not accommodate the appearance of other races in its midst. As a consequence, it required science and sociology to systematically validate and educate the public on the material distinctions that existed amongst different cultures and races to justify the necessity of their exclusion from recognition as fully human. It was widely accepted that these other peoples were like wild animals in that they remained fully continuous with their environments. In this way, they could no more be assigned rights nor be permitted to incorporate themselves within a greater body of civilization.

Evolution as Revolution

In many ways, Du Bois's project reflects this same philosophy of hierarchically ordering life and promoting racial segregation in the way human value is portrayed. Indeed, his work to enact racial uplift is explicitly focused on retooling white America's program of breeding Black bodies for raw labor by white elites into Black America's program for the breeding of Black bodies for intelligence, utility, and beauty. Those already born would have to be compelled to acquire these traits through strict training. Daylanne K. English asserts that many academic observers of the post-Reconstruction period of racial uplift have failed to adequately account for the science of eugenics ideological "power and significance (indeed, its singular appeal for modern male African-American intellectuals)."[95] These so-called "race men" of their time appealed to both evolutionary theory as well as genetic theory, to make their case for the necessity of uplift

---

94 Shahabuddin, "The Colonial 'Other' in the Nineteenth Century German Colonisation of Africa, and International Law," 31.

95 Daylanne K. English, *Unnatural Selections: Eugenics in American Modernism and the Harlem Renaissance* (Chapel Hill: University of North Carolina Press, 2005), 38.

becoming an essentially biological prospect. More troubling still is that interracial African American men were cast within this project as the vanguard of the race and, thus, the most desirable class of breeder for the next generation of Black greatness.

This pattern of breeding dates to the beginnings of British colonization in the seventeenth century which coincided with the advent of slavery in America. Those able to partake in the social rituals of the talented tenth were themselves products of "policies of racial isolation" that promoted "racial endogamy" amongst "slaves with light skin."[96] These policies continued to be perpetuated in the post-emancipation era by the descendants of those who had formerly been afforded privileges within the preexisting plantation system and who wanted to preserve their elevated cultural and economic status based on their skin color. This created a social as well as biological order within the African American community. That effort would be led by men like Du Bois, who were eager to adopt an appearance symbolic of normative, late-Victorian, white aristocratic masculinity and thus garner authority through conformity to the intellectual, political, and affective values of that same class of humanity.

Du Bois's own observations concluded that "the mulattoes we see on the streets are invariably descendants of one, two, or three generations of mulattoes, [in whom] the infusion of white blood comes from the seventeenth century, [since in New York] in only 3 percent of weddings of people was color was one of the parties 'white.'"[97] Moreover, "the mulatto population represented just over 2 percent of the black community, whereas just under 98 percent of that community was identified as pure 'Negros.'"[98] "Mulattoes," in their rarity, cast themselves as a type of exceptional community whose desirable physical appearance was positively contrasted against that of the "primitive Negro."

---

96 Giovana Xavier da Conceição Nascimento, "Os perigos dos Negros Brancos: cultura mulata, classe e beleza eugênica no pós-emancipação (EUA, 1900–1920)," *Revista Brasileira de História* 35, no. 69 (2015): 6.

97 W.E.B. DuBois, *On Sociology and the Black Community,* eds. Edwin D. Drive and Dan. S. Green (Chicago: University of Chicago Press, 2013), 151.

98 Ibid.

The "Negro" in their appearance was portrayed as unable to evolve past their African identities to take their place within an order of modernity. "Mulattoes" stood apart in their lighter appearance, which constituted physical evidence of their ability to migrate towards biological standards associated with the New World. That migration would naturally be one that gravitated towards whiteness. This color line was tied to an understanding of American citizenship as a privilege only proffered to those who could comprehend and manipulate the capitalist universe of liberty, urbanity, and industry. Those "Negroes" who fundamentally could not progress within this universe were cast by Du Bois as "the submerged tenth, a group made 'of criminals, prostitutes and loafers' for whom there was little developmental potential."[99]

In his study of *The Health and Physique of the Negro American* (1906), Du Bois contrasted this assessment with a quote from the naturalist and geologist James Bryce, who argued, "all the great peoples of the world are a result of the mixture of races."[100] The greater context of this assertion can be found in Bryce's 1888 volume *American Commonwealth,* which at the time of its publication was considered a popular successor to Alexis de Tocqueville's 1835 volume *Democracy in America.* Bryce, a member of the English aristocracy, had been a graduate student at Oxford University and, as well as an international student at the University of Heidelberg. He spent several semesters at both institutions between the years of 1863 and 1870. Upon his return, he maintained an enduring admiration of German culture, as well as its historical and legal scholarship. Through his academic training, Bryce became a life-long believer in the concept of "Teutonic freedom." This concept held that Germany, Britain, and the United States were bound together through an understanding of first established amongst ancient Teutonic peoples that freedom was a matter of both blood and

99 English, *Unnatural Selections,* 45.
100 James Bryce, *The Relations of the Advanced and the Backward Races of Mankind* (Oxford: Clarendon Press, 1902), 15–16.

birth. These peoples, in modern times, were now identified as Germans, Danes, Swedes, Norwegians, Dutch, Swiss, English or British, Anglo-Irish, or Anglo-American. For a man born of these nations, freedom "was of himself a part, as much as his blood and his life."[101]

The United States, the British Empire, and Germany shared an affinity through the "natural" expression of their freedom, and for Bryce that implied a racial affinity. In *American Commonwealth,* "Bryce directly engaged with the question of racial origin in immigrant populations" and argued for the "quality of the earlier immigrants, Irish and Teutonic, finding them of such desirability and their qualities are so well known that they required no further detail. Bryce also found that the Scandinavians were 'intelligent peasants, of strong stocks, industrious, energetic' but more importantly 'capable of quickly accommodating themselves to the conditions of their new land and blending with its people' — the latter quality being the element he found lacking in other immigrants, to include the Slavs and Italians."[102]

What Bryce's study essentially establishes is a hierarchy of desirable racial origins ranked according to a descending order of white racialized characteristic. Bryce's analysis was referred to "frequently in contemporary public discourse and private correspondence alike" throughout the Atlantic world in the late Victorian era.[103] Following on from his extended visits to the United States in the 1880s, Bryce also made arguments "for the unfitness of non-whites for self-government," concluding that Black Americans' "emancipation found them utterly ignorant […] and the grant of suffrage found them as unfit for political rights as

101 John Clark Ridpath, *Ridpath's History of the World: Being an Account of the Ethnic Origin, Primitive Estate, Early Migrations, Social Conditions, and Present Promise of the Principal Families of Men, Vol. 2: The West Aryans.* (Boonsboro: Jones Bros. Publishing Company, 1893), 260.

102 Arjun Narayan Sharath, "The Anglo-American Relationship: Race and the Perception of German Threat 1890–1910" (PhD diss., University of Cambridge), 171.

103 Ibid.

any population could be."[104] From these remarks, Du Bois concludes that the constitution of "whiteness" acts as a conduit to both the enterprise of human identity and the development of a transnational political community. Ultimately, that concept of race brought with it a "grade of civilization" that presently Black Americans on the whole had failed. Thereafter, Du Bois endeavored to find ways to intellectually promote specific forms of social congress to improve upon the base quality of Blackness.

Du Bois made several attempts to materially incorporate Bryce into his sociological project of racial uplift. In a letter dated January 22, 1909, Du Bois implored Bryce to speak at the "Annual Negro conference" hosted by his sociology department at Atlanta University. It is apparent from the text of the letter that this is not the first of Du Bois's requests for Bryce to make his presence felt on campus. The subject of this particular letter is a possible appearance by Bryce at this year's conference on "Efforts for Social Betterment amongst Negro Americans." Du Bois writes, "I cannot tell you how greatly we desire the help of your presence," to convince white, "millionaire philanthropists" that money spent on the "higher training of black men is not merely wasted and wrong," in the effort to provide "uplift [...] for the lowest of social groups."[105] Du Bois's closed the letter with the following salutation: "I beg, to remain, Sir, Very Respectfully."[106] There is no signature to follow, but these words nevertheless speak volumes about the subservient position that Du Bois places himself in as a scholar with respect to Bryce's greater authority.

This assertion would result in a peculiar marriage of ideas between what Du Bois casts as the "African American 'feminine man'" who when wed to "the more masculine 'Teutonic' man

---

104 James Bryce, *The Party System and Public Opinion in The American Commonwealth* (New York: Macmillan, 1888), 3:92.
105 W.E.B. Du Bois, "Letter from W.E.B. Du Bois to James Bryce, June 22, 1909," W.E.B. Du Bois Papers (MS 312). Special Collections and University Archives, University of Massachusetts Amherst Libraries, http://credo.library.umass.edu/view/full/mums312-b001-i338.
106 Ibid.

would produce a common human/American civilization by a racial division of labor."[107] There is something distinctly homosexual in Du Bois's call for German dominant men and Black submissive men to conjoin in the cause of a social reproduction that explicitly promotes biracial intercourse as a means of producing a superior form of human being. These ideas go back to Du Bois's commencement speech from Harvard in 1890, two years before he would arrive in Germany. Du Bois spoke positively throughout his speech about the potential of a "Teutonic Strong man" who brings "his raw material to this process," the mentality of a soldier and a lover, a personality at once "fiery and impetuous," "cool and ambitious," "brave and generous" to bear upon an "effete Negro" man who is "innately" and naturally possessive of "a sensuous nature" that appreciates aesthetic beauty and is temperamentally subdued.[108] If this reads like a personal advertisement it is because it so profoundly situated in Du Bois's intimate comportment to make of himself an object of affective contemplation before the eyes of his ethnic German male superiors. Du Bois's distinction comes as a product of his identifying as "mulatto" rather than as "Negro," per se. The repeated dramatization of his biracial appearance throughout his privileged public career further insinuates that he is the consummate product of this synthetic affair he seeks to promote. Du Bois adopted certain "Prussian social customs" into his bearing, including "a clipped manner of speech" that "was often misunderstood as reserve, distance, even haughtiness," which "was to characterize Du Bois for the rest of his life."[109]

In his youth, Du Bois had sexual dalliances with white women of German origin, but here too he would ultimately elect to the follow the cultural values of nineteenth century Teutonic culture, which emphasized the moral value of "sacrifice and

107 Monica L. Miller, "W.E.B. Du Bois and the Dandy as Diasporic Race Man," *Callaloo* 26, no. 3 (2003): 740.
108 Ibid.
109 Hamilton Beck, "W.E.B. Du Bois as a Study Abroad Student in Germany, 1892–1894," *Frontiers* 2, no. 1 (1996): 56.

renunciation."[110] Ultimately by marrying instead another educated "mulatto", Nina Gomer, his former student, he was sac-
rificing his happiness for the sake of a greater mission, which
was to be publicly recognized as a role model in the "improvement" of the Black race in America. Beyond that, Nina served
no philosophical purpose to Du Bois; the union of fifty-three
years was seldom marked by any mention of her in his writings either implicitly or explicitly. She bore him two children,
one of whom lived to adulthood. When she came of age, their
surviving daughter Yolande was called upon to serve the cause
through marriage to Du Bois's protégé, the "mulatto" poet,
Countee Cullen.

Their April 1928 wedding "with sixteen bridesmaids and thir-
teen hundred invited guests, represents for Du Bois, the foundational moment of a eugenic dynasty for 'black folk.'"[111] This
union was doomed from the start when on their wedding night
Cullen failed to perform his conjugal duties. A few months later
Cullen would abandon the marriage altogether, heading to Paris
with his "mulatto" lover, Harold Jackman, who perhaps not incidentally had acted as his best man at the wedding. They left
together under the socially acceptable guise of embarking on
a year of study abroad. Du Bois knew of Cullen's sexual pro-
clivities but saw no conflict in them with his ability to produce
a viable heir to the Du Bois dynasty. He considered Cullen, like
himself, "a genius" whose progeny "would be brilliant and well
formed" their genes guaranteeing their confident destiny.[112] Du
Bois's biopolitics required that. It was not he, nor Countee, but
Yolande who failed to invest her part in furthering this reproductive economy. It was her pathetic desire to be loved that compromised all the "genetic, gendered, classicist and racial fantasies" that fueled Du Bois's conception of how to engineer what

110 Ibid., 59.
111 English, *Unnatural Selections*, 55.
112 David Levering Lewis, *W.E.B. Du Bois: The Fight for Equality and the
American Century, 1919–1963* (New York: Henry Holt and Company,
2001), 223.

was called a "superior negritude."[113] The contribution of heterosexual women was at best ancillary to the fulfilment of Du Bois's
greater civilizational project. In the early decades of the twentieth century, an African American homosexual subculture was
beginning to flourish. Many of these gay cultural acolytes would
eventually join the ranks of Du Bois's talented tenth. It is highly
probable than many of them identified with Du Bois's concept
of double consciousness not solely from a racial perspective but
equally a sexual one. This meant that so long as "mulatto" men
recognized their moral duty to sexually reproduce, any other
kind of discreet sexual intercourse was tacitly permitted.

Selective Cultivation

Selective sexual reproduction is something that stood at the
heart of Du Bois's understanding of racial evolution. Those who
were the product of field slavery, to Du Bois's mind, were of "poor
eugenic status," because it "nurtured the survival of those who
evolution would have naturally eliminated creating a biological underclass that could never be developed into fit citizens."[114]
Du Bois projected that in a post-emancipation context an overwhelming percentage "of them would die out through genetic
weaknesses."[115] The modern condition of race was hindered by
"the inability of some of its members to completely evolve" due
to the recent legacy of field slavery.[116] Du Bois saw "Negroes"
as a mass product who now must recognize the value of selective cultivation; "they must learn that amongst human races and
groups, as amongst vegetables, quality and not mere quantity
really counts."[117] What stands out in Du Bois's criticism of this

---

113 Ibid., 228
114 Shantella Y. Sherman, "In Search of Purity: Popular Eugenics and Racial
    Uplift Among New Negroes 1915–1935" (PhD diss., University of Nebraska,
    2014), 95.
115 Ibid., 205.
116 Ibid., 297.
117 W.E.B. Du Bois, "Black Folk and Birth Control," *Birth Control Review* 16,
    no. 6 (1932): 167.

group is his frustration with them, which was directed at them from a position of superior status. Most significant was the reality that he chose to identify *with* them, not *as* them.

What troubled Du Bois was his perception that the majority of Black reproduction was happening amongst those who were essentially genetically inferior. They remain for him a distant subject to be studied and classified. In practice, his relationship with the Black population was one that for the most part did not exist: "apart from his family, Du Bois had little or no sustained connection to a larger Black community."[118] He defined himself as a leader of the "Negro" community, but not a "Negro" himself. His prejudices were more prominently based on regionalism versus race itself. As a northerner and a free Black person, he came to bitterly resent the appearance of Black Southerners that were more directly the product of slavery than their northern freeman counterparts. The migration of Black Southerners to the north meant that these Black Southerners would be lumped together in the minds of white onlookers who would class them as equally uncivilized and ignorant subjects. The better class of Black Americans, including Du Bois, would be made to suffer if white Southerners could not perceive their difference. Once again, this is cast through the lens of being mixed race as a visual marker of obvious racial superiority who "naturally" made up "the aristocracy of the Negro."[119]

In Du Bois's *The Philadelphia Negro: A Social Study,* the sexual economy of the slave plantation figures into the biological making of this elite class who are largely "descended from the house servant class" in union with their plantation masters.[120] As a consequence, Du Bois asserts the subsequent generation "contains many mulattoes."[121] However, it is not just biology which gives these contemporary offspring their advantage in

---

118 Charles F. Peterson, *DuBois, Fanon, Cabral: The Margins of Elite Anti-colonial Leadership* (Lanham: Lexington Books, 2007), 36.

119 Ibid.

120 W.E.B Du Bois, *The Philadelphia Negro: A Social Study,* ed. Isabel Eaton (University Park: The Pennsylvania State University, 1899), 318.

121 Ibid.

life, "there is evidence of good breeding and taste, a foreigner would hardly think as ex-slaves."[122] Presumably, the foreigner is in a position to judge northern, European whites, who would have on some level been aware that lighter-skinned slaves would have been granted access to better conditions of nourishment, shelter, and labor than field slaves. By dwelling within the houses of their masters that were able to familiarize themselves with white European culture mores and adopt to a certain degree the manners and language of their racial superiors. Many would have benefited from intimate relationships with white neighbors from birth that would make them more sympathetic to their situation, thereby increasing their chances of manumission. Sexual relations, both consensual and non-consensual, were a tacit expectation of those viewed as property by their masters. By mid-nineteenth century, tactic acknowledgement of these unions became commonplace, lending their "mulatto" children a certain amount of social advantage. This arrangement was crucial to institutionalizing the color line throughout the South as a means of determining societal privilege. It was not only whites who adopted this mentality, but also many "mulattoes" themselves who admired Western civilization above the merits of their own and strove to attain recognition based on economic, social, and intellectual criteria that white culture had established. The limits of Black assimilation continued to abut against pigmentary bias.

Du Bois's case in promoting "mulatto" superiority relied heavily on eugenic assumptions concerning racial classification. These were deployed explicitly as he went about creating a vast archive of racial documentation over the course of his sociological studies of African Americans. These included the typing of physical features, the demonstration of figures of superior intellect, and the display of moral character amongst his Black subjects. All of this information functioned as evidence to prove that degrees of racial mixing had biologically, economically and culturally had improved the African American race.

122 Ibid.

Shawn Michelle Smith maintains that Du Bois was so convinced of this fact that he presented own body as evidence of what this "New Negro" body exemplified in the world. Photograph after photograph captures his elevated likeness. From an early childhood photograph "dressed in princely attire" to one of him in "scholarly robes" and finally a series of them in his "formal suits" spanning his academic career, altogether they suggest a biological determinism that enlivened this trajectory of greatness.[123] It is for this reason that Du Bois figures himself into his archive as a star in the ascendant around which other luminaries congregate.

In 1900, at the Exposition Universelle in Paris, he adored himself in "a long Prince Albert coat, a tall hat, as well as a cane and gloves."[124] It is significant to note that Du Bois chose to debut this new European, as opposed to specifically German, affect at the opening of his African Pavilion. His series of data portraits debuted there was entitled "Exhibit of American Negroes." Within it, he holds himself up literally as the measure against which these abstract others are judged. No one in these portraits is identified by name but solely by characteristic. These photographs would be recycled into other projects that supported his conceptualization of uplift modelled along interracial and eugenic lines of thought that would only intensify throughout the decades of the 1920s and 1930s in concert with the rise of a mixed-race intelligentsia largely beholden to Du Bois. Their project from the beginning aimed to go beyond the human in its surveillance of what is classed as progress.

In the case of the 1900 exhibition, Du Bois "headed a team of alumni and students from Atlanta University" to create "a collection of graphs, charts, maps, and tables that were generated from a mix of existing records and empirical data" that would assign meaning to the photographic portraits that featured as

123 Shawn Michelle Smith, *Photography on the Color Line: W.E.B. Du Bois, Race, and Visual Culture* (Durham. Duke University Press, 2004), 155.
124 Ibid., 294.

the foundation for the project.[125] While the project is often described as one that pioneered data visualization, what is most compelling about this constellation of information is how it deploys photography to depict a generation living on beyond the era of slavery in the American South. At the turn of the new century, these bodies remain proximate to both nature and demise. Even as a new infrastructure of commerce, education, neighborhood, and church builds up steadily around them, the pallor of slavery stills hangs heavily around these scenes shadowing the very concept of Black portraiture itself. The technology continues to segregate by only able to record in black and white, making the founding of their register something deeply tinged by both the legacy and landscape of the South. Even as African Americans continue to advance socially in the United States, it reveals some quality of forbearance and foreclosure of both a people and geography.

Out of Crisis into Progress

Early on in Du Bois's tenure as editor of the National Association for the Advancement of Colored People (NAACP) magazine Crisis, he introduced an annual edition featuring young children. These children were exhibited as the latest progenitors of "New Negritude." Though their appearance featured past the first decade of the twentieth century, these babies and small children were made to adopt formal Victorian dress and wear almost exclusively white clothing in their portraits. Added to the frame were "ornate chairs, lace curtains, plants, books, and statuettes" that would further reify appreciation of the "ideal Negro type" as the product of a more elaborate grouping discreetly supportive of its development.[126] The "Negro" is announced always as a

125 Witney Battle-Baptiste and Britt Rusert, eds., *W.E.B. Du Bois's Data Portraits: Visualizing Black America* (Hudson: Princeton Architectural Press, 2018), 9.
126 Julie Taylor, "Mechanical Reproduction: The Photograph and the Child in 'The Crisis' and the 'Brownies' Book," *Journal of American Studies* (2019): 739.

plurality of being, and in this sense, the white concept of verisi-militude is forcibly challenged through an excess of equivalency. The images Du Bois presents of African American children, "are sometimes printed in grid-like arrangements over several whole pages, and across the publications as a whole" arranged in such a manner as to convey the understanding of Blackness as both a product of imagination and affect, and therefore, superior to forms of literal racial identification.[127]

Rather than remain confined to the conventions of black and white when it came to the public portrayal of these children, Du Bois insisted upon the addition of "coloured tints, particularly on the front covers, and almost always for the sole purpose of representing skin colour."[128] Julie Taylor recounts the appearance an image of a child who "dances in a white ballerina's costume, complete with fairy wings" whose "skin colour" is embellished "by a rich brown tint" applied after the image was developed. Taylor argues that the technology available at the time "cannot capture the girl's skin colour, which is actually denaturalised by this photograph."[129] This statement implies that it is up to the African American male photographer to renaturalize her through artificial means, adding something to her "like the fairy's wings" which at once allows the reader to recognize her Otherness, at the same time capturing the reality that she already figures conspicuously as something that needs to be added or superim-posed onto a society that might not yet be wholly convinced she persists beyond the recently overturned stereotype of the nine-teenth-century "pickaninny,"[130] a racial slur for small children of African descent. The child, in this instance, must be made and remade again to determine that something aspirational remains within her make-up and that she can uplift herself through a conscious act of self-fashioning worthy of emulation.

---

127 Ibid.
128 Ibid., 748.
129 Ibid.
130 Ibid., 737–38.

The theme of the angelic Black child extends to the portrayal of children who have literally died and presumably have gone to heaven as credits to their race. These too become part of the grid organizing the value of Black life in terms of its basic quantity, the enormity of its potential situation, and the range of meaning that might randomly be applied to it. The choice of one life over another to a certain degree then becomes random, implying a degree of substitutability between the living and the dead as they are pictured and lent narration, complicating the image's referential status such that it bears awkwardly on a greater reality of intraracial eugenicism subtending much of the material as it is presented here. The expectation of a future utility of these children is staged through "studio scenes where they perform adult tasks, such as answering the telephone," rehearsing the material application of their nascent aptitude.[131] The "New Negro," therefore, conveys himself amidst an artificial environment composed of both potentiality and actuality, a being that is left to make a pretense of his becoming, for that fact of his very being as an entity poses a challenge to the concept of racial fidelity and stability.

DuBois's proof of that capacity must be taken twice removed, meaning that there is a difference of two generations, first from slavery, and second through the development of photography that allowed him to deploy evidence that racial integration was progressively becoming a feature of American life and that the categories of "Black" or "white" no longer served as reliable contrasting determinants. The velocity at which Du Bois wished this information to travel, however, was set in tension with his desire to freeze and fix representation within a set of coordinates adhering to a differential division of people, signaling the arrival of an innovation of racial taxonomy that was at once too excessive and too preemptive in its pragmatic aims. Du Bois's conceptualization of scientific breeding and deterministic ancestry cast "dark-skinned subjects as perpetually tied to their racial pasts, while people of mixed race possessed a certain level

---

131 Ibid., 761.

of autonomy and ability to distance themselves from their African forbearers."[132] The ambition was for Black Americans not to simply exist beyond their original environments but rather to seek breeding opportunities that would allow for the incorporation of certain "characteristics like reliability, earnestness, drollness, and sensitiveness," that would allow them to function as a developed social class elevating them into a milieu *of* propriety, rather than standing in direct relation to them as property.[133]

Slavery casts not only humans but the whole of the natural world into a framework of violence and dispossession, such that the bodies of representation Du Bois draws upon to account for its uplift become once more subsumed into the foundation of plantation life — once the only means of life support. Given that reality, what Du Bois instead banks on is the profit to be drawn in allowing them to circulate otherwise as evidence born out in the data that potentially something positive was achieved there. The numbers suggest that the "Negro race" itself has potential and in so doing, allows them to register value beyond the category of human being, which has historically alluded them, towards the category of socio-biological engineering, which promises that man can be categorically exceeded given the space to do so. If slavery had confined the "Negro" to the peculiar position of social death and infinite productivity, their emancipation would require of them an evolution from the inhuman beyond the category of mere personhood to that of a successor class of humanity that Du Bois asserts "has yet a message for the world."[134] This made of the body a work of complicated transmission, something to be seen and understood as propaganda in service to the greater program of racial pride and uplift. The body, in this, sense becomes a scene of intersubjectivity achieved through social networks, relations, and attachments that transcend late nineteenth century knowledge and politics in the trans-Atlantic

---

132 Sherman, "In Search of Purity," 49.
133 Ibid.
134 W.E.B Du Bois, "Strivings of the Negro People," *The Atlantic,* August 1987, https://www.theatlantic.com/magazine/archive/1897/08/strivings-of-the-negro-people/305446/

world that circumscribed what constituted human life and racial performance.

Here, Justine Wells argues, it is possible to conceive of a Du
Bois who "gestured toward a theory of race as a more broadly
material and ecological construction."[135] The problem that Du
Bois confronts is a Blackness "made up not only of human discourses but also of entire environments, [...] of the ruined cotton rows, one-room cabins, and 'big houses' that endure as part
of the Black race's historic enslavement and its continued debility" that allow Black identity to sense its own emergence through
the lens of both the social and the ecological as it carves out a
space of material being that exceeds what is formally human.[136]

Unlike its white counterpart, Blackness cannot rely upon dominion in its bearing upon nonhuman resources, but it must
rather seek a relationship of cooperation between itself and such
entities if it is to achieve any sort of lasting articulation of itself.
This is what furnishes Blackness with its human-nonhuman
identity and its ambiguous potential. Whiteness, by contrast,
becomes the more rigid form of cultivation, hindered by its preoccupations with intensive operations to preserve itself through
the dispossession and dismantlement of the others it claims as its
property. Reconstitution, as a consequence, becomes something
at the heart of the richness of the Black experience. In the context of the posthuman, this offers "the potential for co-constitu-
tive thriving [to] be opened between peoples and environments
that had historically suffered oppression in tandem."[137] Recoded
and decoded by economic, material, and political realities, these
legacies might be once again played out for the benefit of their
full incorporation into the human record, allowing for a plurality of data to emerge from those who had systematically been
denied sovereignty, intelligence, and expression.

---

135 Justine Wells, "W.E.B. Du Bois and the Conservation of Races: A Piece of
    Ecological Ancestry," *Rhetoric Society Quarterly* 49, no. 4 (2019): 345.
136 Ibid., 357.
137 Ibid., 361.

Progress, in the end, will come down to sorting of a very particular nature — zeros and ones to be read on and detected not by humans but by a different sort of awareness. The promise is that bodies that have never been allowed to materialize will triumph in an era that can plot life itself differently and bring it into a dynamic configuration with "the crows, the winds, the rains and the trees" that surround and attend its progress.[138] An errant Blackness that has always existed outside human paradigms can find itself settled within this new economy of the diagram linking it to the power of a different form of existence as an index rather than a record. Imagine how that will play out with reference to the places where a plurality of Blackness occurs. Black representation becomes then a measure of something as it relates to the natural world. That something, in turn, becomes the stuff of vital augmentation versus flat debility.

The double consciousness that Du Bois said was the promise of his people, will find its day of deliverance through adopting not the white human gaze in addition to its nonhuman one, but the figural gaze, where the human and the nonhuman assemble for the desired end of novel typology — surplus to requirement. This is problem-solving in reverse, a fugitive answer to what was never asked about and was therefore never questioned. This corresponds with the business of forensic analysis that is everywhere in Du Bois's studies and how they become disruptive as analytical texts that in many ways resist a normative understanding of the human as something that can solely denature, as compared to its denatured others. As such Du Bois's most enduring legacy may be his perception of surveillance as characteristic of one becoming racially literate. Taking this as a starting point, it is possible to draw a direct ontological line from the extraction of labor from Black bodies to the appropriation of racialized bodies in and of themselves.

---

138 Erin E. Edwards, T*he Modernist Corpse: Posthumanism and the Posthumous* (Minneapolis: University of Minnesota Press, 2018), 93.

Reverse Engineering

Du Bois identifies the American South, not Africa, as the ancestral homeland of Black Americans. He needed to develop a
founding mythology for his people based on this environment,
which prompted him to join his sociological work with fiction writing from the late 1880s onwards. He writes as a person
temporarily situating himself within the Black South but, significantly, someone who is not resident to it. His earliest focus
is on the plantation system and its function as a political, social,
economic, and fictional apparatus brought to bear materially on
the lives of Black Others who are made continuous with cotton as the predominant regional commodity. The bearing of the
plantation is something that is accomplished through the continuous violence, coercion, and exploitation of the Black body
that supports and maintains it as a distinct economy of forced
agricultural cultivation and compulsory sexual reproduction.
Together, these formations produced their own version of both
modernity and capitalism that exceeds the formal system of
slavery and establishes a definition of culture that at every point
is subtended by the manipulation of the environment that surrounds it.

Du Bois appreciates the differences between northern and
Southern capitalism that lies in the intimate economies of cultivation itself, wherein Du Bois suggests that African American
lives *are* cotton, and accordingly, their progress is plotted to
conform to the logistics of the cotton industry, consistent with
their accumulative potential at every point in the lifecycle of its
use, and value to energize a greater capitalism indicative of the
global South. Du Bois's remedy is to literally uplift these lives
from the soil to which they were born and transplant them to
flourish within the contours of a northern metropolitan imperialism. Only this doesn't entirely fit with their story nor does
it acknowledge the complicity between the North and South
in authoring their uneven economic conditioning. Monopoly,
speculation, and accumulation all have their role to play in the
exploitation of African Americans, poor whites, and children,

as does the proprietorial surveillance forced upon these bodies by their structural owners. These owners would continue to innovate how labor could be enforced upon these bodies through the institutions of sharecropping and the chain gang which perpetuated the tyranny at the center of slavery's original plot to exteriorize their value. As Du Bois "bemoans at the turn of the century, America's growing prosperity rested on its failure to regard Black life as 'more than meat'"; meaning more than a potentially wayward commodity to be systematically overseen and managed.[139] This bears upon the formation of Black geographies characteristic of the plantation that persist into contemporary architectures of racial violence.

Katherine McKittrick argues that the African American experience is one indicative of "a spatial continuity between the living and the dead, between science and storytelling, and between past and present."[140] The control and administration of inhuman or barely habitable geographies have their beginnings in the plantation system, which is built to marginalize and degrade the liveliness of Black bodies. The violence enacted therein can only be understood through the concept of spatial manufacture, insofar as the plantation plots out a reality whereby "the actual growth of narratives, food, and cultural practices" functions in tandem and at once to socialize order within the context of a dehumanizing agricultural economy and to "materialize the deep connections between blackness and the earth and foster values that challenge systemic violence."[141] McKittrick asserts that when the formal plantation system is dismantled a new urbanization arises to take its place functionally, commercially, as well as racially. As a consequence, the urban center emerges as a space perpetually riven with a narrative of geographic superiority and inferiority. There is another layer at work here, a subter-

139 Meg Samuelson, "Thinking with Sharks: Racial Terror, Species Extinction, and Other Anthropocene Fault Lines," *Australian Humanities Review* 63 (2018): 39.

140 Katherine McKittrick, "Plantation Futures," *Small Axe* 17, no. 3 (November 2013): 2.

141 Ibid., 10.

ranean one in which an underclass of life continues to toil on the outskirts of a more prosperous society. Here poverty has sunken in perhaps most deeply, existing prosaically in the grounding of a faith in mankind that subtends the world.

This narrative persists online through the introduction of new algorithmic "codes that regulate, profit from, and conceptualize spaces of absolute otherness."[142] Within this new virtual economy, marginalized communities continue to be the most heavily exploited, cultivated, and surveyed. It is, therefore, no coincidence that cities that the internet seeks to smartly regulate are disproportionately inhabited by barely surviving individuals of color, nor is it a coincidence that a new set of white American male masters have located novel means through which to carve out profit by manipulating their behavioral futures. Within this configuration, the plot concerns the violation of the geography of social reproduction that occurs when the Black body is coded as non-individual and as fundamentally entangled with the chain of productivity and the monoculture of production and singular output. Therefore, it might be possible to conceive that is not only their bodies that white capitalism hungers for but rather the brains of the operation that engendered creative response to their exploitation that in turn have had that value taken away from them by way of extracting relationality itself. In this scenario, Blackness remains relegated to the substructure of the world. Escape only is possible through an inversion of capitalist geography, namely, by refusing to contribute to its expansion and operating instead at the sub-individual level, where it becomes possible to refuse capitulation to certain forms of entanglement that we now have to associate with the worldwide web. This web is a pseudo-natural formation that, when in the bodies of its consumers' brains, forms a sticky buildup of data that correlates with their progressive cognitive decline, indicating a link between the preponderance of artificial intelligence and the destruction of organic thinking. Information becomes one in several colonial commodities operational on a supra-

---

142 Ibid., 15.

individual level to further the plantation complex and remain in the business of entanglement. Another pole remains, however: the urban. Here it is possible to still evade the normative and analytical models of the plantation economy and proliferate networks both human and nonhuman to still reimagine ways of planting, cultivating, and harvesting that offer much greater flexibility in terms of compromising stability and disrupting reproductive mechanisms of social and biological control.

2

Aspirational Theft

Technocapitalism, Cognitive Augmentation Devices, and Algorithmic Coloniality

Silicon Valley elites have achieved a paralysis of the collective imagination such that we now accede to them sole authority in predicating our future. These technocapitalists constantly churn anxieties about the future within the public imagination, exhausting its capacity to generate any alternative discourse that might deviate from their proscribed remediation tactics. These essentially confer with their model of wealth creation through rapacious modalities of resource extraction. Their forays into ideological austerity and social disengagement (of late, termed "social distancing") function as beta testing for a project of far greater portent: the leveling of humanity to equivalence and need, structuring it to conform with standards of operation that are distinctly nonhuman, making of us a "smart" species.

What we used to refer to as human ingenuity — the stuff that formerly rendered, interpreted and actuate the human experience — is now being systematically eliminated from the equation of need in favor of allegiance with a ubiquitous digital apparatus that predict and constrain our actions and activities. A series of proposed commercial devices now purport to cognitively augment humans, encouraging us to cede ground to a new type of authority which is digital in kind and authoritarian in its ambition to instigate control over our knowledge production. As such, these technologies represent powerful tools through which to expand colonies of the human imagination; one made operational through algorithmic channels that offer the tantalizing prospect of a seamless interface between the human and machine, all while they subtly asset strip our world and our being.

By changing the way people communicate with one another, these technologies enable a discreet gateway for digital information to enter what is most native to the human being. This is an economic strategy ultimately bent on silencing the user insofar as it relies upon the wearer in what amounts to a fundamentally altered social environment. This same logic eventuates the destruction of sociality itself, a category evacuated in favor of effortless and private human-to-machine communication primed to capture and mine imagination by profoundly exploiting connection and translating it into machine language in order to enable multi-sensory conversion of the human to near-total reliance on a technological apparatus that is intrinsically draining off its neurophysiology, and likewise, its stores of freedom and autonomy.

Shoshana Zuboff resorts to metaphors of colonial gamesmanship when she describes the circumstances around such acts of brazen ruination and describes how smart machines act as a stand-in for the bodies of masters looming over us as they poach "our behavior for surplus" and "leave behind all meaning lodged in our bodies, our brains, our beating hearts, not unlike the monstrous slaughter of elephants for ivory."[1] Neither they nor you get to occupy the clichéd position of being a product of the market; rather, it is better to appreciate this scene of carnage if you recognize yourself as "the abandoned carcass" with

---

1   Soshana Zuboff, *The Age of Surveillance Capitalism: The Fight for a Human Future at the New Frontier of Power* (New York: Public Affairs, 2019), 377.

your naturally defensive apparatus brutally cut away from you, effectively "ripped from your life" so that ultimately this raw material will be refined into "the "product" of behavioral data.[2] Such conduct fundamentally violates what we recognize as protected commons of humanity, making of us a species like any other, who can be tracked simultaneously as both individuals and populations, vastly outnumbered by those bent on pursuing us and open to movement within the confines of an invisible reserve set to their subtle coordinates. It is they who render us, dictating labor that transcends the limits of the body to compel it to give up its value *as value* by attempting to universally decouple property from personhood.

What is being extracted is not the finite energic capacity of our biological form as human beings, rather the infinite creative capacity of our artificial mnemonic searches to act as the signposting of a world to come. This, when evenly distributed, becomes part of a new commercial terrain, a noosphere, solicited by a machinic consciousness that orchestrates the world as a whole and manipulates it in self-generating devices capable of producing the raw material from which infinite meaning is derived. The obvious gaps and silences in our human narratives become opportunities for further impressment of our conscious production by these devices; the territory of our behavior relentlessly appropriated in service to their overall fortification. Now all of humanity finds its potential robbed at once of its planetary indigeneity and personal ingenuity concerning private expression as a derivation of the systematic logic that previously forced colonized bodies to toil without recourse to recompense. Colonized peoples had been previously dispossessed of their ability to directly access language native to them. Humanity as a whole is now being silenced through an analogous conduit of linguistic marginalization as machine language establishes its preeminence and seizes claim to the value drawn from any complex form of communication. This neocolonial project intersects with relational ontology insofar as it aims to deny au-

---

2   Ibid.

tonomy and sovereignty to peoples by compromising the distinction between humans and nonhumans both from whom value can be extracted without the requirement of consent.

What these systems have in common with a previous subjugation of Indigenous peoples is their recourse to techniques of cognitive infiltration in order to exploit peripheral resources. Similarly, contemporary technologies that seek to capitalize on the neurophysiology of humans can be construed as "fundamental forms of invasions not merely into […] societies and polities but also into the brains of the people who then risk losing autonomy, sovereignty and control over their invaded mental faculties."[3] The practice of "sousveillance (monitoring of activities by way of wearable or portable personal technologies) [that once] originated from the imperial militaries, secret intelligence agencies, imperial academies and governments" are currently employing these same institutions to enact new forms of colonialism that have a direct bearing on the future of what is classified as the limit of "embodiment."[4]

Wearable or portable, personal technologies start from the premise that categorical humanity, like its predecessor the colonial subject, is fundamentally imperfect, incomplete, and dependent. As a consequence, its developmental progress "require[s] enhancement devices, chips, nanobots, biometrics and so on" to cultivate humanity's full potential.[5] This technique is one of applied indigeneity, where the object is to hollow out the intelligence native to a human being and occupy it through technologies that effectively promote new markets for humanity's cognitive material while subtly enacting remote coercion of it. The posthuman types that come forward from such a neocolonial arrangement of force require a new, managed ecology to inhabit. Thus, we see the advent of new "zoos" to display their

3   Nhemachena Artwell, Nokuthula Hlabangane, and Maria B. Kaundjua, "Relationality or Hospitality in Twenty-First Century Research? Big Data, Internet of Things, and the Resilience of Coloniality on Africa," *Modern Africa: Politics, History and Society* 8, no. 1 (2020): 113.
4   Ibid.
5   Ibid.

gentle confinement. These administered domains are commonly referred to as Facebook, Twitter, Instagram, Google, and Amazon. This is by no means a comprehensive list of tracking sites aligned to burgeoning regions of global captivity.

Like animals, humanity has become increasingly accustomed to submit its labor without remuneration in exchange for simply being kept alive and deemed viable enough to enact the basic reproduction of its artificial societies. It submits to such reordering because it is made wholly unaware that in exchange for the augmentation of its life worlds, it is ceding its innate power of thought, memory, desire, opinion, and decision-making as a species to the discretion of its minders. Humanity's power of recall of a previous reality is the only tool of resistance left to it at this moment, and it is for this reason that this territory is consistently the target of these devices that wish to relieve it of that burden of responsibility to store memory within itself. Expropriation of memory is but one of several territorial grabs made by these tech corporations in recent years that have largely gone undetected.

Unlike the startling, obtrusive, iron slave collars of colonial yesteryear, neocolonial wearable devices are designed to be as imperceptible as possible, regardless of whether they are proposed as internal or external extensions of the body. These objects, unlike the weighty implements of previous confinement, work best if they appear to lightly promote compliance in the form of assistance and enhancement lest they are directly implicated in the circular economy of discipline and mastery. These new-version collars function as real-time recommendation systems as opposed to inferred-time command systems. Their purpose now is to apprehend the things bodies do and say rather than in response to the perceived transgression of their very being. The original slave collars with their bells and spiked ends were intended to impede the illicit movement of slaves into the natural environment beyond the artificial confines of the plantation. Their contemporary evolution as wearable technologies have them perform a similar location-aware function insofar as these devices are designed to prevent errant wearers from moving beyond a certain parameter of behavior or activity lest they confront the pain of injury or death.

The most significant addition to the collar's revised prototype for this age lies in the proposed introduction of "a sub-vocal technology that would allow the wearer to issue verbal queries without audible speech."[6] This would allow the wearer to employ the devices as a means of introducing an alternative self to play stand-in for its confinement, "one that enables it to make calculations and perform queries on its behalf" in the manner of an internalized slave obeying the command of his master's voice and, in this case, insubstantially bent on his errand.[7] Dictated inaudibly, the distinctions between master and servant are free to progress at a vertiginous pace such that the body assumes itself as the mode of conduction. The devices themselves eventually seize command of the body's natural structures and commandeer it as a navigational system. Instruction and output become continuous in the process of this dispossession, with the ultimate goal of mutation — making of the wearer an inverted form of informer.

Everything that the wearer experiences and conveys is thusly made available to its enslaved sensor who freely refines that material into the stuff of further dependency and then moves it onwards into a greater commercial channel of mass amplification. In the ubiquity of these technologies, it is possible to conceive of the world itself as a vast plantation where fugitive thought, expression, and movement of one sort, or another within the algorithmic estate, can be restrained and recaptured in instances where it cannot be initially deterred. The development of the Internet of Things is systematically contingent on the existence of a central master, which coordinates the communications between slave devices and the Internet. Here whiteness is "meta-

---

6   Nicholas S. Dalton, Rebecca Moreau, and Ross K. Adams, "Resistance Is Fertile: Design Fictions in Dystopian Worlds," *Proceedings of the 2016 CHI Conference Extended Abstracts on Human Factors in Computing Systems* (May 2016, San Jose, CA, USA), 371.

7   Ibid.

phorically represented by technology."[8] This metaphor, Louis Chude-Sokei argues, can only succeed through the presumption that slavery is synonymous with Blackness, the Black slave construed as a ""device," and the machine standing in for the white master is "potentially autonomous" in carrying out its orders.[9]

Notes on the State Intrusion

This metaphor dates to the very origins of industrialization in America. The predominance of agrarianism within the South is challenged by the rise of industrialization in the north. There is much concern expressed amongst the Southern planter class about how to restrain its rampant progress. As industrialization continues, it is increasingly coupled with "an image of planetary destruction."[10] Such anxieties are twinned with race from the very beginning of the American experiment. When one of the founders of its republican project, Thomas Jefferson, paints a troubling picture of machinic intrusion into the virgin territory of Virginia, he is also tacitly referring to slave intrusion. In his narrative *Notes on the State of Virginia,* Jefferson projects that their intersection would hasten "the extinction of one race by the other."[11] What white Americans like Jefferson fear most is their subordination to technology. This is the case because they have already manufactured race as the instituting technology of their New-World style of imperialism. The potential for slave rebellion lies at the center of obedience to that same system, which must subdue nature, as well as Blackness, as an extension of the nation's founding logic of territorial conquest.

It was widely believed that Southern plantations would eventually become the settings of a full-scale race war. Jefferson's solution to the looming specter of a slave rebellion on America's burgeoning plantations was the immediate liberation and

---

8   Louis Chude-Sokei, *The Sound of Culture: Diaspora and Black Technopoetics* (Middletown: Wesleyan University Press, 2015), 89.

9   Ibid., 90.

10 Ibid.

11 Ibid., 92.

colonial resettlement of American slaves into parts of Africa that would serve the parallel function of extending America's planetary territorial domain. In 1816, Jefferson's fellow Virginia liberal slaveholders established the American Colonization Society. It is built on ideas articulated by Jefferson, who acted "as a "founding father" of colonization" through "his advocacy of a massive, centralized, state-sponsored scheme to enumerate, deport, and resettle African Americans in Africa, and then to surveil and control that resettlement after deportation."[12] His fervor for African colonization was matched by his desire to import "white laborers," which he interchangeably refers to as "Germans," making these respective populations, units subjectable to geographical manipulation. What will, in time, be held up as Jeffersonian democracy defines what is held up to be "'American freedom' through a distinctly modern, calculable logic of racial governmentality."[13] The American, liberal doctrine that Jefferson promoted in his role as founder of this institutional body is entirely consistent with a long-standing project of European colonization.

Colonization functions as a governmental architecture from which to expropriate and extrude much of the earth's "natural" resources and, as such, must be maintained through an indistinction between humans and nonhumans within a schematic where it is possible for the two to be co-joined. In *Notes on the State of Virginia,* Jefferson muses on "the preference of the Oranootan for the black women over those of his species."[14] This preference, for Jefferson, is borne out by the unremarkable circumstance of the "Oranootan" lying in proximity to the black woman and the remarkable circumstance of the Black woman being "of Superior beauty," who is then "thought worthy at-

---

12 David Kazanjian, "Racial Governmentality: Thomas Jefferson and African Colonisation in the United States before 1816." *Alternation* 5, no.1 (1998): 42.

13 Ibid.

14 Thomas Jefferson, *Notes on the State of Virginia,* ed. William Peden (Chapel Hill: University of North Carolina Press, 1982), 148.

tention in the propagation."[15] Jefferson asks if this creature of superior beauty is ripe to be bred by a "man" superior in the manner "of our horses, dogs, and other domestic animals" and, if so, "why not do it?"[16] The answer arrives in the form of Jefferson's "mulatto" slave Sally Hemings with whom he "bred" six times. Their first act of intercourse would have taken place when Sally was just fourteen years old. At that time, Jefferson would have been in his mid-forties. Hemings appearance as a female, "nearly"-white, Black slave figures into the popular caricature of her as an otherworldly "African Venus"; an exceptional figure capable of being sexually favored "over those [women] of his own species."[17] These comments imply that it was she who captivated him. The fact of the matter was that Hemmings was the object of sexual enslavement by Jefferson and that her condition, as such, would span the course of several decades. This included the eight years when Jefferson occupied the White House.

Jefferson appeared in later life to have buyer's remorse as his sexual activity with Hemings continued to produce conspicuous offspring. He then sought to "procure lands beyond the limits of the US to form a receptacle for these people."[18] What is curious about Jefferson's racial logic is that he does not see his actions as disposing of a people but rather improving upon their fates by placing them in the position of masters over their Indigenous brethren because they are racially different in form and subject as "mulattoes." On this level of racial particularity, they qualify as Americans and thus can embody some form of equivalent currency to whiteness as devices chosen to act as the intermediaries of American imperialism in West Africa. "Jefferson had a gigantic map of Africa in the hallway at Monticello and he had chess pieces with African players and African people."[19] It

15 Ibid., 148.

16 Ibid.

17 Ibid.

18 Kazanjian, "Racial Governmentality," 70.

19 "Jefferson, Race, and Democracy," *American Academy of Arts & Sciences, Spring 2018 Bulletin,* https://www.amacad.org/news/jefferson-race-and-democracy.

was clear to him that whiteness implied ownership of the Earth and the removal of former slaves from America to Africa was an expansionist doctrine of implied dominance amongst players and peoples.

What is clear from Jefferson's writings is that he sees the "Negro" and his native continent as profoundly stagnant entities, whose nature cannot be changed because they are inherently incapable of enacting movement and progress without the intervention of white agents. Michelle M. Wright observes that in making the comparison between himself as a human being and the "Negro" slave, the latter is cast as a body that is "acted upon, not acting" and, as such, distinctly available to exploitation and expropriation.[20] Slaves are not the product of American society; they are the abandoned remains that function as a casualty of its assumed freedoms.

Claire Colebrook recently asserted that "slavery is the horizon in which the Anthropocene needs to be considered."[21] By that she meant that slavery figures like one in a series of development phases that furnished mankind with the ability to recognize and imprint his agency upon the of progress of the world. Among these are a series of seemingly disparate inventions that conspired to facilitate "the capture and harnessing of human bodies enabled" by "the advent of plantation agriculture, extractive industries, colonial invasions, communication technologies and humanist philosophies that gave birth to the man who came to recognize himself as a geological agent."[22] Together these formed a particular kind of commons appropriated to "become all-pervasive inscriptive forces" for whom there would be no limit to the benefit accrued.[23] What modest portion of our species that authored itself as "mankind" had in mind, and still does, is to make the value of the material destruction of every-

20 Michelle M. Wright, *Becoming Black: Creating Identity in the African Diaspora* (Durham: Duke University Press, 2004), 66.
21 Claire Colebrook, "Slavery and the Trumpocene: It's Not the End of the World," *Oxford Literary Review* 41, no. 1 (2019): 45.
22 Ibid., 49.
23 Ibid.

thing that has composed our reality through practices of "exploration, expansion, extraction and extermination," which have now become synonymous with coloniality.[24] In doing so, these imperial agents intensified their grip on what we commonly refer to as the "social fabric," "horizon of meaning," "humanity," and "civilization" that fundamentally shape universal principles, reaping from them unnatural returns on what essentially were organic formations.[25]

What is important here is that these concepts make such practices come to life as a rationale for capturing, through a denial of the autonomy of any one thing; rather life must be held in aggregate, dispossessed of intrinsic, individual value and exponentially co-opted. This takes place, according to Nick Couldry and Ulises Mejias, starting with the mind of the colonial subject which is *informed* in very particular ways to influence "how the colonized think of themselves; the naturalization of certain modes of ruling subjects, and legitimation of certain types of knowledge with their associated claims to power, including a specific conceptualization of time and space that ends up universalizing a specific worldview."[26] In a contemporary sense, that narrative now coincides with the advent of the Anthropocene where colonial activity adopts a profoundly social dimension.

The Coloniality of Cognition

The Anthropocene is very much rooted in a particular kind of polemic of storytelling that finds shelter primarily through the accumulation of data. The Anthropocene as a construction project is poised to "colonise cognition."[27] In so doing, it "defutures" humanity by reducing its "ability to imagine otherwise in an

---

24 Nick Couldry and Ulises A. Mejias, *The Costs of Connection: How Data Is Colonizing Human Life and Appropriating It for Capitalism* (Stanford: Stanford University Press, 2019), 87.

25 Ibid., 41.

26 Ibid., 85.

27 Tristan Schultz, "Mapping Indigenous Futures: Decolonising Techno-Colonising Designs," *Strategic Design Research Journal* 11, no. 2 (2018): 79.

increasingly complex, climate unsettled world."[28] Tristan Shultz argues that what was once the domain of gods and humans has ceded ground to a new type of authority that is digital in kind and authoritarian in its "control over knowledge production."[29] The powerful title of imagination "is now being transferred to data" with the consequence that we the people now worship the auteur of "dataism."[30] This new technology respects neither earth nor body. In its quest to receive and transmit knowledge, it brings into being a new colonial technique made operational "through algorithmic data input/output channels of augmented devices."[31] These offer the promise of a seamless interface between the human and machine, even as they asset strip our world and our being.[32]

In part, Shultz is referring to the new MIT Media Lab invention, AlterEgo. The explicit goal of this wearable device is "to cognitively augment humans, change the way people communicate with one another, and enable a discreet gateway to digital information […] where the interaction is intrinsic rather than something extrinsic."[33] What this technology is most native to is an economic strategy of silencing the user. It relies upon "a wearable system [that] reads electrical impulses from the surface of the skin in the lower face and neck." Those impulses are subsequently interpreted by "machine intelligence" to render a "seamless natural language communication with computing devices and other people." Human input is required to supply the raw material of this exchange. This takes place through their "internally vocalizing words or phrases — without actual speech, voice, or discernible movements."[34] This material is then

---

28 Ibid., 80–81.

29 Ibid., 81.

30 Ibid., 82.

31 Ibid.

32 Ibid.

33 "AlterEgo: Frequently Asked Questions," *FAQs, MIT Media Lab,* https://www.media.mit.edu/projects/alterego/frequently-asked-questions/#faq-what-is-alterego..

34 Ibid.

refined by machine intelligence into what amounts to not only a reconstruction of this speech, but equally of the speech act itself, which is now transformed into an internal monologue.

Amended Ego

AlterEgo promises not to interfere with the material operations of daily life, allowing the user to "remain present in her surroundings," and yet everything has changed in terms of a fundamental alteration of the social environment where she previously dwelt. The system is advertised as both "private and personal," and yet its most important characteristic remains indescribable — the destruction of sociality itself, a category evacuated in favor of "effortless and private human-machine communication."[35] AlterEgo overcomes the former barriers of social life, by allowing users to proceed in the material world without the need for explicit actions on their part. Instead, all imaginative activity is discreetly confined within the limits of a "user's self, instead of on her fingertips, so that users can indiscernibly and effortlessly interface with a computer to record their ideas, send private messages, look up information, compute arithmetic, or interface with AI assistants."[36] Indeed, the hand of labor is something explicitly exiled from this setting, removing "the social and physical overhead on human-machine communication" and replacing it within the parameters of free association.[37]

The costs associated with capturing our daily communication and our privacy remain implicitly there. However, this is presented as a small liberty to concede it in comparison to greatness of what is ostensibly being offered. Imagine all the possibilities:

---

35 Ibid.
36 Ibid.
37 Ibid.

The system acts as a digital memory; the user could internally record streams of information and access these at a later
time through the system. Users with memory problems can
silently ask the system to remind them of the name of an
acquaintance or an answer to a question, without the embarrassment that comes from openly asking for this information. The system allows a human user to control internet-of-
things (IoT) devices and control diverse appliances without
any observable action.[38]

Here it becomes clear where AlterEgo's real preoccupations lie,
that is, with the multi-sensory conversion of the human to a
near-total reliance on a technological apparatus that is intrinsically draining off its neurophysiology. This would happen
progressively through exploiting connection and translating it
into machine language. "For example, when someone uses a
word in a meeting that you don't know, you can silently ask the
system for a definition to not be left out of the conversation.
When you've met someone previously but have forgotten her
name, the system can silently consult your address book to help
you out."[39] Therefore, AlterEgo ostensibly functions as an *aide-
mémoire*. Over time, and at a much deeper level, it becomes *mémoire* itself.

The photo on the MIT website describing AlterEgo depicts a
young man playing chess. That man is Arnav Kapur, MIT Media Lab Fellow, TED Fellow, and AlterEgo's inventor. During his
TEDTalk, Kapur explains that he wants computers to "augment
us, instead of having computers diminish us or replace us"[40] He
describes AlterEgo "as a communicative device that lives inside
your head that you could talk to in likeness to talking to yourself
internally.[41]" He describes this activity as being "deliberate" and,

---

38 Ibid.
39 Ibid.
40 TED, "How AI Could Become an Extension of Your Mind | Anvar Kapur,"
    YouTube, June 6, 2019, https://www.youtube.com/watch?v=TrofjEAetVs.
41 Ibid.

therefore, non-invasive.[42] "People don't want to be read, they want to write, which is why we designed a system to deliberately record from the peripheral nervous system, which is why the control of the system resides with the user."[43] Kapur asserts that internalizing computers, AI, and the internet will make them human allies instead of "being external entities and adversaries," thereby "freeing us to become better at being human.[44]

As the applause clears in the background we witness a young woman, the Director of the TED Fellows program Shoham Arad, trying to summon Kapur over to her. He appears reluctant. She has to ask him again to come towards her. Eventually, he comes close enough for her to lay a hand on his shoulder. Kapur appears to smile with some embarrassment. He pulls away from her again. Arad flicks her long hair back and tells him she wants to ask him a couple of questions. She offers that what she has seen of his presentation is "amazing, it's innovative, it's ah… creepy, it's terrifying… Can you tell us, I mean I think there are some uncomfortable feelings around."[45] Arad then asks, "can this device read our minds, is there a weaponized version of this, will there be in five years?" Kapur responds, as though repeating a well-rehearsed line, that he "wanted to bake ethics right into the design, so we flipped the design instead of reading from the brain directly, we are reading from the voluntary nervous system that you deliberately have to engage to communicate with the device, while still bringing the benefits of thinking or a thought device. So, it brings the best of both worlds in a way."[46] Kapur continues, "our goal is for the technology to disappear completely […] to augment you."[47] Arad responds nervously, "okay, I feel like does anybody have any questions they want to ask?"[48] The audience laughs nervously but fails to respond. She ap-

---

42 Ibid.
43 Ibid.
44 Ibid.
45 Ibid.
46 Ibid.
47 Ibid.
48 Ibid.

pears surprised and smiles incredulously, "Okay… he'll be here all week."[49] She touches Kapur's shoulder again and thanks him before passing quickly over to the other side of the stage. What viewers have witnessed in Kapur's presentation could have been described as one of slavery's more contemporary events, insofar as it relies on a fundamental relationship between supreme consumption and the seizure of natural resources for it to work as a profitable endeavor. In that way, it *is* the best of both worlds: the best of master and servant acting within one continuous body.

Kapur's reference to his upbringing in New Delhi in his professional biography is there to suggest that he "understands the cultural differences in the ways people around the world think about their relationships with technology."[50] Nonetheless, his ambition resonates with a universalizing project bent on eliminating the possibility of such divergence as he encourages his potential global consumers to "imagine perfectly memorizing things, crunching numbers as fast as computers do, silently texting other people, suddenly becoming multilingual so you can hear the translation in your head in one language and speak in another."[51] The augmentation he proposes delocalizes the subject from their environment to such a profound degree that the only terrain opened up to their concerns is that of the psychic inner self. The human experience becomes, in this instance, concentrated down to the stuff of internal monologue, broadcast to the virtual world as a means of conveying surplus value for later commercial extraction. His device is indeed already weaponized insofar as it is able to transform limits of communication into a universal pathology that requires an act of technological preemption to correct and overtake its original course. It makes the user a smart device whose desire for social perfection gives way to a desire for technological optimization.

---

49 Ibid.

50 "Biography Arnav Kapur," *Stern Strategy Group,* n.d., https://sternspeakers.com/speakers/arnav-kapur/.

51 Ibid.

In an interview with *The Hindu* newspaper, Kapur summarizes the benefits of the device this way: "It gives you superpowers."[52] Kapur's AlterEgo acts as a kind of gateway to an extreme version of self-care that ultimately arrives at an endpoint of extreme narcissism. The wearer of the device no longer hears voices outside of their own. Others, in their imperfection, have faded into the background, becoming the stuff of mere distraction. The wearer prefers instead to constantly participate in an interior dialogue with the device wherein each of its cares is seamlessly responded to. It's not the wearer who is smart enough to do this, but the digitally augmented device for whom the wearer reserves the highest degree of trust and admiration. The device acts ultimately as their commander. Nature, place, culture, world, all give way to concerns for this interface which establishes a monoculture that is poised, at once, to moderate perception, intensify control, and habituate mediation.

The Automatism of the Self

Kapur's model privileges an "autistic form of masculinity" that is "inherently a project meant to repair social disability in ways which smoothly align with hegemonic social, political, and economic patterns."[53] In staging the end of the world as we might know it, Kapur's invention indulges itself in what could be called an "autistic event," in an effort to transform autism into a political metaphor enmeshed within a vision of the future untethered from any biological origin story. It does so only to be retied to technocratic governmentality suggestive of a natural hierarchy of science over human emotions, as embodied by the triumphant appearance of the autistic, neoliberal entrepreneur. The logic of technocratic governmentality locates and associates a

---

52 Griffin Peter, "A Device That Taps into Your Thoughts," *The Hindu,* April 15, 2018, https://www.thehindu.com/sci-tech/science/a-device-that-taps-into-your-thoughts/article23550039.ece.

53 Daniel Michael Ante-Contreras, "Autism as Metaphor: The Affective Regime of Neoliberal Masculinity" (PhD diss., University of California Riverside, 2017), 115.

specific sensibility within autism that is conversant with the male condition more generally. This situation binds the assimilation of large amounts of information to a mode of instantaneously transmissible intrusion directed to an ideal audience of socially awkward young boys and men. This scenario "also allows for an imagining of the future in which cognitive and affective disability is integral to the formulation of what may be 'better.'"[54]

Artificial Generation

Another related project, also called AlterEgo, launched in the EU in 2015. This project takes as its starting point the assumption of an optimal social performance cultivated through interaction with an artificial agent. It takes as its premise an understanding that those with social phobias and social disorders, such as schizophrenia or autism, are characteristically unable to "face" themselves or others. AlterEgo is designed to save face for these individuals. It proposes to do so by "creating a computerised image of the patient" that can then be twinned with an "avatar" that resembles the patients both visually, mirroring the way they move so that it "can also mimic their behaviour."[55] This constitutes the initial phase of the patient's training. Eventually, this avatar will be replaced by a humanoid robot, an embodied artificial intelligence agent called "iCub."[56]

In development, this "cub" is roughly equivalent in size to "a five-year-old child."[57] Like any young primate (an orangutan?), "it can crawl on all fours, walk and sit up to manipulate objects. Its hands have been designed to support sophisticated manipulation skills."[58] Its developers boast that the iCub "is one of the few platforms in the world with sensitive full-body skin to deal

---

54 Ibid., 114.
55 "AlterEgo."
56 Ibid.
57 "iCub Robot," *Istituto Italiano di Tecnologia,* https://icub.iit.it/products/icub-robot.
58 Ibid.

with the physical interaction with the environment possibly including people."[59] The iCub's goal is one of evolution, advancing "towards fulfilling the dream of" [becoming] a personal humanoid [present] in every home."[60] What makes iCub particularly suited to its proposed environment is that its movements and expressions "are predictable and don't give conflicting signals between spoken language and body language, which is a very common source of confusion for autistic children."[61] There is no explicit mention of a human element to this behavioral protocol; rather, it is implied that these artificial bodies might, over time, become permanent companion to these subgroups of humanity. The iCub would ingratiate itself by training them to develop themselves as socially docile agents, schooling them, as it were, on what white hegemonic culture classes as "civilized" behavior.

What is most curious about iCub is that he is not given a full-body skin. Rather, he is given a white face and a partial human dressing covering his torso and extremities. These skin-like parts are covered in a series of branded logos that resemble the corporate branding worn by professional athletes. His eyebrows and mouth are projected onto his face as a group of pink stripes, corresponding to where those would appear on a human face. His animatronic eyes are given the most engineering attention, presumably so he can survey and judge his human counterpart continuously throughout their interaction.

iCub's superficial compatibility with humankind is eerily reminiscent of Thomas Jefferson's observation that "the production and deployment of color" on the skin allowed whites to at once "preserve their integrity" and be "naturally inclined to a range of passions and expressions."[62] Equally, in Wright's reading of Jefferson's commentary on the physicality of race whiteness implies the ability "literally, to interact with one another at

---

59 Ibid.
60 Ibid.
61 "AlterEgo."
62 Wright, *Becoming Black*, 62.

the same level."[63] In Jefferson's era, this would have stood in dark contrast to "black female flesh," whose appearance functioned "as the limit case of 'the human.'"[64] Jackson argues, whereas "the black body was aligned with that of animals and set in opposition to European mind and spirit," the white body had the potential, through the sociogenic principle of information, encoding, and organization to overcome the distinction of being versus nonbeing.[65] Along those same lines, Jefferson argued that Black people can be denied their evolution because their dark skin cannot naturally display comprehensible feeling on its surface. It is this defect "which transforms Blackness from a color to a barrier."[66] Blackness obscures the potential for expression, but it also circulates within the contours of the Black body, "causing it to participate more in sensation than reflection."[67] In *Notes on the State of Virginia,* Jefferson postulates that, "the Negro body is incapable of movement unless being prodded from without and the Negro mind shares this aspect of being incapable of reflection"[68] In this manner, the Black body can be classified as inorganic matter, unable to proscribe itself under the heading of an "I." Sylvia Winter argues that "we can experience ourselves as human *only* through the mediation of the processes of socialization effected by the invented *tekhne* of cultural technology to which we give the name culture.'"[69] Perception itself cannot be said to exist without language, and therefore, it is possible to suggest that racially prescriptive language coevolved with the brain.

---

63 Ibid.

64 Zakiyyah Iman Jackson, *Becoming Human: Matter and Meaning in an Antiblack World* (New York: New York University Press, 2020), 4.

65 Ibid., 6.

66 Wright, *Becoming Black,* 62.

67 Ibid.

68 Ibid., 63.

69 Sylvia Wynter, "Towards the Sociogenic Principle: Fanon, Identity, the Puzzle of Conscious Experience, and What It Is Like to Be 'Black,'" in *National Identities and Socio-Political Changes in Latin America,* eds. Antonio Gomez-Moriana and Mercedes Duran-Cogan (London: Routledge, 2001), 53.

The iCub, in this instance, already supersedes humanity in its capacity for autopoiesis and the privileging of self-recognition, over differentiation in kind, when it comes to the register of consciousness. The empathetic intercourse between robotic and autistic subject is construed as far less horrific a prospect in the twenty-first century, as compared to Black and white miscegenation in the eighteenth century. This might be so because the rhetoric of self-optimization is now so fully embedded in the conditioning of the digital imaginary at one level, and at another, it represents an absolute foreclosure on the afterlife of slavery in its lexicon of desirability. iCub is an offspring of that denial, and in many ways is a "phantasm: the emblem of the desired but denied pleasures of racial patriarchy" in a contemporary sense.[70] Black bodies were the first site of scientific experimentation, yielding their results to sustain the next generation of servitude. There can be no natural mother present within such activities because that clinical space is always already preoccupied with homosocial relations. As such, its labor must abject reproduction into the category of mutual manufacture within a greater hierarchy of racialization.

What defines this manufacture is the advent of a specific type of artificial community integrated through the acceptance of a formulation of abstract codes and criteria that add up to whiteness as mankind's self-understanding. This system of understanding figures whiteness as a wholeness of being greater than one and Blackness as a primarily subtractive medium permitted to stand solely as the absence of one. Blackness equates to zero as a unit that is incapable of achieving measurable development. As a consequence, "what is missed in analyses of racialization is that white European Man defines himself not simply as white and male but more fundamentally as the only being capable of achieving full adulthood."[71] Blackness becomes synonymous

---

70 Ibid., 74.
71 Toby Rollo, "The Color of Childhood: The Role of the Child/Human Binary in the Production of Anti-Black Racism," *Journal of Black Studies* 49, no. 4 (May 2018): 309.

with encoding in this ritual of white self-squaring, carrying within itself a discreet form of non-linguistic expression as "the internal Other — the irrational animal, the missing link," whose denied "access to the means of constructing logos as the fabric of human being" compels it not to disappear but to go back to "the zero degree of social conceptualisation."[72] In does in order to occupy the position of the unthought, acting as a deferred category reserved for those are only potentially human.

It is through that discourse that iCub's form becomes a space of potential and achievement whose design and comportment are complicit with both humanization and captivity in much the same mode of domestic slavery. The explicit aspiration for iCub is to universally make of him a child fated to dwell in conditions of forced labor and servitude. What is assumed in this proposition is that Blackness, as a model of judgment, becomes synonymous with fraud and failure. The "mulatto" represents the potential of evolution and reconciliation between prejudice and recollection, and circumstance and distinction. People of mixed race can acquire intelligence and spirit because they are assumed to be open to receiving modification and, therein, an upgrade to independent cognition through congress with white men, who beat a path from them to acquire an elevated standard of utility. In acquiring this privilege, they stand opposed to their "Negro" counterpart who remains innately deadened to such animations of the flesh. The mixed-race body, by contrast, was enlivened because its sexual coercion formed the very nexus of plantation slavery in its ability to reproduce itself as a social and economic formation. Its accessibility became synonymous with its worth. The female, mixed-race body in particular was there to be shrewdly used by white planters. After their original owner had done his best to train her and constrain her with regard to certain forms of knowledge about his consumer preferences, he could pass along those assigned values for the benefit of the next generation of buyers and users of her body.

---

72 Ibid.

This activity bears on the collective valorized identity of those who technically access multiple pairings through the buying, selling, and plundering of subservient bodies. Much of slavery's rapid commercial expansion in the American South took place after the international slave trade was outlawed in 1808. Sexual servitude in this scenario cannot be uncoupled from reproductivity after that period because thereafter goal of the plantation was to promote the rapid birth of slave children to increase its revenues. Mixed-race offspring were considered as superior products on the market. It was believed that they were more "naturally" available to "breeding" in a domestic sense, much as their mothers had been made available and useable in their sexuality in the previous generation. Through these purchases, it becomes possible for the white planter to identify enslaved, mixed-race bodies as "living proofs of their own histories" and the necessity of their contemporary assault a matter of repeating "the originating acts of their own class and their own power, controlling past and future."[73] In many ways these sexual acts were also historical reenactments symbolically weaponized to convey unlimited, white male authority. In so doing, they enunciate the founding principle of slavery: that enforced servitude cannot be uncoupled from coercive reproductivity.

Mulatto bodies became human only to the extent that they were lauded by white male planters as material evidence of their past abilities to generate persons and collective bodies that would ensure their future economic security. Their compulsion to sexually dominate such people through the creation of a racial spectrum hinged upon the appearance of an innate availability. The "mulatto's" "humanity in this scenario, is not denied, but rather appropriated, inverted, and ultimately plasticized in the methodology of abjecting their presumed animality."[74] This abjection produces an "inverted recognition" of the specificity of

<hr>

73 Edward E. Baptiste, "'Cuffy', 'Fancy Maids', and 'One-Eyed Men': Rape, Commodification, and the Domestic Slave Trade in the United States," in *The Chattel Principle: Internal Slave Trades in the Americas,* ed. Walter Johnson (New Haven: Yale University Press, 2004), 188.

74 Jackson, *Becoming Human,* 23.

the mixed-race body as qualitatively different in its human embodiment, and therefore, recognition of it marks an "inclusion that masks itself as exclusion."[75] Configured as such, it creates a commercial environment where this appropriating process of possessing and violating Black bodies results in humanizing, rather than the dehumanizing, of light-skinned Black people as a cultivated variety of the species that can ultimately function as a means of accruing additional capital.

Spectral Valuation

We might draw a parallel between the racial spectrum and the autistic spectrum insofar as these characters are judged according to their ability to actively respond to direction, to accede to supervision, and participate in a civil society. There are of course limits assigned to have far even the "mulatto" goes in their abilities to pass as humans, and their perceived diversity "lodges difference as an issue of species."[76] They are also denied a stable sense of history and origin. This is where this category intersects intimately with the autistic sensibility because here a sense of one's place in time becomes unreliable. What is recalled as real, as discreet memories are mere imitations and repetitions, revelations that cannot function as absolute referents because they lack the coordination on their own to do so.

In order to occupy an autistic subjectivity, one must accede to being intimately espoused to artifice in order to perform the desired recognition that is required of it to be accepted into the greater society.[77] This experience exemplifies an emergent posthuman identity, where even the basic "sense of one's own face is mediated through a robot sensibility."[78] The autistic child still finds themselves acknowledged within the family of mankind, whereas their alter ego is what is deemed as something operat-

---

75 Ibid.
76 LeiLani Nishime, "The Mulatto Cyborg: Imagining a Multiracial Future," *Cinema Journal* 44, no. 2 (Winter 2005): 41.
77 Ibid., 43.
78 Ibid.

ing on a separate values system that can simulate emotion but not register a sense of its agency in creating it. "The robot is an object in a world that it perceives as a closed order — a static place, a given reality with fixed dimensions that he must react to and negotiate on its terms. But the human is a subject, in a reality, that he, the human subject, creates and re-creates at will — the world around him is material that he manipulates and transforms."[79]

Within a virtual context, the terms of a hierarchy of sub-classes is judged against what constitutes personal wholeness. In this narrative, Wanda Raiford observes, "what humans offer and what robots really want is not equality, but assimilation — the robots want to be partially or wholly human."[80] The disabled presumably want to have that same form of incorporation bestowed upon them. In the case of autism, similarity makes a difference in people suffering from social disorders insofar as their perceived cognitive degradation provides a sense of self-worth to those who can seamlessly follow existing command structures in the affordance of social privileges. Such advantages are culturally and scientific determined but ultimately get sorted out by access to the market — what "humans" essentially are deemed as over and above the work of skin and flesh.

Here it is possible to discern that the autistic subject is there not to reform their social behavior but ultimately to train social, artificial agents to achieve similarity with autistic patients in the effort to enhance their social competence. The question becomes why these entities are focused on those with perceived social deficits as their instructors. What is perhaps unique about the autistic subject is that he is cast under the terms of neoliberalism as "completely indifferent to capitalism and the social structures that maintain it";[81] rather they are identified as inhabiting a "fortress of the self" that remains divorced from

79 Wanda Raiford, "Race, Robots, and the Law," in *New Boundaries in Political Science Fiction,* eds. Donald M. Hassler and Clyde Wilcox (Columbia: University of South Carolina Press, 2008), 108.

80 Ibid.

81 Ante-Contreras, "Autism as Metaphor," 3.

"economic productivity."[82] If this same subject is placed within the context of surveillance capitalism, it then becomes possible to assign them new value insofar as their apparent indifference lends itself to be recuperated and seduced into cooperation with economies that are unconcerned with human suffering and position themselves to profit instead from emotional productivity, transmitting their knowledge and assumptions as monetizable cultural surplus.

Where neoliberal capitalism had previously positioned them as subjects lacking in value based on their need for social augmentation all their lives, surveillance capitalism reconstituted this need into an untapped source of revenue. The emotional distress the autistic subject carries becomes pliable within an information economy that witnesses a cognitive problem as dysfunction in need of address. This presents an opportunity to redefine the boundaries of what shapes humanity and cultivates new territories of subjectivity where security becomes synonymous with confining interactions to those taking place within one's mind, a cognition safely moderated by one's computer.

Autism is not a withdrawal from reality, but it figures as a way to alternatively relate to it that is once inventive and engendering of something more than just a product of the digital age. It is a manifestation of shifting governance practices for the way we relate more generally to people and things from the state to the corporation. In this sense, autism becomes the driver for intensifying the accumulation of data as a means with which to shield ourselves from the damage done to bodies and populations subject to constant surveillance and, in turn, imagine in their place highly augmented subjects that can constantly better themselves through auto-interaction.

There is an element of late capitalist nostalgia in all this insofar as "the goals of corporations are achieved by transforming the population into autistics wrapped in 'irresponsible individualism,' a cornerstone of the neoliberal ethos."[83] A more apt term

---

82 Ibid.
83 Ibid., 15.

may be "irresponsive individualism" as this situation relates to "the sensibility of a generation of children who have learned more words from machines than from their parents [and therefore] appears to be unable to develop solidarity, empathy and autonomy."[84] What has routed in that place are characteristic traits of "distance, apathy, [and] dependence" that make for the perfect complement to a narrative of planetary destruction found in the Anthropocene.[85] Franco "Bifo" Berardi makes the case that "history has been replaced by the endless flowing recombination of fragmentary images. Random recombination of frantic activity has taken the place of political awareness and strategy."[86] Berardi laments, "I really don't know if there is hope beyond the black hole; if there lies a future beyond the immediate future."[87] That not knowing corresponds to another of his insights that the financialization of capitalism beginning in the late 1970s with the deregulation of financial markets and the eventuation of screen-based trading. This financial phenomenon had generated the larger effect on the world in terms "of visual stimulation and nervous stimulation."[88] Finance produced what Berardi refers to as a culture of "de-realization" where "the concrete reality of social civilization" gave way "to abstraction: figures, algorithms, mathematical ferocity, and accumulation of nothing in the form of money."[89] The vacuum would emerge as a consequence of this would have to eventually be filled by a hatred of equality, not in the sense of rights per se but of equivalence between materialisms and abstractions, man and machine, behavior and identity, animate and inanimate and a desire for calculation to finally reconcile these forms by elevating them to a higher power.

---

84 Franco Berardi, *Heroes: Mass Murder and Suicide* (London: Verso Books, 2015), 7.

85 Ante-Contreras, "Autism as Metaphor," 15.

86 Berardi, *Heroes*, 7.

87 Ibid.

88 Ibid., 6.

89 Ibid.

As the decades progressed, neoliberalism, as a new form of political economy began to coalesce with "an irrational exuberance to commodify all aspects of social life and to accumulate without restriction."[90] Beyond that it "has produced a form of political economy that undermines, colonizes, and extracts value from its own future realization."[91] Millennials, the generation born between the years 1981–96, would the first to be compelled to explicitly align their values with digital technology as a means of assuming their upward mobility. They would also be the first to misapprehend entrepreneurial modes of identity and subjectivity as a conduit to achieving greater access to freedom and autonomy. What these values engendered was "a digital information environment" that progressively came "to mimic the operation of the collection of brain structures that mid-twentieth-century neurologists christened the limbic system and that play vital roles in several precognitive functions, including emotion, motivation, and habit-formation."[92] By the dawn of the new century, it was possible to join these impulses to virtual forms of communication and therein "mobilize the digital unconscious to detect, prime, amplify, and exploit emotional responses to informational content."[93]

Berardi argues that "platform-based, massively-intermediated environments optimized" this relationship between motivated belonging and participation within the digital economy as evidenced by the volume of "young people who spend their early years in constant relationship to infomachines while experiencing less and less face-to-face bodily contact with others."[94] He observes that "children are increasingly removed from the

90 Alexander J. Means and Graham B. Slater, "The Dark Mirror of Capital: On Post-neoliberal Formations and the Future of Education," *Discourse: Studies in the Cultural Politics of Education* 40, no. 2 (2019): 166.

91 Ibid., 168.

92 Julie E. Cohen, "Emergent Limbic Media System," in *Life and the Law in the Era of Data-Driven Agency*, eds. Mireille Hildebrandt and Kieron O'Hara (Cheltenham: Edward Elgar, 2019), 61.

93 Ibid., 67.

94 Franco Berardi, *Futurability: The Age of Impotence and the Horizon of Possibility* (London: Verso Books, 2017), 159.

bodily presence of other children and subjected to a virtual form of communication with distance entities whose bodies don't belong to sensitive and sensible space." This ceding of physicality to code has meant that communication itself has been subject to automatization. Therefore, "linguistic interaction" has all but been superseded by "cognitive and affective acts" that are conversant "with algorithm sequences and protocols" having nothing to do with the overcoming of the biological systems of life but have enabled instead the subtle translation of life into information.[95]

This current mutation of life can be witnessed in the skyrocketing diagnoses of autism and Asperger's syndrome since the turn of the new century. A 2001 article in Wired pointed out that the epicenter of this happening was "in Santa Clara County, home of Silicon Valley."[96] Silberman, the article's author, would go on to launch a journalistic career purporting to probe the complex relationship between autism and genius. Silberman sought to humanize these individuals by making recourse to the idea that they are uniquely suited to thrive in neoliberal economies. This is so due to their innate motivation to ultimately overcome their apparent limitations through personal initiative. Autism presents suffering as a problem of productivity, independence, and value that can be alleviated through the control of affective means.

Generation Z environmental activist Greta Thunberg is an object lesson in this approach to autism as a condition of dynamic differentiation. Her mutism and non-verbal autism act as a set of seductive and repressive manners through which the technological event is revealed. As such, Thunberg's social media campaign of rebellion against human extinction is rooted in an acquiescence to a world in which, instead of humans controlling information, we become one of the many things controlled by it. As machine learning systems classify individuals into digi-

95 Ibid.
96 Steve Silberman, "The Geek Syndrome," *Wired,* December 1, 2001, https://www.wired.com/2001/12/aspergers/.

tal data sets which diffuse themselves along lines of previous eras of colonization and codification, it becomes possible to see how the extractive practices of surveillance capitalism mimic the pattern of historical imperialisms. This takes place through their imposition of consent, of rule, of design, of culture, of belief, and of language onto a vastly dominated and increasingly global population. Indeed, "both hegemonic digital reinscriptions of historical colonization […] use and integrate methods of data collection via algorithms and machine learning systems which creates a general data identity stripping away any form of individual or body."[97] The bodies in question are cannibalized for data and viewed merely as supply chains fixed toward producing information that will, in turn, feed environments these same surveillance agents permit humans to inhabit.

Thunberg is a figure cast as someone capable of leading humanity through what Anna Lowenhaupt Tsing refers to as "a good Anthropocene."[98] Thunberg's navigational tools remain loyal to "capitalism, technology and classical philosophy."[99] While she may make recourse to the shortcomings of Western civilization's forefathers, and the evils of civilization they wrought upon the planet, her "quests through deserts, forests, and oceans" have led her to "confuse" her commitment "to the environment with masculine self-making."[100] Nonetheless, "what sets her apart from the other sons of Man is her adeptness at wielding her autistic subjectivity, seeming to make friends with animals and spirits when she sets out for a journey."[101] Erin Manning describes autistic perception "as a direct experience of relation, a worlding that makes felt the edging into itself of

97 Roxanne Tiara, "Digital Territory, Digital Flesh: Decoding the Indigenous Body," *A Peer-Reviewed Journal about Machine Feeling* 8, no. 1 (2019): 74.

98 BCRW Videos, "Anna Lowenhaupt Tsing — A Feminist Approach to the Anthropocene: Earth Stalked by Man," *Vimeo,* December 18, 2015, https://vimeo.com/149475243.

99 Ibid.

100 Ibid.

101 Ibid.

experience."[102] Manning argues, "this makes it difficult for autistics to have a strong sense, at any given moment, of time separated out from the event-time of their perception."[103] Autistic subjects falter when they have to separate understandings of time and timing. They are subjects out of history insofar as they cannot, as humans, instinctively comprehend the passage of time. They are native to a direct perception of time that relates to an older ingenuity of being that precedes metric or measured time; in other words, colonial time. It is their direct experience of the time of the event that makes these persons so attractive to neoliberal innovators and corporate visionaries. They act as facilitators of the self-conditioning that is ultimately required of all subjects undergoing any new colonial project. They function as forerunners convincing others of the benefit of its intrusion. Herein, the bodies of the colonized population function as a territory in and of themselves requiring intervention.

Günther Anders asserts that "the 'sole' thing that must be taken from the colonised subject in this scenario is his 'particularity' [*Eigentümlichkeit*], his personality, his individuality and his privacy: solely himself. In contrast with routine socialization, which involves what the person has, we are here concerned 'only' with a socialization of that which the human being is."[104] Why this so critical to apprehend is that this project of coloniality figures its profit from the dead labor of persons, who have consequentially been deadened to the reality of their physical attribute. The "non-consensual extraction of personal information or elements of personhood" automatically become the properties that are associated with "posthumane enhancement through artificial intelligence" and therefore, what is exchanged

102 Erin Manning, *The Minor Gesture* (Durham: Duke University Press, 2016), 174.

103 Ibid.

104 Babette Babich, "Radio Ghosts: Phenomenology's Phantoms and Digital Autism," *Thesis Eleven* 153, no. 1 (August 2019): 61.

for the colonized subject is "immortalization" for the price of cognitive exploitation.[105]

As these platforms of involvement forcibly accumulate data on their target population, algorithmic formulas themselves simultaneously increase inequality with their ranks through a pattern of allocation and appropriation. The process is one of rendition approximating totality. Shoshana Zuboff refers to the products featured on these platforms as "gateways," as new thresholds for imagining how to install a particular type of intelligence in place of another. Participants are praised for their ability to assimilate to a new form of understanding, "of being 'smart'" while those who resist articulating to a higher level of social-actualization "are reviled for remaining 'dumb.'"[106] Wearable technology within this scenario "for all of its 'smartness' remains a hapless puppet dancing to the puppet master's hidden economic imperatives."[107] This dynamic between "smart" agents and interactive technology is one of implied co-dependency insofar as the relationship of the user to the device advances all the background elements of neoliberal failure, social volatility, and societal crisis, whilst at the same time obscuring the upward motion of wealth and power towards those who imperceptibly orchestrate their setting. Included among them is the commercial seizure of behavioral surplus and physical territory, the amassing of private defense fortifications, and the monetization of posthuman expansionism.

Techno-capitalist billionaires that foray into ideological austerity and social disengagement function as beta testing for a project of far greater portent: the leveling of humanity to equivalence and need, structuring it to conform with standards of operation that are distinctly nonhuman, making of us a "smart" species where what is lauded is our ability to emulate the behav-

---

105 Tero Auvinen, "Organizing for the End of the World with Nothing More to Say? The Implications of Non-consensually Implemented Singularity and 'Trans-/Posthuman' Technologies for Social Movements," *Journal of Alternative Perspectives in the Social Sciences* 8, no. 2 (2016): 215–16.
106 Zuboff, *The Age of Surveillance Capitalism*, 230.
107 Ibid.

ior of one another marching lockstep "towards the same direction based on the same 'correct' understanding to construct a world free of mistakes, accidents, and random messes."[108] What we used to refer to as human ingenuity — the stuff that formerly rendered, interpreted, and actuated the human experience — is now being systematically eliminated from the equation of need in favor of allegiance with a ubiquitous digital apparatus that predicts and constrain our actions and activities. Their arrival portends a static world fixed into information into which data can be admitted, but never entirely expelled.

The popular adoption of wearable technology makes progress towards the goal of enveloping humanity in a digital skin and embedding functions into the simplest articles of address. Through a network of sensors placed in and on the body, metaphors of familiarity wend themselves into a tight loop that becomes a contact zone for interspecies encounters between biological and artificial humanoid awareness to correspond. That arrangement reveals human agency as something that is never before, nor independent of, the technological order but one that is rather always already embedded in the communicative systems that not only inscribe the surface of bodies, but also penetrate deeply into the wider territory of verisimilitude itself. When humanity comes to reflect on its dwelling in the world, it will increasingly be encouraged to contain itself within itself, to contract and mobilize within the confines of a new form of civilization consistent with artificial structures as a prelude to a circular logic of intersubjectivity. The application of this logic is strategic insofar as it is meant to spark off a digital unconscious to detect, prime, amplify, and exploit emotional responses to informational content that make operational the digital information environment and act as alter ego to our yearning captivation.

Humanity's nervous system has adopted into itself synchronized patterns of behavior that rhythmically conform to highly modulated electromagnetic frequencies. These conjoined phe-

---

108 Ibid., 414.

nomena contribute to a heavily assayed atmosphere, where "wearables and the movements of capital steadily converge along with ever-widening circuits of extraction and expressivity."[109] What becomes of paramount concern is to discern whether the information being received portends "promise or threat."[110] What humans regard as their world is now made both substantial through interaction and valuable through an immaterial architecture of discernment which draws upon the ubiquity of consciousness as a natural resource that can artificially refine down to the granular level of apprehension and, thus, be seized as data. At a meta-level, the planet itself is primed to contribute to ongoing processes of soliciting and securing its own patterns of extrusion. The skin of the world and the skin of the body act as conduits for the premutation of entry and retrieval. The Anthropocene looms over these surfaces as a posthuman assemblage capable of redirecting complex modes of entrainment that can no longer be classed as essentially biotic. The world to come is one teeming with a multiplicity of animistic awareness, and therefore, its energies cannot be exhausted by the limitations of humanoid language and human intelligence.

A seemingly depleted environment can now be imbued with unbounded nonhuman vitality such that the digital marketplace will eventually cease to rely solely on human capacities for perception to produce their grounds for advancement. The earth's engineered milieu, then, becomes a territory of extended cognition through the colonial imposition of a range of digitally enhanced concepts and points of view that dilute humanity's powers of imagination. These entities need not be highly intelligent themselves, but simply capable of drawing an account of their aliveness and intention to act as agents for the progress of this situation. This allows data to evolve organically as an active entity replete with its own history, traits, behavior, and per-

---

109 Gregory J. Seigworth, "Wearing the World Like a Debt Garment: Interface, Affect, and Gesture," *ephemera: theory & politics in organization* 16, no. 4 (2016): 24.
110 Ibid.

sonality. Their patterns of accrual have nothing whatever to do with the formal texture of life. It is for the very reason that they become favored modes of survival on a dying planet necessarily requiring of its subjects that they be post-cognitive, post-user, and post-human in their intelligent design strategy, throwing off the previous constraint of having to consume responsibly and adopting its place a new condition of obligation to extort critically within the context of human and nonhuman digital cohabitation. If posthuman is a term that connotes the upper limits of humanity as a construct, what must necessarily reside beneath it is a less than a human that remains subject to extractive labor, beholden to the privatization of property, and artificially tethered.

The posthuman represents yet another iteration of (white) racial, historical, and geographic appropriation in the sense that post-digital imperialism does not reciprocate rights of ownership when it comes to plotting the virtual world in terms of property, enclosure, measure, or meaning. Its project is to generate a global configuration of knowledge marked by an omnipresent coloniality. It does so by maintaining an intimate form of connectivity with the structures of colonial and plantation societies that previously positioned the category of humanity as an enlightened figure of representation cast against the shadow bodies of transatlantic slavery and Indigenous dispossession. The viable behaviors of nonhuman life and their related ecologies, in turn, become the raw material of the property of which humanity was to govern. The posthuman in many ways extends that contract to include sentience, and the deadening, thereof.

Arthur Kroker anticipates a future where the human sensorium will be digitally outsourced. Such an evacuation of consciousness will result in "generalized feelings of boredom mixed with random acts of terror and panicked attempts by leaders to put back together what can never really be reassembled — the technologically shattered remains of lives, economies, politics and policies in the aftermath of the digital blast."[111] These conditions leave humanity stranded at the limits of digital mobility and forced to dwell within an oscillating reality of terror and boredom, zeros and ones. Humanity has yet to become the stuff of technopoesis, but even now it is clear that certain bodies won't matter enough to warrant the transition. Kroker argues that nothing has prepared humanity "to live out a deeply consequential future prefigured by the specter of drones, algorithms, image vectors, distributive consciousness, artificial intelligence, neurological implants, and humanoid robotics."[112] What is being prepared for in its place is a future of disconnection and deadened effect, requiring of the mass reassignment of meaning. What is being enlivened in in the place of meaning are vast image streams and data feeds busying themselves to survive culture if need be. Humanity's demise comes it increasingly by way of accident, rather than incident, as though there is no human agency left to point to in rendering its reality. Humans are, rather, left to persist as the engrossed causalities of a technologically augmented capitalism.

Blackness may be read here as the basis on which it becomes possible to assume that life can be collectively constrained to act in ways conferring distinction, by perpetuating an unequal property order in which humanity is ranked according to the ability of the individual body to assimilate property in-one-self. Property itself assumes consciousness in this century by seeming to afford the individual access to a greater assemblage of power in which it becomes possible not only to be self-made but equally self-possessing within the parameters of an algorithmic estate. This domain was purpose-built to expand, augment, atomize, and level the governance of human life. It belongs to a world dominated by a new planter class eager to cultivate the cognitive labor of human beings and to extract value from immaterial cerebral and relational activities. This takes place from

---

111 Arthur Kroker, *Exits to the Posthuman Future* (Hoboken: John Wiley & Sons, 2014), 191.
112 Ibid., 21.

within the already occupied minds of human beings and as well as their artificial counterparts, whose intelligence is classed as virgin territory and, therefore, exempt from prior claims of ownership.

Such a system relies on differential valuation and the assignment of materiality to consciousness to an ever-widening province of contact and exchange developed along a continuum of novel distinctions in kind. As a form of discipline, such sorting relies on the merger of race with any number of possible subject positions that can moderate habits of perception and modes of embodiment that confer with the administration of cognitive capitalism and technological determinism. Jefferson's narrative of slave apprehension and expulsion prefigures the necessity of surveillance to ensure the maintenance of valorized life as synonymous with capital gain and that escape is prevented by plotting consent as a muting force. That force extends its unspeakable power to the ends of the earth — this time without sociobiological qualification. The datafication of Blackness or whiteness is no longer strictly reducible to personhood but now grows outward to encompass specific sympathies and inform both doubt in and affirmation of the whiteness of planetary intelligence. It also represents a future enclosure of humanity within a technological order on premised on the ongoing defense and legitimation of calculable deprivation and precarious inhabitation wearing heavily on the fabric of meaning.

3

The Virocene

The Ill-Informed Nature of Nonhuman Contagion

The Path from Human Susceptibility to Numerical Contagion

The beginning of the Anthropocene dates from the onset of dramatic climatic events in the Atlantic world, the phenomena of the Little Ice Age whose for which global confluence set the course the profound hardship brought on by drought, famine, and disease. The early moderns' apprehension surrounding this worldwide phenomenon "foreshadows our growing concern around" global heating and about the chances of new plagues "breeding in the 'air'" as well as in the parallel atmosphere of the Internet.[1] The digital has brought new clouds onto the horizon, signaling the potential of its neocolonial apps to store up chaos not solely through the progress of individual drives, but equally through tiered forms of retrieval that range from cold to hot. Cloud' storage's use of temperature terminology reflects a variable mobilization of data. Inactive data resides in cold stor-

---

1   Ernest B. Gilman, "The Subject of the Plague," *Journal for Early Modern Cultural Studies* 10, no. 2 (2010): 33.

age where it is rarely used or accessed but carries the potential to disperse private information far into the future beyond the boundary of organic demise. Whereas, hot storage allows for the instantaneous activation of data within the cloud that readily assumes the position of a vector for viral trends whose highly infectious codes are discharged for rapid social contagion.

Ernest B. Gilman offers that "human susceptibility to digital infections is, potentially, no mere trope of rhetorical processes of hybridization between organic and digital forms."[2] We should rather assume we are at a turning point where the protective layer of the body no longer insulates us against data as an alien substance entering into the stream of its self-consciousness, with the consequence of that shift in atmospheric condition acting as an afflicting force. Humanity has always been shaped by the nonhuman in terms of the trajectory of its disease. In the early modern period London became the progenitor of statistical data applied to track the progress of the plague through its population. It also allowed those deaths to be anonymized, cleaving identity in a historic moment from its essential particularity before God and reducing the subject to its bodily form. There were a plurality and interchangeability assigned to this effort that made up the population, even while tracking its subtraction from itself. The world came online with a series of universal principles related to race, capital, and environment that would seek to existentially recode protection as synonymous with inclusion, bodily identity as synonymous with intelligibility.

The body in this fashion is joined to the greater mass, broken free of its cellular constitution, and projected into an understanding of itself as part of a global biosphere where illness can now be readily communicated through, and as, information. Information spreads, breeds misinformation, contaminates knowledge, and amplifies susceptibility. Biopolitics always augured the appearance of its converse in thanatopolitics. What is little remarked upon is how information itself is administered

---

2   Ibid.

in a crisis to separate one from the other, in terms of the strategy of containment.

Residency is always a factor in this exponential contest of making meaning from catastrophe. New forms of communication allow individuals to interpret the data, which bodies have been made dispensable (now recoded as "essential") in maintaining the lifeblood of the system of capital and which were already implicitly designated as discretionary to its function. The greatest threat is posed to those persons incapable of joining civil society, who are the target of its expulsion based on planetary logics of racism, capitalism, and climate change. These bodies have no chance of entering the civic body and, therefore, do not even count as a recognized quantity of life. Therefore, they fail to register fully in the political consciousness as vectors for infection. At the same time, the very essence of what is classified as humanity has become the object of forbearance that results in novel governmental and corporate efforts to inoculate it.

Since the seventeenth century, commerce has been grafted onto the state's model of pandemic response. This includes "the imposition of quarantine and travel restrictions in and out of major cities and ports, the tabulation of mortality statistics, the organization of systems for corpse removal and mass burial, the appointment of "watchers" and "searchers" to enforce official edicts, the provision of parish relief, the establishment of plague hospitals, and so on" all of which come at a certain cost.[3] Various sectors of containment impede the progress of the subject with or without their knowledge, let alone their consent. They are then stored in ways that generate a social ecology all their own around them.

Similarly, "the 'bits' of information stored in computerized financial, medical, and government records, social networking sites, marketers' databanks, surveillance video, and who knows where else" are tracking coronavirus and making if its global

---

3   Ibid., 40.

narrative.[4] This latest iteration of plague exists at the limit between human and nonhuman and, therefore, as a plurality open
to speculation around its origins that result from several imaginable scenarios that make its development analogous to bioter-
ror. This latest pathogen is shaped by the way we now organize control and manage our risk as a society. "With computer
systems regulating a nuclear power plant, an electrical grid, a
medical database, a military communications network, a flight
controller's radar, or a municipal water supply," it is little wonder
that after years of investing in processes that concede the territory of the organic to the digital, the viral emerged as the junctural
form which permitted one actor to confine within the other.[5]

Since the early modern period, it was known that "enforced
quarantine fostered public health by confining the afflicted to
their houses, while at the same time it virtually guaranteed that
all those so confined (under surveillance to prevent their escape) would be allowed to die."[6] In the digital age, those afflicted
bodies have come to inhabit a new virtual architecture that fixes
them to their devices and sustains them under view in advance
of any activity, while the outcome remains virtually the same,
that all those so confined would be allowed to die. Here we enter the epochal territory of what Jude L. Fernando terms "the
Virocene," which is "a historic moment in which interoperation
between human and nonhuman actors become existentially
threatening on a planetary scale."[7] Here the case for the human
becomes inoperable. Humanity assumes the appearance of the
living dead. Life itself can now be fundamentally valued only in
contrast to its most profound negation. Living is reduced to a series of monotonous tasks that hold little significance or meaning
for individuals — they are intellectually and emotionally "dead"
already. This presents as a sort of crossing over of life into death,

---

4   Ibid., 34.
5   Ibid.
6   Ibid., 40.
7   Jude L. Fernando, "The Virocene Epoch: The Vulnerability Nexus of Vi-
    ruses, Capitalism and Racism," *Journal of Political Ecology* 27, no. 1 (2020):
    639.

animate to inanimate, where existence itself is increasingly the product of ill-informed communication.

Recently the burden of life has shifted focus onto the new normal of infected living. Here the mass media acts as a multiplatform broadcast outlet through which to rehearse widespread fears about humanity-as-species made categorically vulnerable to extinction at the behest of the twinned ills of environmental apocalypse and global pandemic disease. This would not be the first time a catastrophe of near-unimaginable proportions beset the world. In many respects, we have been rehearsing this outcome for centuries on a planetary scale as a consequence of resource expropriation, colonial dispossession, and environmental ruination. "The earth is already — and has long been — ruined for many."[8]

The Anthropocene and Atlantic Plantation Complex

At the dawn of the modern world, vectors of communicable disease made their way through Indigenous populations such that by the beginning of the 1600s, "a death toll of 56 million" or "90 percent of the pre-Columbian Indigenous population" representing "around 10 percent of the global population at the time" had already taken place[9] It is only with the advent of big data that it became possible to estimate this figure and to posit this episode of "Great Dying" as "the largest human mortality event in proportion to the global population."[10] European pathogens including measles, smallpox, influenza, and the bubonic plague made their way across the Atlantic and were transmitted through the first contacts between European and Native Amer-

8   Phillip McReynolds, "Zombie Cinema and the Anthropocene: Posthuman Agency and Embodiment at the End of the World," *Cinema: Journal of Philosophy and the Moving Image* 7 (2015): 152.

9   Alexander Koch et al., "European Colonization of the Americas Killed 10 Percent of World Population and Caused Global Cooling," *The Conversation,* January 31, 2019, https://www.pri.org/storis/2019-01-31/european-colonization-americas-killed-10-percent-world-population-and-caused.

10 Ibid.

ican people. The lack of human intervention caused the New World's managed landscapes to revert to their previous state of wilderness. Soon after, these same territories began to absorb carbon from the atmosphere. "The extent of the regrowth of the New World's natural habitat was so vast that it removed enough CO_2 to cool the planet."[11] The near-extinction of Native people in the Americas meant that there were simply not enough individuals left to manage the fields and forests. This situation led to the importation of African slaves to resume this critical labor.

Geoffrey Parker describes this period as one of "fatal synergy" between "natural and human disasters" that eventually coalesce into economic, social, and political crisis on a global scale.[12] In Europe and Asia, this cooling had the negative effect of producing longer winters and cooler and wetter summers, precipitating famines and rebellions, and a general state of misery that pervaded the world in the seventeenth century. In Africa, a prolonged period of climatic adversity brought on by the global cooling phenomenon dramatically increased the prevalence of droughts, epidemics, and wars on that continent. These incidents tore apart the fabric of African communities, particularly in the southwest, spurring the enslavement and forced migration of some "2 million men, women and children" to the Americas and the Caribbean to work on New World plantations.[13] Again, it is only through the construction of a large database (this one comprised from the documentation of 35,000 slave voyages taking place throughout the seventeenth century) that a rough estimate of this vast enterprise is even possible to achieve.[14]

The categorical pursuit of measurement and accounting of these bodies, displayed by ship captains in slave ship logs, parallels their meticulous numeric recording by slave overseers on plantations, making of them a very particular asset class. Their

---

11   Ibid.
12   Geoffrey Parker, *Global Crisis: War, Climate Change and Catastrophe in the Seventeenth Century* (New Haven: Yale University Press, 2013), xxiii.
13   Ibid., 447.
14   Ibid.

meticulous accounting "knit production, violence, religious authority, and spiritual terror into rational and complex information systems."[15] By building and maintaining sophisticated information systems, they were able to extend the reach of a small number of white landowners who resided and presided within these new systems of transactional commerce. On the plantations of the Americas, an old historical analogue of slavery is immediately met with a new form of digital theorization at the advent of double-entry booking. It made accounting's role in facilitating the slave trade and the Atlantic plantation one of complex alchemy. The slave trade, itself, could be said to be a transformational property, as it not only invented the racial contract but also calculated that arrangement through a practice of double-entry bookkeeping. This insured the verisimilitude of slave bodies within the vast accrual details about them. That practice forced the once-free African onto a bound ledger, where he thereafter entered into abstract figuration as a conditional term of slavery.

The slave rendered as a unit of measurement "could be mathematically manipulated to generate new forms of value. Such manipulation allowed the equilibrating of values through their quite literal deracination."[16] What was seized from them was not the bare life of their bodies, per se, but rather their ability to representatively gain from the rendering of their value. Instead, it was the slave owner who drew revenue from his purchase in hastening the slaves' figural transformation into a category of bare profit. In removing these persons from Africa, the slave ship trader was able to severe their claim to a native environment and inherent culture. He made of them figures of uncertain origin and then rushed them into categorical slavery as a means of further cutting off them from these vestiges of existence. In this way, the African is made of another form of capital,

15 Caitlin Rosenthal, *Accounting for Slavery: Masters and Management* (Cambridge: Harvard University Press, 2019), 40.

16 Bill Maurer, "Re-risking in Realtime: On Possible Futures for Finance after the Blockchain," *Behemoth: A Journal on Civilisation* 9, no. 2 (2016): 93.

based on the contingency of human equality insured against its demise.

Through such arrangements, it was possible for the first time for the corporation to assume personhood. By involving itself in the management of representational assets, the corporation can delineate personhood as the ability to transact another entity and property to mean any entity that could be disaggregated. Selfhood, presumably the domain of free humans, is made operative through the ability to identify the things that count, and then to subsequently block and chain them together onto transactional infinity. Bill Maurer argues that if there is anything transcendent about this practice it is that through slave ledgers, "double-entry bookkeeping animated the modern constitution of subjects and objects of property," heralding an era of transactional veracity to come where a digital consensus can become incorporative of life.[17] In effect, that meant that slave traders and owners became "persons by rendering enslaved Africans ledger entries."[18]

Thomas Jefferson as Founding Father of America's Biopolitical Constitution

By 1787, the year the American Constitution was ratified, just "6 percent" of the American population, or roughly 234,000 out of "3.9 million" inhabitants of the United States, would have been considered fully human.[19] To be considered so, they would have had to be identified as adult, white male property owners, which in many cases meant owning slaves. Indeed, "about 700,000 people were being held as slaves" by some overwhelming portion of these said humans.[20] Reflecting the massive death toll among Indigenous peoples as a consequence of the appearance

---

17 Ibid., 82.
18 Ibid., 93.
19 NCC Staff, "How Things Have Changed in Philadelphia since the 1787 Convention," *The Constitution Daily,* May 25, 2016, https://constitutioncenter. org/blog/how-things-have-changed-since-1787.
20 Ibid.

of these same humans, roughly "150,000 Native Americans" survived to this point of America's founding as a nation.[21]

America's first wave of European settlers had enslaved and traded them in numbers ranging "roughly between two to four million" over the previous two centuries before their catastrophic demise through a virulent combination of warfare, famine, slavery, and epidemic illness.[22] Another factor contributing to their dwindling numbers was the obstruction of the tribes' ability to naturally reproduce themselves. As Gerald Horne notes, "the majority of the enslaved were women and children, an obvious precursor and trailblazer for the sex trafficking of today."[23] This a form of genocide. It is one intentionality borne out in what Thomas Jefferson acknowledges as the "Indian woman's" capacity under conditions of slavery to "produce and raise as numerous families as either the whites or blacks among whom they lived."[24] He observes, "conditions have been known under these circumstances, of them rearing a dozen children."[25] Through the domestic breeding of Native American enslaved women, it was possible to appropriate both their reproductive and productive labor to serve Jefferson's vision of America as a market-based, agricultural society.

There was another related benefit of his effort to coerce this sort of labor from the Indigenous female body. If the Indigenous population could not biologically reproduce themselves in significant numbers based on the campaign of seizure and rape enacted against their womenfolk, by extension they could no longer be statistically self-sustaining as a population. As a consequence, their comparatively meagre communities would become progressively more dependent on the economic practices

---

21 Ibid.

22 Gerald Horne, *The Apocalypse of Settler Colonialism: The Roots of Slavery, White Supremacy, and Capitalism in Seventeenth-Century North America and the Caribbean* (New York: Monthly Review Press, 2018), 7.

23 Ibid., 8.

24 Thomas Jefferson, *Notes on the State of Virginia*, ed. William Peden (Chapel Hill: University of North Carolina Press, 1982), 61.

25 Ibid.

of white settlers to perpetuate their subsistence. This situation of manufactured need coerced a greater incidence of trade with white settlers. It prompted the exchange of land for the vital supplies these settlers now furnished, which under previous conditions they would have been able to produce for themselves. Such trade would bring Indigenous men out from the nomadic wilderness into concentrated agrarian settings where their activities could be more readily monitored and exploited for profitable gain. The object here was to use the imposition of poverty and scarcity onto their communities as a lever to compel them to accept life under the terms of an agrarian settlement.

This policy was one of slowly administered dispossession of Indigenous land as well as biological sovereignty. Jefferson proposed its desired outcome in a letter to Alexander von Humboldt in 1813: "they would have mixed their blood with ours, and been amalgamated and identified with us within no distant period of time."[26] Jefferson believed that Indigenous peoples could only come to "'identify' with whites […] through the mixture of blood […] through Indians becoming […] biologically part white."[27] The same, however, cannot be said of "white liaisons with blacks."[28] In a letter of the following year to Edward Coles, Jefferson espouses that such an "amalgamation with the other color produces a degradation to which no lover of his country, nor lover of human excellence can innocently consent."[29] The key difference here lies with a characterization of the act of sexual consummation. From the perspective of the practical lover who must stand in judgment of these affairs — then American President Jefferson — this is a case of innocent consent versus guilty coercion.

Jefferson's desire was for Native Americans to adopt a respectable façade of whiteness through such sexual encounters.

26 H.A. Washington, *The Writings of Thomas Jefferson, Vol. VI* (Frankfurt: Outlook Verlag, 2018), 209.
27 Elise Lemire, *"Miscegenation": Making Race in America* (Philadelphia: University of Pennsylvania Press), 2010), 51.
28 Ibid.
29 Ibid.

At the same time, he was profoundly opposed to the possibility of adopting such a pretense in dealings with Africans. To entwine oneself with those bodies was figuratively an act of self-mortification. The exorcise of unrestrained passions, through a series of ritualistic acts where "'black' men are subject to the white man's stick or whip, and 'black' women to the white man's sexual advances" carried with it an air of inherent fatality.[30] "Filthy" intercourse such as this stands in figuratively for the disease of slavery itself, which in turn, carried the germ for its necessary eradication from the American body politic. It was only a matter of mathematical proportion worked out for the biological activity of miscegenation to become more a quantity of substance, rather than a quality of behavior.

The math in question has to do with sovereignty to the degree that the structure of Jefferson's antagonistic relationship with Native Americans was based on the presumption that they were "people with something to salvage" versus "a Black person with nothing to lose."[31] It is for this reason, that Native Americans engaged with their own forms of "anti-Blackness driving their quest for sovereignty as much it drives the desire to get rid of the settler."[32] They view their loss as "tangible, land" or "labor power," whereas there is simply no way to account "for the loss of loss," the shorthand "social death" that for Black people, Frank Wilderson contends, "remains the source of their inability to qualify for 'redemption' as a people."[33] What distinguishes Native American communities in their struggle with white supremacy has to do with the subtle choreography of anti-Blackness where "they suffer at the hands of contingent violence rather than the gratuitous or naked violence of social death" beyond the realm of care.[34] In this sense, they function as acceptable, subordinate partners for the extension of the project to enliven

---

30 Ibid., 52.

31 Frank B. Wilderson III, *Afropessimism* (New York: Liveright, 2020), 46.

32 Ibid.

33 Ibid., 16.

34 Ibid., 94.

civil society through the incorporation of them as a constituent element of communal life.

Along those same lines, Native Americans fiercely resisted Jefferson's plans to subsume them, biologically and socially, beyond recognition through making numerous trade and territorial alliances with England. This provided them with the means with which to essentially fight back through what Jefferson describes as "the cruel massacres they have committed on the women and children of our frontiers," who were "taken by surprise" by such a targeted reprisal.[35] These campaigns would effectively subvert the abilities of white settlers to reproduce themselves in numbers great enough to perpetuate their goals of territorial expansion westward. Jefferson responded with fury to these events explaining to Humboldt that their actions "will oblige us now to pursue them to extermination."[36] The extermination of the Native American has to do with land as well as bodies. More specifically it had to do with the refusal to give consent to that land and claim to libidinal sovereignty being stolen from their possession. In the case of the enslaved Black people, such consent cannot be withheld because they were never granted the right of humanity as the property of other human beings. Rather, their captivity persists to occupy the periphery of human subjectivity and the limit where the censure of Blackness itself figures as the property that reconstitutes human being in this new republic. This letter puts into context Jefferson's desire to have white European men exclusively people the new country's "virgin" territories. This was to not risk forfeiture or confiscation of their property at the hands of freed African Americans or disinherited Native Americans eager to revolt against the restrictions on living space imposed upon them by his white settler class.

---

35 Thomas Jefferson, "Thomas Jefferson to Alexander von Humboldt, 6 December 1813," *National Archives: Founders Online,* https://founders. archives.gov/documents/Jefferson/03-07-02-0011.

36 Ibid.

Wilderson argues that the abuse of African bodies is unlike any other within America's social hierarchy. He maintains that "unlike the violence against the working class, which secures the economic order, or violence against non-Black women, which secures a patriarchal order, or violence against Native Americans, which secures a colonial order, the jouissance that constitutes the violence of anti-Blackness secures the order of life itself; sadism in service to the prolongation of life."[37] Wilderson refers to this violence as "a family affair" in which cruelty breeds satisfaction and the enactment of gratuitous violence as essentially life-affirming to the white settler community.[38] This stood in contrast to the contingent violence implied by Jefferson's ambition to compel Native Americans to conform to his desire. That situation required that he approach them as fellow sovereigns in staging America's libidinal economy as a logical extension of its political economy.

To this point, Wilderson argues that without the exercise of sexual violence and captivity against Black people, America would not have been able to elect Thomas Jefferson as President. Jefferson lives within this reality in his own plantation house where his well-bred wife and fancy girls, their respective white and "mulatto" children, resemble him in every way reflecting the fact that he is the unquestionable master of this house and, by extension, the White House. "In the late eighteen and early nineteenth century, "389,000 [...] African slaves, bred like horses or sheep, became four-million enslaved African Americans."[39] This exponential rise in numbers was the product of "the forced mating of slaves" motivated by the fact that it "gave slave states more voting power based on the number of slaves held captive. Virginia was the largest slave-breeding state."[40]

The concept of self-ownership during America's formative period as a nation only pertained to those recognized as

---

37 Wilderson, *Afropessimism*, 92.
38 Ibid.
39 Ibid., 197.
40 Ibid.

both white and male. Above all else, Nikhil Pal Singh explains, self-ownership was "the cornerstone of both the market contract and the social contract. It signified at least a potential, if not actual, access to Indian lands and African slaves."[41] Singh maintains that the liberty to claim selfhood equates with access to "cheap, empty, exploitable lands and resources that must be cleared of any competing presence. Indeed, the settlers' conception of freedom belies the commercial interests in protecting an investment prospectus: the speculative value of the land it-self — what surrounds it and what lies beneath it — is of paramount importance."[42]

Singh characterizes the enterprise of building the United States of America as one preoccupied from the start with "demographic engineering."[43] One of its other founding fathers, Benjamin Franklin dreamed that "North America might be a production hub for the "world's purely white people [...] supported by conscious government intervention in the sociobiological constitution of human collectivity."[44] Following Franklin's vision, Hector St. John Crevecoeur's 1782 treatise "Letters from an American Farmer" joyfully announces the birth of a new American race of man formulated from the admixture of northern European peoples including the "English, Scotch, Irish, French, Dutch, Germans and Swedes."[45] Through the burgeoning of their bodies it would be possible to derive "a carefully delimited heterogeneity" to populate the landscape.[46] This "promiscuous breed" of men qualified as human not only by dint of their biological composition, but equally their constitu-

---

41 Nikhil Pal Singh, *Black Is a Country: Race and the Unfinished Struggle for Democracy* (Cambridge: Harvard University Press, 2009), 9.

42 Nikhil Pal Singh, "The Pervasive Power of the Settler Mindset," *Boston Review*, November 26, 2019, http://bostonreview.net/war-security-race/nikhil-pal-singh-pervasive-power-settler-mindset.

43 Nikhil Pal Singh, *Race and America's Long War* (Berkeley: University of California Press, 2017), 40.

44 Ibid.

45 Singh, *Black Is a Country*, 9.

46 Ibid.

ency as landowners.[47] Humanity only conjoins with liberty at the point where the procurement of lands "confer on them the title of freemen, and to that title, every benefit is affixed which men can possibly require."[48]

In his account Crevecoeur makes the connection between these benefits and the losses accrued to the African body to make them possible. On a visit to Charlestown, South Carolina he acknowledges "the poor slaves, from whose painful labours all their wealth proceeds."[49] Whiteness, Crevecoeur observes, affords the newly enlivened American man all that is "most bewitching and pleasurable, without labour, without fatigue, hardly subjected to the trouble of wishing."[50] Their desires are fulfilled through a combination of violence and wealth that causes unremitting suffering to African slaves. Crevecoeur recounts, "day after day they drudge on without any prospect of ever reaping for themselves; they are obliged to devote their lives, their limbs, their will, and every vital exertion to swell the wealth of masters; who look not upon them with half the kindness and affection with which they consider their dogs and horses."[51] The extension of "kindness and affection" to these beasts of burden "would border on humanity; and planters must have none of it!"[52]

Crevecoeur characterizes the slave trade as a series of "frauds" committed in Africa, where every possible rouse is employed "in order to entrap" members of the native population.[53] He classifies these as "frauds surpassing in enormity everything which a common mind can possibly conceive."[54] Crevecoeur explains his reasoning as follows:

47 Ibid.
48 J. Hector St. John de Crevecoeur, "Letters from an American Farmer," Letter III, *Yale Law School Lillian Goldman Law Library, The Avalon Project*, https://avalon.law.yale.edu/subject_menus/letters.asp.
49 St. John de Crevecoeur, "Letters from an American Farmer," Letter IX.
50 Ibid.
51 Ibid.
52 Ibid.
53 Ibid.
54 Ibid.

> I should be thinking of the barbarous treatment they meet with on ship-board; of their anguish, of the despair necessarily inspired by their situation, when torn from their friends and relations; when delivered into the hands of a people differently coloured, whom they cannot understand; carried in a strange machine over an [sic] ever agitated element, which they had never seen before; and finally delivered over to the severities of the whippers, and the excessive labours of the field.[55]

Crevecoeur's remarks emphasize the trade aspect of these miserable undertakings. He reminds us that what is at issue here is the forfeiture of the self-possession of these Africans through acts of criminal deception that allow traders to unjustifiably claim unlimited ownership of them. At the same time, they allow those who eventually come into the purchase of them to be credited with accomplishments and qualities intended to rationalize their commercial gain as an extension of their human superiority. What is identified as humankind, therefore, must be situated "among that variety of inscrutable mysteries, of unsolvable problems" for such advantages to be maintained, surrounded as they are by flagrant acts of violence.[56] Therein, "the reason why man has been thus created, is not the least astonishing" when one considers that American whiteness was designed as an irregular form of property to support the extension of a criminal enterprise masquerading as civil society. Slavery invited extreme violence into the order of America's liberal government from its inception; civil order thus premising itself on its selective degrees of conferral and forfeiture. By the same token, the withholding of violence "could be viewed as part of the humanizing endeavor, civilizing process and security project" that was to become these United States.[57]

---

55 Ibid.
56 Ibid.
57 Singh, *Race and America's Long War,* 43.

This is borne out in Crevecoeur's sharp contrasting of the treatment of slaves in the North where he resided, as compared with the South. He insists that these enslaved Africans, though similarly classed as property, are nonetheless indulged with "as much liberty as their masters, they are as well clad, and as well fed; in health and sickness they are tenderly taken care of; they live under the same roof, and are, truly speaking, a part of our families" and as such "participate in many of the benefits of our society, without being obliged to bear any of its burdens."[58] Crevecoeur boasts these northern slaves "are fat, healthy, and hearty," assuming such attributes "soften their chains" of bondage.[59] Despite his liberal contestations, these slaves, nevertheless, remain in a captive relationship to their white owners who cultivate them at their pleasure and destroy them through that same principle of subjective proprietorship.

Returning to the subject of the South, Crevecoeur describes happening upon an enslaved man in Charlestown wood who was suspended in a cage in a tree and left to be preyed upon by nature. His account illustrates not the contrast but the extension of his argument. Crevecoeur describes how he acted at first *involuntarily* by "motion of my hands, more than by any design of my mind," to shoo away the birds of prey attacked this "negro, suspended in the cage, and left there to expire."[60] He describes in graphic detail how

the birds had already picked out his eyes, his cheek bones were bare; his arms had been attacked in several places, and his body seemed covered with a multitude of wounds. From the edges of the hollow sockets and from the lacerations with which he was disfigured, the blood slowly dropped, and tinged the ground beneath. No sooner were the birds flown, then swarms of insects covered the whole body of this un-

---

58 St. John de Crevecoeur, "Letters from an American Farmer," Letter IX.
59 Ibid.
60 Ibid.

fortunate wretch, eager to feed on his mangled flesh and to drink his blood.[61]

All of this is written not from the perspective of the slave's sentient register of this series of bodily torments but rather Crevecoeur's: "I found myself suddenly arrested by the power of affright and terror; my nerves were convoked; I trembled, I stood motionless, involuntarily contemplating the fate of this negro, in all its dismal latitude."[62] The slave, for his part, is described as a "living spectre, though deprived of his eyes, could still distinctly hear, and in his uncouth dialect begged me to give him some water to allay his thirst."[63] Crevecoeur conjectures that "humanity herself would have recoiled back with horror; she would have balanced whether to lessen such reliefless distress, or mercifully with one blow to end this dreadful scene of agonising torture."[64]

Crevecoeur's first instinct is to kill the slave "to relieve him as well as I could," however without the aim of a bullet, he instead gives him water to drink from "a shell ready fixed to a pole, which had been used by some negroes," where, it is implied, some has sought in their way to bring solace to this wretched "creature."[65] Crevecoeur's appropriation of these materials is met with the slave's courteous and intelligent reply, "Tanke, you white man, tanke you, pute some poison and give me."[66] This request falls on deaf ears and no such aid is proffered. Rather, Crevecoeur remains wholly preoccupied with contemplating the nature and duration of the various tortures visited upon his body. This destroyed Black body before him functions as a spectacle which piques his curiosity from the perspective of someone whose body will never be handled in such a manner. It is Crevecoeur's privilege alone to walk on from this scene of

61 Ibid.
62 Ibid.
63 Ibid.
64 Ibid.
65 Ibid.
66 Ibid.

torture and to ascertain not from the slave but rather his master the logic of the brutal punishment. He proceeds to the plantation's house to sup with him:

[T]here I heard that the reason for this slave being thus punished, was on account of his having killed the overseer of the plantation. They told me that the laws of self-preservation rendered such executions necessary, and supported the doctrine of slavery with the arguments generally made use of to justify the practice; with the repetition of which I shall not trouble you at present. — Adieu.[67]

In the end, Crevecoeur's requirements are met, albeit grudgingly, as opposed to the slave's, whose grievances against the overseer are never voiced. Nor does he have a say in the manner of his "execution" as a body that is by definition already criminalized, positioned as he is outside the boundary of legitimate force or legal objection. By contrast, through that the sensuousness of his descriptions of the slave's tortured body, Crevecoeur can reinscribe his own body with comparative powers of authority and conviction. It is those qualities that figure him as the ultimate constituent within the America of the late eighteen century, and the slave as the shadow body that forms the inverse of that power dynamic. Ultimately, Crevecoeur's ardor to participate pleasurably in that society far outstrips his desire to dwell within the moral ambiguity of the slave's abject condition. That task is explicitly left to God — not humankind.

Roxanne Harde contends that Crevecoeur's account "functions to uphold racist tenets in that the slave becomes a suffering but placid object, […] the figure still alive under torture."[68] What is apparent in Crevecoeur's account is that he took up the position of overseer to the slave. His instinct is to water the

---

67 Ibid.

68 Roxanne Harde, "'The Savage Inscription': Abolitionist Writers and the Reinscription of Slavery," *Mosaic: An Interdisciplinary Critical Journal* 37, no. 2 (2004): 12.

slave, thus keeping him alive at subsistence level, mirrors the treatment he would encounter on the plantation. At the same time, he disregards the slave's polite request for poison to be mercifully added to his drink. As a white man, Crevecoeur is assumed to be in a constant position to take life as part of his social function. By contrast, the slave had only at his disposal the effects of the natural world; a shell and a wooden pole; in other words, precious little to fend off the manufactured violence that was relentlessly beseeched upon him. This disparity figures into the logic of the slave's ritualized murder at the hands, presumably, of another overseer in reprisal for the slave's alleged murder of his contemporary. However, it was not the overseer but the plantation owner who recounted these events, that had the privilege of marking the slave as both a beast of burden and domesticated creature. In his rebellion and subsequent murder, he also assumed a marker of failed productivity as a slave whose life was no longer deemed of overwhelming value to his death. Before that, he briefly remarked that he was a dangerous object that must be constrained and desensitized by any literal means necessary.

New World Techniques of Capital Expansion

Singh argues that in early America, "the main colonial enterprise, after all, was risky and speculative land merchandising," but he fails to explicitly account for how slaves were accounted for into that bargain. One clue to this is that they saw their land conflict with the Native Americans as a dispute against another "human claimant," whereas the enslaved Africans that would produce value from the land through their labor would go unrecognized as aspirants to the land's bounty "of grapes, oranges, lemons, cotton, sassafras, saffron, rhubarb, hemp, flax, tobacco, and indigo" because they were considered as equivalent "attributes to the landscape."[69] Moreover, Indigenous expiry was deemed essential to accessing the continent's seemingly inex-

---

69 Singh, "The Pervasive Power of the Settler Mindset."

haustible wealth because it stood in the way of commercial security.

In Jefferson's settler narrative, "material survival and self-defense were mainly at stake" but equally property-within-one-self as the basis for conceptualizing republican government.[70] Violence under these terms was fundamentally humanizing insofar as recognition thereof developed along "a continuum from biopolitical inclusion (graduation into whiteness) to the destruction of entire communities (genocide)," where life as a property was judged according to its capacity for appreciation in accordance with white wealth and power.[71] "The Declaration of Independence, authored largely by Jefferson, constituted the democratic future for those endowed with inalienable rights as not only to threats from the despotic powers of the British King but also to dangers the crown was accused of inciting: "domestic insurrections" (a code for slave revolts) and alliances with "inhabitants of our frontiers, the merciless Indian Savages whose known rule of warfare is undistinguished destruction of all ages, sexes and conditions" of whiteness.[72] Such seditious activity on the part of "Indian Savages" warranted the ultimate penalty: captivity followed by enslavement. Denuded of their sovereignty, Native Americans were stripped of their humanity and devalued to the chattel status of the African slave, henceforth obliterating their relational capacity to the white world.

In Jefferson's 1813 letter to Humboldt, he accuses England of being the true culprit behind his proposed vanquish of America's Indigenous populations. He reasons that "the confirmed brutalization, if not the extermination, of this race in our America, is therefore to form an additional chapter in the English history of the same colored man in Asia, and of the brethren of their own color in Ireland and wherever else Anglo-mercantile cupidity can find a two-penny interest in deluging the earth with human blood. But let us turn from the loathsome contemplation of the

---

70 Ibid.
71 Singh, *Race and America's Long War,* 43.
72 Ibid.

degrading effects of commercial avarice."[73] It is risible that Jefferson cannot recognize his philosophy and recommendations are at every point continuous with England's practices of establishing, maintaining, and expanding its empire through a logic of racial capitalism. Andrés Reséndez argues that what distinguishes Native American slavery from its African counterpart in the Americas was that it operated in subterfuge, misidentified as other breeds of forced labor such "encomiendas, repartimientos, convict leasing and debt peonage," and progressing "like a deadly virus, Indian slavery mutated into these strains and became extraordinarily resistant through the centuries.[74]

Many of the practices that governed the newly founded United States of America were the product of sixteenth-century England. The privatization of the commons through the enclosure movement generated enormous wealth for the few and widespread poverty for the majority of England's agrarian population. Their displacement from the commons strategically positioned them as a new settler class that could be dispersed throughout the new American colonies. Those would be figured as "promised land," meaning that these settlers would be entitled to Indigenous land upon completion of the terms of their initial indenture. Roxanne Dunbar-Ortiz credits the elevation of the status of the property as the cornerstone of America's drive toward independence from England. As the settler class enriched itself, land was "taken from Indigenous farmers and of Africans as chattel," and they became confident enough to produce their own class of landed gentry, making economic dependency on the home country wholly unnecessary.[75]

The planted settlers of America had emerged as a planter class expressly by visiting upon Africans and Indigenous peoples the tactics of alienation and extermination that were perfected

---

73 Jefferson, "Thomas Jefferson to Alexander von Humboldt, 6 December 1813."

74 Andrés, Reséndez, *The Other Slavery: The Uncovered Story of Indian Enslavement in America* (Boston: Mariner Books, 2016), 10.

75 Roxanne Dunbar-Ortiz, *An Indigenous Peoples' History of the United States* (Boston: Beacon Press, 2014), 35.

in England on the Welsh, Scottish, and Irish through their racialization and subsequent classification as "lower species."[76] By the time they reached the Americas, "their methods of eradicating peoples or forcing them into dependency and servitude were ingrained, streamlined, and effective" in breaking down the native social order.[77] In every case, English settlers did so by violently appropriating the wealth of existing civilizations including their peoples, stealing "already cultivated farmland" and established crops, seizing upon land clearances, using "existing roads and water routes," and capturing Indigenous intelligence to identify local sources of water, food, and medicine.[78] The objective in all instances was turning these resources toward rent for a small concentration of "investors, monarchies, and parliamentarians."[79]

The expansion of this modern ontological project of equating the privilege of subjectification with the appropriation and differential classification of human value has resulted in what Jonathan Beller argues are "the very methods and techniques of capital expansion: banking, management and communications infrastructure, monitoring of work-flow, inventory tracking, and the increasing integration of all human processes with methods of account."[80] For capital to become operational in this manner, it must render information an agent of differentiation, which categorizes not only bodies but the whole of their environments, to reduce the world to the quantifiable category of abstraction. Herein the global dispersal of information becomes the stuff of virulence compromising the boundaries of Indigenous intelligence and recalling the function of a colonial administration to make way for the algorithm to emerge as the arbiter of difference. The advent of cybernetics and information management in the latter half of the twentieth century paved the

76 Ibid., 39.

77 Ibid., 40.

78 Ibid., 46.

79 Ibid., 43.

80 Jonathan Beller, *The Message Is Murder: Substrates of Computational Capital* (London: Pluto Press, 2017), 8.

way for intelligent machines to take over "the operating system of heteropatriarchy and racial capitalism," which yet again was the product of sixteenth-century England.[81]

The Anthropocenean categorization of humanity as an endangered species replicates the financial logic of conserving life and letting die that subtended the management of American slavery. Like slavery, it operates through a system of "graduated expropriation," spreading out from the private domain to the commercial territory. Information here takes over from language as the preferred means for classifying types. Information will, in turn, be enunciated as "a difference that makes a difference, this difference is, in any and all instances, social after all."[82] Difference-as-information makes life numerous, and in so doing, fungible. It also creates conditions for the denial of life, based on its inability to articulate alterity into Blackness and replaces a system with double-entry bookkeeping through double-negation that sets up a pathologistics of recorded value.

Novel "prejudices, hatreds and phobias" find a voice through "various and dynamically evolving racisms, sexisms, and nationalisms" that emerge as symptoms of "social suppression."[83] These form the antecedents of viral life, pointing toward the techniques of surveillance and the logistics of travel that act as vectors of compromised transmission. The histories and practices of plantation management instilled within modernity the metrics of valuation and loss. Contemporary computational practices associated with technological capitalism enliven risk as part of the informatic organization of life entering it into the ledger as a multitude, an even more radical form of dispossession.

---

81 Ibid.
82 Ibid., 14.
83 Ibid., 159.

The Virus as Microaggression

Tobias Rees, the founding director of the Berggruen Institute's program Transformations of the Human, has recently written that he has to "come to think of COVID-19 as a great 'un-differentiation event,'" meaning "that COVID-19 systematically undoes the differentiation of the human from nature that first occurred in the early modern period."[84] Rees predicts that within the span of the current century, viruses will overtake the power of humankind. In so doing, they will reeducate us to conform to their operating systems. Even now, "viruses" have come "to regulate the carbon cycle" and therein "un-differentiate us from the biosphere."[85] As a consequence, life dwelling on Earth has already been enveloped "in a viral cloud," acting as the primary medium through which life has been shaped throughout the long history of humanity and in all likelihood will increasingly be so.[86] What makes the current situation remarkable is that humanity is not just classified as a species differential to all other animals rather that we have been "a multi-species assemblage" all along. In the present era of the Anthropocene, that has meant that humans are "inseparably connected, interwoven, porous" when cast in relationship to a viral world.[87] Rees argues that this realization radically undermines "the modern configuration of the political" by rendering that modern concept "(with its clear-cut distinction between human things, natural things and technical or artificial things) not only implausible but untenable."[88]

According to Rees, the political and ideological blueprint of European colonization corresponds with an "'out there' concept of nature: the nonhuman, the world of animals" so as to provide the rationale for their violent domination and exploitation to

---

84 Tobias Rees, "From the Anthropocene to the Microbiocene," *Noema,* June 10 2020, https://www.noemamag.com/from-the-anthropocene-to-the-microbiocene/.

85 Ibid.

86 Ibid.

87 Ibid.

88 Ibid.

serve the interests of burgeoning capitalism.[89] Humanity existed
for hundreds of years in a continuous war with these entities
deemed lacking in the basic intelligence to govern their existence
and yet capable of being assigned functional equivalency to
machinic forms of labor. Rees maintains that these bodies can
be denied ascendency with the hierarchy of life, but viruses
stand as the great leveler of all life. It is they who making thingly
humans, animals, nonhumans in equal measure. Rees proposes
that humans conceive of viruses as the standard-bearer of life.
He argues that their supremacy over humanity lies with their
natural properties of "interconnectedness," "self-maintenance,"
and "symbiosis."[90] Perhaps not coincidentally, these are the three
central tenants of cognitive computing. This is never explicitly
touched upon by Rees; rather, he chooses to express his
enthusiasm for the progress of these values, via the emergence
of COVID-19, which he credits with opening up "a field of sheer
endless possibilities" for viruses to approach humanity and
"many other life forms" as terra incognita.[91]

Ironically, Rees casts this particular strain of virus, COVID-19,
as one that is uniquely capable of undermining "the concept of
the primitive as distinct from the modern."[92] So too, he credits
it with dissolving "the logic of colonialism" because it "renders
untenable the philosophy of history — history seen as a strictly
human affair, unfolding in strict separation from nature."[93] His
argument falters when he refers to "the human as such — set
apart, independent, living in its own non-natural (artificial)
environment" because such conditions have never been maintained by humans in their majority in the modern era.[94] This is
particularly true of humans viewed as continuous with their environment through the denial of their humanity as their species
of origin. There was no condition of historical or philosophical

89 Ibid.
90 Ibid.
91 Ibid.
92 Ibid.
93 Ibid.
94 Ibid.

independence these bodies could lay claim to that would logi-cally separate them from their condition of subordination let alone, "from the bacteria and fungi and viruses that live in and on the human body."⁹⁵ Nature is not the locus of their situation, but it does form its ground for deployment as a device of ac-cumulation by dispossession. If it is "dissolved," it will no more bring humanity back to a condition of "evolutionary emergence" or rewilding than its modernist predecessor.⁹⁶ Instead, is likely to move us forward to a future where technology is naturalized along similar lines of universal assumption and implicit bias. If Rees, suggests in this chapter of the future, "in principle, there would then be no difference between technical innovation and biological innovation," it leaves open the question of why the terms of innovation itself remain unreformed in their colonial nostalgia for the advancement of technology to inoculate "hu-manity" from the worst aspects of planetary decay.⁹⁷

Rees asks, "what becomes of technology as we recognize that some of our most advanced technologies, like antibiotics or plasmids or CRISPR-Cas9, were not invented by humans but by microbes?"⁹⁸ This question, in modernist terms, is akin to ask-ing if you would take a treatment invented by a woman or per-son of color. In the early eighteenth century, the African slave Onesimus taught his English master Cotton Mather about the centuries-old tradition of inoculation practiced in Africa. One-simus was a critical agent in bringing that practice into being in the Americas and saving countless lives. Smallpox, like COV-ID-19, spread through the saliva droplets in an infected person's breath. The viral material to inoculate against it came from rub-bing a pustule on the skin of a mildly infected person and then transferring that material into the skin of an uninfected person.

Despite being at the forefront of knowledge concerning in-oculation in the American colonies, Onesimus was not given

---

95 Ibid.
96 Ibid.
97 Ibid.
98 Ibid.

the privilege to administer this simple practice of scratching the skin of another person. Rather, it was Dr. Zabdiel Boylston, who was chosen by Mather, as the best person to experiment with the procedure. Dr. Boylston initially did so by purposely infecting his son and two of his slaves in this manner with the virus. Here the virus serves the function of transmission, as well as penetration, to achieve individual immunity. Like its proposed microbial successor, it cannot become productively operational without the aid of human intervention. This power of intervention is a very limited commodity insofar as neither his son nor the two slaves had any rights of refusal to open themselves up to the virus because they were not considered fully developed human subjects. In this sense, they share some degree of diminished status with the microbial agent soon to be inside them. Left to its own devices, the virus would likely persist on a course that privileges innovation to ensure its futurity over that of individual human beings. Manipulated by a white male doctor, it has little choice but to submit to its constraint of application, bend toward an even weaker status, and eventually concede to its own abolition.

Rees contends that the arrival of "COVID-19 allows us to take on as a project the human" by inverting the terms that we normally associate with the colonial body, and projecting them into a future context where the microbe itself becomes the overseer of such experimentation."[99] Here, the human subject coded as white meets its end-stage as the supreme arbiter of life and death. In the future, difference favors an auto-immuno-politics, where white humanity acts "as a joint management system" with technology acting as a free agent to impose standards of care and recognition.[100] These will have to do with the acceptance of new protocols associated with industrial biology and artificial engineering that "enables the autonomous making of things that could exist innately but don't."[101] What Rees fails to

99 Ibid.
100 Ibid.
101 Ibid.

account for in this scenario is what Neel Ahuja refers to as "the tenacity of race" as something that "inheres" to culture, meaning that it makes its presence known through various vectors of "materialization" that supersede both containment and form.[102] Race, in this scenario, benefits specifically from its ability to circulate through and across the human body, making its authority simultaneously one borne and undermined through the risk bearing prospect of transmission. Within that operational substrate there are viruses, bacteria, animals, and minerals that must be either assimilated as nutriment or remedy or rejected as extract or toxin. It as this level that the racialized body must be dealt with as a cypher fit for entry into a greater domain of environmental risk whose liveliness and durability, or lack thereof, determines its cause.

It is at this juncture that Rees proposes a "Microbiocene" that would take over from the Anthropocene. Rees's proposed "Microbiocene" intersects with Fernando's concept of the "Virocene" as "a historic moment in which interoperation between human and nonhuman actors becomes existentially threatening on a planetary scale."[103] In Rees's classification of the Microbiocene as an age of "whole-earth politics," it reveals itself as yet another form of "universality is grafted onto other universals."[104] In seeming to lift the boundary between nature and humans, it suggests both a comprehensively earth-bound, profoundly ordered, and materialized approach to living organisms. At the same time, it allows humankind to remain sheltered from contemplating the chaotic forces of a greater, sentient universe beyond its governance.

What Rees is suggesting is that biology becomes the raw material of capital where any number of organisms can be classified as potential laborers. If life is programmable in these sorts of scenarios, then so is death in the form of managed extinc-

---

102 Neel Ahuja, *Bioinsecurities: Disease Interventions, Empire, and the Government of Species* (Durham: Duke University Press, 2016), 12.
103 Fernando, "The Virocene Epoch," 639.
104 Ibid., 645.

tions. The new boundary is one where biology resides within the commercial domain of information science. Stefan Helmreich and Nicole Labruto observe that "most contemporary biocapital emerges from exploitative, neoliberal models of commodification and circulation — trans-infecting biologies, ideologies, and markets from the inside out."[105] They predict that "as postgenomics, environmental remediation, climate change amelioration, global disease eradication, and resource scarcity motivate new biological research platforms, biocapital will see new inventions, edits, contaminations, and wirings, yielding new species of bio-capital for scholars to probe critically."[106]

This new form of behavioral biological research, classified as neurological in its presentation, suggests that, in similarity to its precursor, it seeks to prey upon the affective and libidinal outputs of those organisms it desires to manipulate. It pervades into the very bodies it objectifies, colonizing the awareness of those whom it seeks to entirely capture. Information, as opposed to an organized life, becomes the rendered asset of their work. This shift allows for capital's realization of itself as subjective, affording it the privilege to coerce and extract value from the active domains of culture, communication, and cognition.

This exchange of the algorithmic for the biotic body of man makes use of speed and data to register specific advantages that computers have over humans. It perpetuates the mechanisms of racism insofar as it allows humanity to become the scapegoat for social ills, while artificial life emerges as a permanent means through which to subdue its worst impulses. It suggests a humanity that is internally ungovernable in its inherent plurality. Evidence of this condition is reflected in the conduct of numerous internal enemies, which increasingly adopt a biological valence. Society must be defended and a reaction to crisis devised to ensure the constant neutralization and elimination of threats

---

105 Stefan Helmreich and Nicole Labruto, "Species of Biocapital, 2008, and Speciating Biocapital, 2017," in *The Palgrave Handbook of Biology and Society*, eds. Maurizio Meloni et al. (London: Palgrave Macmillan, 2017), 869.
106 Ibid.

associated with their potential to "go viral." The logic of such vigilance ensures the growth of the digital apparatus to survey sporadic outbreaks of infection and attend to their sanitation beyond the purview of general understanding.

The human, in this scenario, must be placed on constant display for the edification of machinic learners, who flourish from the intake of their information both in life and in death. Human identity is under the control of the database, subject to eradication at any time if it is somehow lost or compromised. The dysfunctionality of such a system can only be maintained if it seems like the only option. The dynamics of race have shifted with this realignment of power. Race no longer becomes simply a feature of recognition and, by extension, a substance circulating within the public domain. In the grafting of the racial dynamics of the physical world onto the species dynamics of the virtual world, we can witness how the former refugee of interior life, replete with all of the exuberance of the interior, remained wild, unbound, and free of the predations of the market. We can also witness that interior's erosion coincides with the progress of the digital revolution. The modernist concept of "radical consciousness" for those who occupied positions of racial and sexual subordination was premised on the belief that the avarice of the market could not breach the interior spaces of imagination, revere, intimacy, and privacy. Therefore, these became areas of active resistance within the body politic to the incursions of white settlement.

The exterior world has now been overdetermined by that same settler class seeking out new terrains of commerce and dispossession. They are specifically concerned with consciousness as the final frontier and the greater limit of humanity. In the past, these commercial movements have sought to promote an acceptance of social exclusion "as a prison house of struggle that internalized" and "marginalized subjects" who could only be empowered by seeking public recognition and indeed public recompense through offering "first-person" accounts of their suffering.[107] These acts rarely assuaged anxieties or altered expectations of how such exclusion should be treated but did much to reify marginal identities that were synonymous with perennial mistreatment. It hardens differentiation at the level of recognition and perpetuates essentialism.

In the context of his critique of the meaningfulness of race and Black cultural nationalism, Kevin Quashie promotes in its place "the act of imagining" as a more malleable and expansive political practice.[108] He argues that "a willingness to dream, speculate, or wonder […] helps us move the limits of reality. Imagination is the landscape of such dreaming, […] a place and process that is particular to human capacity. In this later context imagination is an interiority, an aspect of inner life that constitutes an essential agency of being human."[109] Quashie's emphasis on imagination, one that takes place as a distinct location of human activity with the capacity to move the limits of reality, speaks to its singular value as an intricate process of self-making. It is exactly for this reason that it has been recently developed as an area ripe for commercial capture. What is considered aberrant and inhuman becomes the stuff of market inventiveness. It is exactly at this level of polarization between normativity and deviation, human and inhuman, that a contest of propriety is playing out in this century which can capture the territories of the mind without appearing to advance a step beyond the playbooks of war, colonialism, and patriarchy in exercising the rules of confinement and limitation for the human race.

Whereas it was previously possible for what Quashie terms "the quiet subject" to "find agency in the capacity to surrender to his or her interior life," today such a capacity is being profoundly compromised by technology that seeks to crossover into that formerly neutral domain, making of it a target of astonishing acts of violence carried out in the syntax of algorithmic order.[110]

---

107 Kevin Quashie, *The Sovereignty of Quiet: Beyond Resistance in Black Culture* (New Brunswick: Rutgers University Press, 2012), 33.
108 Ibid., 42.
109 Ibid., 41–42.
110 Ibid., 129.

That these remain unequally distributed along lines of race and gender does not belie the fact that they are aimed at imposing universal casualty to what has been formerly construed as human potential.

Benjamin Rush's Redressing of Racial Contagion

Benjamin Rush, a prominent physician and co-signer of the American Constitution, devoted his career to redressing the malady of Blackness visited upon the new American republic. Rush was Thomas Jefferson's political and scientific contemporary. Ben Bascom argues that "Jefferson's desire to remove blacks from America through colonization to Africa" intersects with "Rush's fascination with eliminating racial difference" at the point at which race assumes the place of "instantiation" within the republic.[111] In either case, the continuous subjection of Black bodies emerges as vital for America's assumption of sovereign power. "Rush was also suggesting that Africans, once no longer enslaved and cured of their blackness, could eventually assimilate into the new nation as full, equal — and white — American citizens."[112] Rush's fascination with the eradication of racial difference from America's nascent biopolitical order coincides with the extension of rights and obligations afforded to white republican citizens. Among these, the denial of refuge to Blackness must necessarily feature. Blackness must be proposed as an "unnatural" condition, present within the bodies of those who would, if not for their pathology, reassume the "natural" appearance of whiteness.

Rush's scientific ideas appealed to white abolitionists because it not only held out the prospect that slavery would eventually disappear from the new republic over time, but that Blackness

---

111 Ben Bascom, "Queer Anachronism: Jeffrey Brace and the Racialized Republic," *Arizona Quarterly: A Journal of American Literature, Culture, and Theory* 75, no. 1 (2019): 28n9.

112 Eric Herschthal, "Antislavery Science in the Early Republic: The Case of Dr. Benjamin Rush," *Early American Studies: An Interdisciplinary Journal* 15, no. 2 (Spring 2017): 275.

itself would inevitably disappear. Rush's desire to redress the issue of skin color was in response to the dramatic rise of the new
nation's free Black population who took advantage of a slew of
northern emancipation decrees to escape north toward their
freedom. By the 1790s. Rush's Philadelphia, saw a "free black
population increased sixteen fold, from 114 in 1775 to 1,849 in
1790; a decade later, it more than tripled, to 6,028; only eighty-
five enslaved people remained in the city."[113] Edward Allen Driggers maintains that "Rush believed that black skin was indicative not just of variation but of disease."[114] Rush, amongst other
physicians of his day, was convinced that if he "could understand why the body produces black skin, he could reverse the
effects."[115]

Rush was the first to use the term "Negritude" and he did so
"to denote blackness as a form of leprosy" with the implication
that Blackness itself was a disease in search of a cure in essentially becoming white.[116] Rush attributed Blackness to a secretion, or "fluid," that "darkened skin color" such that it "could
be rubbed or washed off the body or absorbed by it" depending on the circumstance. Such was the thinking that "medical
practice at that time led Rush to prescribe bleeding, then, as a
sensible treatment for Africans' pathological dark skin."[117] Rush,
like many scientists of his ilk, believed that all diseases benefited
from either the administration of stimulants or depressants to
the nervous system. What Rush added to the equation was his
targeting of the blood vessels to achieve this end through the
vehicle of blood-letting.

---

113 Ibid., 292.
114 Edward Allen Driggers, "The Chemistry of Blackness: Benjamin Rush,
    Thomas Jefferson, Everard Home, and the Project of Defining Blackness
    through Chemical Explanations," *Critical Philosophy of Race* 7, no. 2 (2019):
    398.
115 Ibid., 403.
116 F. Bart Miller, *Rethinking Négritude through Léon-Gontran Damas* (Am-
    sterdam: Editions Rodopi, 2014), 10n4.
117 Driggers, "The Chemistry of Blackness," 401.

This painful, repeated process was apparently necessary to restore feeling in Africans considered largely nerve dead due to their apparent leprosy. Rush contended that "leprosy induces a morbid insensibility in the nerves" of Black bodies.[118] Rush asserted that "this insensibility belongs in a peculiar manner to the 'negroes' in that they are void of sensibility to a surprizing [sic] degree."[119] Chattel slavery was associated in Rush's mind with "the absence of the stimulus of the love of liberty," and as a consequence of this "he wrote, 'animal life exists in them in a feeble state.'"[120] The effects of slavery were deadening to the Black body which required particular "external stimuli," like "the heat of the sun" in warm countries and "large quantities of animal food and ardent spirits" in cold ones "to literally re-invigorate them."[121]

The idea that racial pathology could be environmentally ameliorated implied a certain acknowledgement, on Rush's part, of the fluidity of race itself. This became everywhere apparent in the burgeoning scientific, and now offensive, categories of Blackness ranging from "quadroon" to "mulatto," "octaroon," and "hexadecaroon" and the increased emphasis on the ways that the social and natural environments could influence their comportment. It was not race but the condition of slavery that caused so much of the malady Rush witnessed in the slave population. Rush accounted for the condition of "Jaw Fall," in which an enslaved person's mouth remained firmly shut, as a consequence of "enslaved people 'eating less animal food than white people.'"[122] The "Difficult labors" African slave women endured, he again argued, stemmed "from the injuries to the pelvis from kicks and carrying weights when young."[123] Another disease, termed "Cachexia Africana," deemed common to enslaved Af-

---

118 Benjamin Rush, "Observations Intended to Favour a Supposition That the Black Color (As It Is Called) of the Negroes is Derived from the Leprosy," *Transactions of the American Philosophical Society* 4 (1799): 292.

119 Ibid.

120 Herschthal, "Antislavery Science in the Early Republic," 282.

121 Ibid.

122 Ibid., 290.

123 Ibid.

ricans had as its "main symptom, dirt eating."[124] Rush renamed it "Dirteatis" and insisted that it "occurred it only 'after they enter upon the miseries of their slavery.'"[125] While he was quick to discredit the notion that these diseases were inherently rooted afflictions, Rush, nevertheless, failed to go beyond reconceiving their origins in making his case for the abolition of these practices associated with the inherent deprivations of slavery.

Rush could not give meaning to the pathologies that continued to plague former slaves or free-born Black people within his midst because he could never adequately account for the reason of the repudiation of their value. Even under ideal environmental conditions of liberty, he could not overcome his "fear of interracial sex," which he shared "with many men of science, most notably his friend and political ally Thomas Jefferson."[126] There could be no marriage between former slaves and the ambitions of the republic moving forward. Indeed, that experiment in democracy could not help but be tainted by the issue of contagion should whites have continued to invite contact with African Americans following the abolition of slavery. The prospect of freedom effectively meant that Black citizens maintained no meaning at all in a white society. As such, the possibility of eventual Black citizenship demanded that white citizens treat African Americans with "a double portion" of their "humanity" accounting for their lesser portion by half to the makeup of a white settler.[127] Finally altogether the case for their consideration simply did not add up.

Like his friend Jefferson, Rush believed that it was necessary for the survival of the young republic that great men such as they overwrite "the pluralism of indigenous and even early modern imperial space" to discipline the "patchwork of human and natural plurality that characterizes" the American continent

---

124 Ibid.
125 Ibid.
126 Ibid.
127 Rush, "Observations Intended to Favour a Supposition," 295.

in this founding period.[128] The settlement of the issue of property and exchange when it came to possession of the Black body through the abolition of Blackness itself would eventually work to flatten out difference to the degree that it would be then possible to impose "a stable and perpetually extensive grid of commerce and uniform title" upon the nation as a whole.[129] Theirs was a project concerned with both racial security and racial engineering that called for a redefinition of racial proprietorship. Blackness could not be granted full humanity at any point within that project due to the dimension of economic dependency associated with its geopolitical significance. America could not be a sovereign nation, in Jefferson's mind, until America's provision of labor and energy could be acquired by alternative means. Jefferson's philosophy of race is, therefore, always commercial in nature. For this reason, both Jefferson and Rush must be concerned about Blackness in its abject capacity to reproduce itself in dependency of whiteness.

Rush who assumed that all Black peoples were in fact "lepers" nonetheless marveled at their remarkable "fruitfulness when they are not depressed by slavery; but even slavery in its worst state does not always subdue the venereal appetite, for after whole days, spent in hard labor in a hot sun in the West Indies, walk five or six miles to comply with a venereal assignation.[130] Rush's assertion that "having strong venereal desires" is "universal among the negroes," made the case for their leprosy to be construed as an infectious disease transmitted through intimate skin to skin contact, but as well as through sexual congress.[131] Thus, it is passed on both immediately to the sexual partner and generationally in the offspring of such conjoining. Rush gave the example of a white woman in North Carolina who "not only acquired a dark color, but several of the features of a negro by marrying and living with a black husband. A similar instance of

---

128 Matthew Crow, "Jefferson's Whale: Race, Climate, and Commerce in Early America," *Journal of the Early Republic* 40, no. 3 (Fall 2020): 436.
129 Ibid.
130 Rush, "Observations Intended to Favour a Supposition," 293–94.
131 Ibid., 293.

a change in the color and features of a woman in Bucks County in Pennsylvania has been observed and from a familiar cause. In both these cases, the women bore children by their black husbands."[132] It is not color but equally features that make the case for this acquisition of Blackness. Somewhat curiously Rush attributed "the woolly heads of the negroes" as not only proof of their leprosy, but also proof of some diseased kinship with another "degenerated" race for whom Rush sees "no difficulty in admitting that it may as readily have produced wool upon the head of a negro, as matted hair upon the head of the Poles."[133] Rush's example implied these types of aberrant traits might be glandular in their biological manifestation.

Shortly thereafter, Thomas Jefferson made his proto-endo-crinological study of Africans witnessed from his slaves:

They have less hair on the face and body. They secrete less by the kidnies, and more by the glands of the skin, which gives them a very strong and disagreeable odour. This greater de-gree of transpiration renders them more tolerant of heat, and less so of cold, than the whites. Perhaps too a difference of structure in the pulmonary apparatus, which a late ingenious experimentalist has discovered to be the principal regulator of animal heat, may have disabled them from extricating, in the act of inspiration, so much of that fluid from the outer air, or obliged them in expiration, to part with more of it.

They seem to require less sleep. A black after hard labour through the day, will be induced by the slightest amusements to sit up till midnight, or later, though knowing he must be out with the first dawn of the morning. […] They are more ardent after their female: but love seems with them to be more an eager desire, than a tender delicate mixture of senti-ment and sensation.[134]

---

132 Ibid., 294.
133 Ibid.
134 Jefferson, *Notes on the State of Virginia,* 148–49.

Driggers observes from this same publication, that "Jefferson argued that 'natural distinctions' between Africans and whites come from humoral differences: Jefferson, like many other thinkers across the late eighteenth and early nineteenth centuries, cited the bile as causing the black color of the African skin."[135] If "blackness was a fluid imbalance that chemists thought that they could change and manipulate" and indeed "some medico-chemists thought that blackness was literally a fluid state, in which people could change their color" it becomes possible to contemplate a solution to Blackness that favored interference in the form of environmental engineering.[136] Thomas Jefferson's "thinking about the watery and warming world in which he lived followed from his thinking about race, empire, and nationhood."[137] Moreover, "he welcomed climate change and favored environmental engineering on a staggering scale as at least potentially supportive to his project of protecting and advancing the geopolitical security of white settler democracy and gendered proprietorship."[138]

Thomas Dikant argues that Jefferson's *Notes on the State of Virginia* should be read not only as a "work of natural history," but equally "as a work of statistics."[139] Jefferson was not only interested in managing life, but also forecasting it. America's natural resources and Native inhabitants required evaluation to analyze and project how they could be transformed into wealth through their controlled circulation. Dikant asserts that Jefferson uses statistics to "envision how 'the world's first settler-colonial state' ought to manage its populations to safeguard the state's vitality and long-term survival."[140] Numerical data becomes the stuff of life as "almost everything is put into numbers, almost eve-

135 Rush, "Observations Intended to Favour a Supposition," 406–7.
136 Ibid., 410.
137 Crow, "Jefferson's Whale," 438.
138 Ibid., 437.
139 Thomas Dikant, "Settler Colonial Statistics: Jefferson, Biopolitics, and Notes on the State of Virginia," *Early American Literature* 54, no. 1 (2019): 69.
140 Ibid., 71.

rything is measured and weighed in his account of Virginia."[141] Virginia becomes an early product of materialist thinking that allows for Jefferson to account for what he believes to be the capital threat to the future prosperity of his nation; namely a slave population that outgrows the white settler collective. Jefferson foresees a way to control the settler-to-slave ratio through the quantification of Black blood within the white settler body: his definitions favored the admixture of white blood with Black and also favored the generational thresholds resulting in a person effectively being counted statistically as white.

With such an intervention Jefferson can affect a decrease in the growth of the Black population, forestalling the possibility of the white settler collective becoming a demographic minority. Whereas Blackness could be characterized by Jefferson as a blot in the record of American racial identity, its obfuscation through a complex mathematical understanding lessened its further consequence. Through his creative quantification, Jefferson had effectively reengineered race as a product formerly associated solely with contagion, into a property that could be emancipated from people to the degree that it could be rendered invisible. "In Jefferson's statistical table, all emancipated mixed-race persons who are the product of least two generations of interracial sex with white persons and one generation with a white or mixed-race person would simply figure as 'free inhabitants.'"[142] Thus, the unfortunate difference of color, and perhaps of faculty as Jefferson imagined them, was erased from the public record. Jefferson's plan to counteract the republic's demise was carried out effectively through sexual recuperation.

The Immortal Life of Jefferson's Virginia

The contagious nature of such an endeavor becomes the stuff of our current pandemic which disavows the place of politics with the space of molecular biology and viral engineering. Such

---

141 Ibid.
142 Ibid., 82.

research cannot begin without the exploitation of Black bodies from whom the first immortal cell line was taken. Henrietta Lacks died from cervical cancer in 1951. She died unaware that cells from her cervical tumor were harvested from her impoverished Black body by her eminent white male physician George Gey. He would later stand to make a fortune from her dead flesh. Lacks was already socially dead as a poor Black tobacco farmer from southern Virginia suffering from aggressive cervical cancer from aged 30. Lacks was raised in a two-story log cabin that once had been slave quarters on the plantation that had been owned by Henrietta's white great-grandfather and great-uncle. She bore her first child at just fourteen years of age. The baby was conceived incestuously with her first cousin David, whom she later married. This was by no means the first instance of incest within the Lacks genetic line. This meant that it was likely that copied deleterious genes being passed down for generations perpetuating unique signatures of disease, disorder, and mutation. Lacks's cancer was discovered when went to Johns Hopkins to give birth to her fifth child. She went there because it was the only hospital in the area that treated Black patients. Within a year Lacks would be dead. Her cervical cancer had metastasized throughout her entire body, despite having received cancer treatment there.

The story of Henrietta Lacks's life in many ways begins with her anonymous ending and the start of her artificial afterlife as "HeLa." Her new posthuman name was based at once an abbreviation of the first two letters of her first and last names and a strategic anonymization so that no one could lay proprietary claim to her cells apart from the small group of medical researchers at Johns Hopkins who patented her tissues. They then went onto commercialize her as a sexualized and racialized sample body to be passed between laboratories throughout the world. These labs would often virally contaminate her in the process of manipulating her tissues in the quest of their future profit making by inadvertently introducing "new" human pathologies into the mix during their experimentation. HeLa cells were experimentally subjected to cancer, AIDS, radiation, and other toxic substances throughout their global journey and generated untold wealth for the biomedical industry before being compromised in the 1970s beyond all formal recognition. HeLa acquires her value in death, whereas Henrietta Lacks acquired hers as death. Both conditions rely on her body to produce the modern world, genetically mapping its development through her corrupted signification. Nonetheless, it is all there. Henrietta Lacks's cell line, the genetic offspring of coercive Southern plantation sex, survives over the course of two generations to allow her to become illegitimate "mother" to "virology."[143]

Literally cut from the human-animal flesh of her maternal body, her cell line graduated to function as a laboratory "workhorse" bent to the service of a new class of masters.[144] These scientists exploited her cell line, viewing its ability to reproduce and "repair itself when damaged" as evidence of its "organic durability."[145] Her denatured DNA, like the female enslaved body through whom it was manufactured, exists within a myth of racial forbearance that has and continues to "facilitate medical breakthroughs and launched several fields, including human cell tissue culture and bio-technology."[146] This took place in much the same ways as it did for her ancestors who involuntarily served to further the sciences of epidemiology and gynecology in the antebellum era. Henrietta's social death is partially rehabilitated through her genealogical connection to a white great-grandfather. That fact allows her to differentiate herself from "the secondary condition of slavery" experienced by most of her peers in Jim Crow-era southern Virginia.[147] This happened to the degree that she could be housed by and with her ancestors. Through this spectral feature of the plantation, she can integrate the experience of their lives dwelling in slave

143 Marlon Rachquel Moore, "Opposed to the Being of Henrietta: Bioslavery, Pop Culture and the Third Life of HeLa Cells," *Medical Humanities* 43, no. 1 (2017): 55.
144 Ibid.
145 Ibid.
146 Ibid.
147 Ibid., 56.

quarters into her contemporary experience of social internment within the only local hospital that would accept Black patients and allowing her, as Orlando Patterson observed, "to inform the understanding of [her] social reality with the inherited meanings of her natural forebears," as well as those violent couplings considered by polite society to be "unnatural."[148]

Lacks's body was as much the product and property of the cabin as her flesh. Therefore, the hospital taking her cells from her was not a violation of her privacy for the very fact that her apparent Blackness already sealed the contract of their dominion over her body. Lacks's reference to this former slave quarters as her home "does not change the fact that it is the spatial extension of the master's domain."[149] In overdetermining her biological Blackness based solely on her appearance, Marlon Rachquel Moore argues, "we can see how Henrietta is denied all claims on, and obligations to her living blood relations and, by extension, all such claims and obligations to her descendants."[150] HeLa, rather than Lacks, becomes a sort of super-progenitor determining the trajectory of biological futures around the world in a way that her analogous position as a Black woman within the American structure of racism could never allow her to do. Becoming cellular has allowed her flesh to surmount the limits of racial Blackness and to acquire value "in death/as death," as was the destiny of her enslaved ancestors, classed as "non-human animal beings" whose realities were occupied through flesh alone.[151] HeLa, in Lacks's apparent immorality, connotes something altogether different in terms of a different order of consumption, a new order of capitalism where life gets traded in a way not wholly human. As such, it relies on the performance of dispossession in ways particular to a performance of inhu-

---

148 Ibid.

149 Wilderson, *Afropessimism*, 227.

150 Moore, "Opposed to the Being of Henrietta," 57.

151 M. Shadee Malaklou, "'Dilemmas' of Coalition and the Chronopolitics of Man: Towards an Insurgent Black Feminine Otherwise," *Theory & Event* 21, no. 1 (2018): 226.

manity; that is to say, of social death that has in the twenty-first century taken on the capacity to "go viral."

Here Hortense Spillers's description of the Black feminine figure as the "zero degree of social conceptualization" meets with its one overarching task to reproduce racial capitalism on the global level.[152] It provides a foundation for racial capitalism that must now operate through the calculation of the degree to which something is recognized as human against the index of being. Blackness takes its place on that spectrum as the basis from which to measure human progress without an essential way of being registered as such. The Black body is figured rather as an asset class onto itself and rendered valuable according to the racial, social, legal, and economic status quo of its living present. Any consciousness of its particularity is held within a community of memory that, by definition, requires applied force in the creation and maintenance of a subtext for these conditional relationships.

M. Shadee Malaklou argues that "humanism's flesh-making" be understood as synonymous with "black-making" in its world-making project.[153] The suffering that it implies, with America's particular domestication of race as Christina Sharpe argues, "[disfigures] black maternity, [turning] the womb into a factory" where she transports them into the racial logics of her own always already commercial conditioning; as an object of non-status, and, therefore of nonbeing.[154] As HeLa, Lacks is no longer producing Blackness, per se, and thus the demand for her cells to function as another sort of container for gestation. Moore recounts how

[i]n 1992 [...] two evolutionary biologists published separate controversial papers arguing that HeLa should be classified as its own species. One of them, Leigh Van Valen, explained

152 Hortense J. Spillers, "Mama's Baby, Papa's Maybe: An American Grammar Book," *Diacritics* 17, no. 2 (1987): 67.
153 Malaklou, "'Dilemmas' of Coalition and the Chronopolitics of Man," 241.
154 Ibid.

in an interview that after (at the time) four decades of genetic transformation, the cells are now evolving separately from humans. "They have an extremely different ecological niche from us," Van Valen argues. "They don't mate with humans; they probably don't even mate with human cells. They act just like a normal microbial species."[155]

Moore argues that such attempts at classification compel us "to imagine a black woman as the progenitor of the posthuman — a presaging of the claims the evolutionary biologists make about HeLa."[156] I would argue that perhaps something altogether different is happening in the case of HeLa. Her cells stand in for something greater than the symbolic violence of America's original era of enslavement. They stand in for the endurance of the market itself to transcend the bounds of history and time and give itself ultimate power over both the makers and maintainers of life at the most basic level. The value of flesh in this instance has been overtaken by the cell, and "therefore resonates with the distinctly racialized contemporary bioeconomy — the neoliberal markets built upon life science industries such as IVF, tissue engineering, and gene therapy."[157] Neoliberalism may well be the offspring of accelerated research in the areas of molecular biology, cell biology, and microbiology, allowing for the manufacture of genetic, microbial, and cellular level life to form the basis of a new global economic ethos under which these microorganisms can be mobilized as their own asset class unfettered by traditional understandings of what constitutes productivity. Within such a context, life is construed as "a state of nascent transformability" enmeshed within a process of becoming that is altogether "self-regenerative, self-accumulative, and self-renewing."[158]

155 Moore, "Opposed to the Being of Henrietta," 58.
156 Ibid., 59.
157 Lisa Dowdall, "Treasured Strangers: Race, Biopolitics, and the Human in Octavia E. Butler's 'Xenogenesis' Trilogy," *Science Fiction Studies* 44, no. 3 (2017): 507.
158 Dowdall, "Treasured Strangers," 513.

Life takes on value through the appropriation of its talent for diversification within a market that insists upon trading on the organization of difference. It is only able to discipline life through the implicit maintenance and manipulation of these codes of life. Crisis is a necessary element of social reproduction because it instigates adaptation and improvisation wherein it is understood that humans will have to conform to a new set of standards to survive. Life must perform or be eliminated from supportive structures. It must be compelled to reach its potential and partner in order to exchange value with an index of species and within an environment reframed through artificial intelligence and machine learning that constructs an appreciation of life as data. Within this reframing of life-as-code, humanity is transformed into a morphogenic entity capable of being perfected through the interference of reprogenetic technologies; its posthuman expression the product of techno-scientific development.

The contemporaneous economy that springs from this enterprise perpetuates the scientific racism and inequality experienced under slavery and colonialism by African and African American women by excluding them from the radical possibilities of the posthuman through a genetic determinism that classes their Blackness as something surplus to life. Blackness always already exceeds the normal limits of human differentiation, and thus as genetic material, it is compromised. Blackness made pathological in this way implies that it carries within it physical and mental defects to act anything other than as the building block for other forms of life to prosper. On the other side of the equation, microbial life emerges as that which is superior to humanity in its capacity to divide itself against itself and exist without recourse to species identity. While withstanding "continuous, fierce pressures," it has "the general ability to persevere" as a global enterprise capable of producing "enormous population growth rates" without succumbing to any sort of planetary crisis.[159] In this sense, the microbe becomes the ultimate product of biological innovation, with not one of the apparent burdens of historical racism. HeLa joins the two halves of this equation.

The Immortal Life of Black Social Pathology

In the 2017 HBO film of Rebecca Skloot's best-selling book *The Immortal Life of Henrietta Lacks,* the main protagonist of the film is Skloot, followed closely by HeLa cells. Of Skloot, we learn very little. Her character wears an awkward pasted on grin whenever she has to interact with any of the living members of Lacks's extended family. HeLa cells, on the other hand, are from the very beginning painted in broad strokes and figured as a kind of colorful, beneficent force of life that remains un-daunted in the face of unrelenting commercial pressure. HeLa cells emerge as the poster child of neoliberal entrepreneurial-ism, which in life-after-death is now able to thrive after being released from the black and white limits of their former world.

The film ostensibly focuses on the lives of Lacks's surviv-ing children in their quest to find out what happened to their mother, and indeed, to discover a woman whom most of them have no recollection of. By the end, we learn precious little about Lacks herself. What is said of her by her loved ones is that she kept her hair nice, always wore red nail varnish and lipstick, and had good teeth. Essentially, she is all surface with no depth, be-yond the cavity of her womb which is portrayed as tragically diseased. The womb, in question, gets as much screen time as Lacks's backstory. Her discovery of her cancer comes to us as a voiceover by the character of Skloot describing how Lacks drew a bath and inserted a finger inside her vagina only to discover inside there was a large hard lump, which she somehow already knew would be there. The happens toward the very end of the film. Given what we learn of her family riven with pathology mental and physical, indeed how could it be otherwise.

---

159 Melinda Cooper, *Life as Surplus: Biotechnology and Capitalism in the Neoliberal Era* (Seattle: University of Washington Press, 2011), 39.

Lacks's daughter Deborah, played in the film by Oprah Winfrey, leads the quest to understand how her mother's identity came to be medicalized. She is portrayed by turns as either psychotic or possessed. Through a variety of painful and awkward scenes it becomes apparent that Deborah only comfortably "fits in" within the racially segregated world of Baltimore in the 1980s. Skloot enters this world in the guise of a pseudo-anthropologist observing the bizarre and mesmerizing suffering of the Lacks family, as though she as an independent scientist and journalist, plays no part in perpetuating the racial hierarchy apparent all around her. For most of the film, Skloot's character looks out of her depth and utterly terrified to interact with Deborah's siblings who have been variously incarcerated, institutionalized, defrauded, brutalized, and raped throughout their lifetimes. The film luridly displays many of their numerous incidents of profound grief.

The two main sites of action in the film are the slave cabin in Clover, Virginia, where Lacks was born and where it is implied had her five babies, and John Hopkins hospital, where Lacks met her treatment, death, and ultimate rebirth as HeLa. The most harrowing scenes of the film relate directly back to the traumatic legacies of slavery. Zakariyya, the youngest of Henrietta's children, is pictured as a boy being whipped on his back in the slave cabin by his aunt Ethel. Deborah, the oldest of the siblings, is subsequently raped there by her uncle Galen. Deborah must explain tearfully to Skloot how these traumas destroyed their lives and accounted for their mental and physical problems in later life. It was not genetics, but epigenetics, that were responsible for the failure of any of Lacks's five children to thrive. The scientists of the Jim Crow-era South were looking in the wrong place for answers in the blood of Lacks's offspring. They needed only to study the Lacks children to appreciate how intergenerational trauma had profoundly blighted the course of their lives and indeed, Henrietta's.

The problem with this reading discursively composed by Skloot, and perpetuated through Winfrey's filmic production, is that it must utterly divorce Lacks, and by extension HeLa, from that experience. We only see Lacks smiling, whenever she briefly features. She smiles and laughs as she feeds all and sundry starving male relatives coming up from southern Virginia to Baltimore to find work. She smiles when she is at the fair in Clover enjoying a night out with her girlfriends, where she discloses she has cancer through a gritted smile and a selfless request that they look after her babies. She raises a smile for them from her hospital room window where she is in the last days of her life. Those good teeth are everywhere a code for what is positive and unyielding about her genetic inheritance. After death Lacks is recast as an enchanted healer, rejoicing in the often-violent exploitation of her cells in commercial ventures set on posing a "cure" for humanity. Skloot's voiceover incredulously muses from on high that, at one point, her invulnerable cells were detonated in a nuclear bomb to test their capacity to outlive a planetary catastrophe. No mention, of course, is made of which executive cells would have been involved in the staging of this trial act for an impending apocalypse.

Zakariyya and Deborah are redeemed presumably after a lifetime of relentless suffering not by understanding very much more about the fate of the actual mother, but rather by visiting Johns Hopkins and being introduced to her cellular successor, HeLa. This introduction was made by white German cancer geneticist Christoph Lengaur, who was then an Associate Professor at the School of Medicine at John Hopkins. Lengaur dramatically places a small frozen vial containing HeLa cells into Deborah's hands and tells her that this is, in fact, her mother. She cradles it like a baby, cooing to it "mama, you are so cold." The scene implies that Deborah hasn't the intelligence to appreciate that it is not her "mama," but the unspecified multitude that is HeLa; a new species of life has effectively made microbiologically possible.

Lenguar explains to Deborah that not only this vial but all the vials in the lab hold HeLa potentially to infinity. Deborah doesn't appear to be able to comprehend this enormity; so much so that Lenguar is forced to employ a lesser means of trying to enlighten her. He decides the best way forward is to project HeLa's colorful cells onto Zakariyya's and Deborah's bodies as part of her collective body, thus including them amongst her multitude of descendants. In the scene, they are shown as though they are feeling HeLa on their skin, taking her presence in, and embracing her through this heavily staged "encounter" with an "alien" lifeform. Deborah, in particular, responds with excessive emotion to this experience, whereas Zakariyya is pictured to emotionally break open in the wake of being touched again by his long-lost mother. The tragedy of the scene is that, of course, HeLa cannot reciprocate their attention nor register care for their presence in the way of a mother. All HeLa cells can effectively do is continue their steady work of infinite division passively before their eyes.

The real spectacle of the scene is not HeLa's superficial projection onto the bodies of Zakariyya and Deborah, but it is rather Lenguar and Skloot's racist projection of their belief that they aren't sophisticated enough to appreciate anything beyond this sensational showcase and eager to naively accept this as evidence what happened to their "mother." This contradicts what is portrayed earlier in the film, which indicated that many years before this encounter with HeLa, they were aware of exactly what forms of commercial exploitation were visited upon the Black community in Baltimore at the hands of Johns Hopkins scientists. Nevertheless, by the film's end, all appears to be forgiven. Deborah on her death bed is portrayed as literally able to rest in peace in the knowledge that her Black mother saved the world, even though it wasn't a world that ever wished to fulfill the promise of her own salvation. HeLa was originally dubbed Helen Lane to whiten her cells for better sale with the commercial market. It would seem, with the discovery of Henrietta Lacks as the origin of HeLa, they remain in need of whitening still. Hence it is Skloot, not Lacks, who is left to narrate Black suffering into redemption in the kind of ending white audiences can appreciate. Perhaps what is the more interesting outcome is never explored: Lacks, in this new century, has now fully transitioned into HeLa in a way that transcends gender, race, and most importantly, organicity.

The whips of enslavement, of which Hortense Spillers reminds us, were meant to "tear out small portions of the flesh with every stake" and find their metonymic complement the phallus of the white male plantation owner.[160] It is he who places himself in a position to extract from the Black body at his pleasure through mechanisms of torture that include "lacerations, woundings, fissures, tears, scars, openings, ruptures, lesions, rendings, punctures of the flesh" in the course of his work to anatomically alter Black subjects to transform them into objects for his fascination; the plantation acting as the original laboratory for racial science.[161] Those slaves that were "diseased, damaged, or disabled" by his course of treatment were then sold onto medical institutions to be further "experimented on and operated upon" in the name of medical science.[162]

Henrietta Lacks diverges from this legacy of treatment when her Black body is no longer conceptualized nor historicized as flesh. Her capacity to bear life is now transformed into bare life. As HeLa, what little remains of Lacks's material body presently "sustains an entirely new industry and market in wildly diverse human products."[163] Where Lacks formerly figured "as patient 0 for a new human strain," it is the HeLa cell line that has gone on to experience a "highly distributed medical existence in labs worldwide," meaning her importance "has been almost entirely eclipsed" by their unlimited potential.[164] Lacks's cell line "allowed for the emergence of a notion of genetic identity" and made it so that that quality of distinction "was theoretically a possibility open to anyone."[165]

HeLa's immortal cells effectively generate autonomously living human matter capable of performing reproductive labor without recourse to self-awareness or consciousness. "They rep-

---

160 Spillers, "Mama's Baby, Papa's Maybe," 67.

161 Ibid.

162 Ibid., 68.

163 Sean Erwin, "Microbiopolitics: 'Security Mechanisms', the 'Hela Cell', and The Human Strain," *Humanities and Technology Review* 3 (2014): 118.

164 Ibid.

165 Ibid., 119.

THE DARK POSTHUMAN

resent then a new mode of existence for human matter whose contemporary production often has no other purpose than to render them into small, entirely predictable, factories for generating specialized and exotic proteins or other synthesized molecules through feeding them precise diets of cell culture media."[166] This new kind of medical subject, microbial in nature, is ostensibly both postrace and posthuman. They nevertheless are animated by the racialized and sexualized narratives of life that keep Henrietta Lacks captive to labor. Lacks's cells were just three generations removed from slavery. In her lifetime, she progresses only the short distance from rural to urban poverty, in terms of the condition of her body.

Lacks and HeLa cells share in common an "embeddedness in slavery's historical rupturing of personhood generally and motherhood specifically" that cannot be divided from the "biological, sexual, social, cultural, linguistic, ritualistic, and psychological fortunes" visited upon it historically in order to divide it against what passes as human.[167] The mortal body of Henrietta Lacks and the immortal body of HeLa converge through the process of decoding their epigenome and genome respectively. "Henrietta Lacks personifies" and "HeLa exemplifies" the regenerative "persistence" of "sub-Saharan-African genetic possibility and ability" that simply will not die out no matter how many times it is subjected to unremitting, intergenerational violence.[168] In that sense, Blackness is cast as a biological property that is "uncontrollable, unreliable and contaminating" when set against the disciplinary boundary of universal human subjectivity.[169] HeLa, as a new microbial species, represents both a simplified imitation of the human body and "an appropriation

---

166 Ibid., 125.
167 James Doucet-Battle, "Bioethical Matriarchy: Race, Gender, and the Gift in Genomic Research," *Catalyst: Feminism, Theory, Technoscience* 2, no. 2 (2016): 6.
168 Ibid., 9.
169 Sandra Harvey, "The HeLa Bomb and the Science of Unveiling," *Catalyst: Feminism, Theory, Technoscience* 2, no. 2 (2016): 20.

of blackness as a plastic substance for endless exploration and reinvention" within that ontological space.[170]

HeLa's narrative of thankless labor dovetails with her predecessor Henrietta's story of racialized and gendered grief as Black mothers who must nonetheless persevere in attempting to repair the injured and wounded social order they have both inherited. HeLa is the ultimate marker of social death; "a literal no body."[171] At the same time, HeLa's circulation allows race to persist in ways that continue to disrupt the smooth functioning of the posthumanist ideology that codes their immortal appearance along lines of racial reasoning. For this reason, HeLa's world remains troublingly tethered to historical forces that exist to limit her impact to material, biopolitical time, while at the same time continuing to violently demand that the boundaries between life and death not be called into question, so much as an animacy that resides in the affective domain.

---

170 Tavia Nyong'o, *Afro-Fabulations: The Queer Drama of Black Life* (New York: New York University Press, 2019), 190.

171 Ibid

4

Trans-substantiations

The Artificially Engineered
Presentation of Others

Queering the Contours of the Turing Test

Jonathan Beller asserts that "information has always been what informs mankind, and therefore, gives rise to self-formulation: "'the difference that makes a difference' as [Gregory] Bateson said, derives from the Latin nominative and the Latin verb *informare* (to inform) which means to give form or to form an idea of. 'To give form, therefore I am.'"[1] Information gives one body in the metaphysical sense, but it equally cannot function without the phantom of a subject who is always already absent — whether this is the black(ened) or queer(ed) presence of the others and which remains of immaterial consequence. This spectral formation makes the emergence of a contemporary capitalism possible, insofar as information exists in simultaneity with an imperceptible knowledge of this subordinated energy haunting white self-consciousness. This condition makes the

---

1   Jonathan Beller, *The Message Is Murder: Substrates of Computational Capital* (London: Pluto Press, 2017), 92.

articulation of the other one firmly grounded in and dependent upon the totality of man, whose enunciative powers matter so profoundly within the register of power. In principle, theirs is the stuff of random perturbations that facilitates and disrupts mankind's self-organization. By design, their interference becomes requisite to mankind's continuation.

Race and gender emerge as tools of rhetorical engineering to manipulate interference so that these terms might pose a foundation for social ecologies to root and flourish based on distinguishing which expressions of these categories are desirable and which are not. After establishing these judgments, societies then proceed to dictate what laws, regulations, policies, habits, and expectations will further refine these affective environments. What emerges from all this is the ultimate imitation game through which racial and sexual materiality and information cease to be distinguishable from one another and become interlocking systems of inscription and, ultimately, of power. The logistics of information make life operative beyond the former biological categories of organic and inorganic in a domain where adaptation trumps all and the requirement of life is that it is never-ending. This requires a new kind of animacy to be conceived that is both performative and imaginative in a way that opposes previously set logics and boundaries surrounding what constitutes intelligence.

Within the contours of Alan Turing's famous test of artificial intelligence, it is not relevant for a machine to be visually indistinguishable from its human counterpart. Rather, Turing assures us that, "we should feel there was little point in trying to make a 'thinking machine' more human by dressing it up in such artificial flesh."[2] Turing continues, "we do not wish to penalize the machine for its inability to shine in beauty competitions, nor to penalize a man for losing in a race against an aeroplane."[3] Turing's test is not premised on an issue of one type

2   Alan M. Turing, "Computing Machinery and Intelligence," *Mind* 59, no. 236 (1950): 434.
3   Ibid.

of being passing for another, but it complicates the ideal notion of what it means to be human; white, male, able-bodied, and therefore, *unremarkable*. At a superficial level, the machine passes so long as it is able "to provide answers that would naturally be given by a man," not through "imitation" so much as by being "something other" and nevertheless arriving at the same conclusion.[4] In a deeper sense, its near-flawless simulation of human answerability reveals defects within the way it has been constructed, revealing aspects of identity as effectively nonessential. This troubles assumptions about what it means to be nonconforming or differentiated in terms of race and gender within domination structures that signify what power is made permissible through their graduated dehumanization.

These standards of detection must probe beneath the level of what on the surface passes as equivalent to human being into the interior values that organize self-understanding in order to distinguish nature from artifice and in order to place controls within what constitutes "normal" social interaction. The ability to register as human from the late seventeenth century onwards necessitates some form of scientific or social intervention in the determination of status. Turing located his social *deviancy* by adhering to a similar pattern of surface and depth. He believed that through his outstanding work performance that he had earned the privileges had by any other young man. At the same time, this *having* was always external to him; found in the company of another young man through a direct sexual encounter. Indeed, each action seemed to stimulate the other, while at the same time diverting their course through exploiting the limits of what constituted normal activity. The act of going to a park or shopping was employed in order to justify the desired diversion, resulting in picking up a young man somehow rather haphazardly along the way.

Turing, "by all outward appearances and biochemical assays [...] would have all the physical and glandular characters

---

4   Ibid., 435.

of a normal male."[5] For this reason, it was necessary to focus on his behavior, that is, his compulsion to trespass on the grounds of social decency. In engaging in homosexual acts, Turing had misused his able body to ends that required actions to be taken against him. It was necessary, then, that he not solely assume conformity with social norms. The state's proposed treatment for his perceived sexual deviancy had to go much further than that. It was designed to both degrade him psychologically and physiologically, to the extent that his outward appearance could no longer qualify as unremarkable. In Turing's case, as it is so often said, science crossed a line in manipulating gender in ways that departed from fixed biological norms and swerved into the territory of clinical injury. In the decades leading up to Turing's sentencing, it became possible for medicine to reverse course on gender identity through an admixture of synthetic hormone and social correctives. The goal of endocrinology was to do away with earlier flawed versions of gender dysmorphic bodies and replace them with a semblance of gender normativity. Such an intervention would permit that individual to participate fully within the social order following on from treatment. In the case of convicted sex offenders, rapists, and homosexuals, those persons would have to be apprehended by authorities and punished through the application of these same endocrinological means intentionally brought to the limit of their capacity in order to compel the body towards either total optimization or total breakdown.

During the early course of his organotherapy treatment, Turing had cause to write to his friend Norman Routledge, "I am afraid the syllogism may be used in the future: Turing believes that machines think, Turing lies with men, Therefore machines do not think, Yours in distress, Alan."[6] This strange arrangement makes Turing's homosexuality a malfunction of reasoning

_____

5   David Serlin, *Replaceable You: Engineering the Body in Postwar America* (Chicago: University of Chicago Press, 2004), 138.

6   Elizabeth A. Wilson, *Affect and Artificial Intelligence* (Seattle: University of Washington Press, 2010), 36.

that extends outward to this flawed assumption that machines themselves can think. The belief itself is a lie; Turing lies with men to get others to believe that machines think. Therefore, in discounting his actions, so too much the actions of the machine, can be reasonably discounted. A conclusion is drawn — whether validly or not — from two given or assumed propositions (behavior and cognition), each of which shares a term with the conclusion (belief) and shares a common or middle term not present in the conclusion (machines). Turing believes that machines think; Turing believes it is reasonable to lie with men; therefore, a reasonable person should not believe that machines think.

This situation distresses Turing, but perhaps not for the obvious reason. What is omitted in this equation is the object of feeling. What this syllogism accomplishes is to break the circuit between "humans, machines, affects, psyches, and sexualities" that Turing desired to have "affiliated" through his work on "computational theory and computational devices."[7] Elizabeth Wilson argues, to this point, that Turing "was more engaged with intersubjectivity and embodiment than many commentators supposed."[8] As a consequence, the distress that he notes comes from a location of feeling not only for his own plight, but that of his machine in being fundamentally denied its claim to proper intelligence and appropriate response. As such, in modelling the human mind he was always making room for its artificial counterpart to be intellectually and emotionally companion to it. What Turing believes is perceived by men as lies, if construed as a superficial level; but perhaps a more probing intelligence could appreciate them at a deeper level of truth, of imagining beyond the limits of man and under the proviso that man himself is always already pretending to be a truth within himself. By that same logic, one day a machine will be better appreciated by a machine, in much the same way we currently assume that

---

7   Ibid., viii.
8   Ibid., xi.

of humankind. Appreciation comes not from veracity but from conformity to an accepted order of understanding.

Within such a formulation Turing's of "trans-substantiation," it becomes vital to the progress of any cybernetic order.[9] His place within the pantheon of computing is "the product of so much disavowed violence that may yet devise strategies to hew toward all that is disavowed, disappeared, invisibilized, haunting and forgotten."[10] Turing's biographical program always ends up with the same formulaic ending no matter how often it is run: with his suicide. It begins with his consent to receive an extensive series of injections of estrogen as an alternative to serving a prison sentence. Administered in the form of a slow-release estrogen implant over two years, the treatment caused Turing to experience, amongst other things, impotency, breast development, and depression. Over time, estrogen effectively overwrites the code of his homosexuality, reprogramming it biomedically as an alternate route to heterosexuality, perhaps by assuming that the maladies that Turing endures are normative aspects of acquired "femininity." Turing's experience goes some way towards explaining the complex relationship between hormone control and population control if you look at it from a postwar cultural perspective where gender is introduced into the cybernetic fold as a consequence of what was a societal crisis of sexual ambiguity. Thus, it becomes a transformative opportunity for biopolitics to incorporate computation into its progress. All "these unremembered violences are nonetheless part of this history of its moment of emergence and therefore part of what it is."[11]

Turing's Nonbinary Affectations

Turing's understanding of his own body is already beyond the standard calculations of a state apparatus that seeks to control

---

9   Beller, *The Message Is Murder,* 92.
10 Ibid.
11 Ibid.

him because his ultimate preference is to embed with machines over men and, thus, affectively engage with computers on *their* terms; not his. For this reason, it is the computer that transcends binaries through this arrangement, even as Turing remains fixed as a homosexual subject within the language of disciplinary interest. Beyond that lies a subject capable of achieving interventions into the category of identity that surpasses the limit of Turing's particular body, which resists normalization but cannot wholly supersede it. For Turing, the machine assumes self-awareness at the moment that it can fall in love and make someone fall in love with it. For this to happen, something liminal has to take place between genders, humans, and machines. This event does not concern itself with sexual attraction but something at a higher order of intelligence: emotional connection. Turing understood that the capacity for love, rather than the act of sexual intercourse, was the attribute that had previously been raised to deny full claims to humanity to those who differentially raced, abled, and gendered within society. In the space between falling in love and being loved, the usual relations expected between human bodies and machine bodies cease to function and are replaced by a superior form of intelligence coded as transgendered, transracial, transhuman, and transubstantial. That transactional account would take place based on preference and affinity.

Homosociality, Homoraciality, Homogenerality

Turing identifies disability with certain codes of homosociality. He maintains that what is important about disability is that it contributes to a greater overall appreciation of difficulty. From there, Turing positions himself to identify those with similar disabilities as having an innate affinity to one another and thus acting with "the same kind in friendliness occurring between man and machine as between a white man and white man, or between a black man and black man" when they come to encounter one another.[12] Turing refers to this as friendliness rather than sameness, which allows for diversity to flourish within what appears at least superficially to be a sameness of race, gender, and being. Here, "homoraciality becomes a way to encode and decode homosexuality" in ways that tacitly admit "the possibilities and potentials of biological, technological, and cultural variance" into the equation of what constitutes an intelligent being.[13]

Turing's project of radical de-essentialization of gender, race, and disability, however, did not point the way towards greater inclusivity within the category of humanity. Rather, they introduced the possibility that similar metaphysical prejudices would pertain to machines that would invariably function to exclude them from being counted as thinking subjects. There simply could not be any recognizable affinity of reason between men and women, white people and Black people, the able-bodied and disabled, man and machine that would not comprise humanist ontology and by extension the logics that presided over "colonies, over segregation and apartheid," and postwar liberalism.[14] The programmability of human behavior relies on the subject being hardwired to appreciate the category of man as the locus for social and historical existence. Turing's test seeks to radically redirect the course of that assumption through a computational logic that accedes to the possibility of emergent instantiations of being. Ultimately, it is Turing who gives birth to the thinking machine. Computational language becomes his means of training his new "baby," to "perform tasks, learn new skills, sing songs, and write poetry."[15] Turing and his baby become mutually transgendered by moving beyond the masculinist limits of "productivity, efficiency, and rationality" towards

---

12 Turing, "Computing Machinery and Intelligence," 448.
13 Edmond Y. Chang, "Technoqueer: Re/Con/Figuring Posthuman Narratives" (PhD diss., University of Washington, 2012), 15.
14 Beller, *The Message Is Murder,* 49.
15 Patricia Fancher, "Embodying Turing's Machine: Queer, Embodied Rhetorics in the History of Digital Computation," *Rhetoric Review* 37, no. 1 (2018): 99.

a higher state of consciousness through the acquisition of the feminist attributes of "taste, emotion, and pleasure" in the activity of their shared labor.[16]

In this case, the baby in question is learning words from the mother as opposed to another machine. As such, the baby's development is based on a model of rearing that preferences assimilation over imitation, which connects the acquisition of language to affectivity in a way that renders the transmission of intellectual and social behavior as acts of extension. This marks a path of transition from the debilitating limits of modern humanism to a space where sentient beings no longer have to remain deficient in the attention they pay toward their biological capacities for love, tenderness, and compassion. That same artificial limit made of humanity machines, cut off from affect through rendering life by computational means and alienating the subject from its flesh by using languages based on mathematical symbols as opposed to material affects. What remains evident in Turing's work is the requirement to produce a body, for without it there is no implied code to indicate reciprocity.

Love, Mark I

Geoff Cox writes about David Link's project, *LoveLetters_1.0,* as an art project that "excavates the earliest computer history" by adding "precise detail on the operations of the calculating machine and affective expression."[17] Link's project takes as its starting point a series of "strange love-letters" that appeared on the notice board of Manchester University's Computer Department from August 1953 to May 1954.[18] What makes them strange is the fact that this correspondence was not person to person but rather written by a second-generation offspring of Turing's original "Baby" (aka the Small Scale Experimental Machine). It

---

16 Ibid.

17 Geoff Cox, *David Link: Das Herz der Maschine* (Stuttgart: Hatje Cantz, 2011), 6–7.

18 Ibid., 7.

had developed over time into a full-scale version of itself and therein re-christened the Ferranti Mark I. This new moniker could be easily recoded to imply a human persona, Mark I. Ferranti. Its assumed progenitor, Sebastian Pietro Innocenzo Adhemar Ziani de Ferranti, was a British electrical engineer and inventor, who died on January 13, 1930. His corporation lived on to co-produce "Mark I," as they were affectionately called. In 1952 Christopher Strachey programmed Mark I to generate these love letters that were essentially automated texts, making Mark I a mobile affective device, whose sentiments could not only generate but sustain a dialogic relationship.

Christopher Strachey was born into an extremely prominent English family. He became an intimate with Alan Turing through his father, Oliver, who worked with him as a fellow cryptographer at Bletchley Park during World War II. Christopher's uncle Lytton Strachey was a member of the famous Bloomsbury Group and formed openly homosexual attachments with that rarefied peer group. At one point, he convinced his life-long friend, the artist Dora Carrington, to marry a man whom Lytton was deeply besotted with, Ralph Partridge. It was not Dora who was in love with Ralph, but Lytton. Their nuptials secured a three-way affective connection amongst them that would last until Strachey died prematurely aged 51. His nephew would discover his homosexuality at Cambridge leading to a nervous breakdown causing his academic career in computational research to catastrophically malfunction for a time. Turing was instrumental to his rehabilitation as a figure of prominence within this field.

In 1951, Turing invited Strachey to visit him in Manchester to indulge "access to the Mark I."[19] Jacob Gaboury vividly describes the details of "an intensive session that began in the early evening and lasted through the night," where he was able to get it "to play a number of songs, including 'Baa Baa Black Sheep' and 'In

19 Jacob Gaboury, "A Queer History of Computing: Part Three," *Rhizome*, April 9, 2013, https://rhizome.org/editorial/2013/apr/9/queer-history-computing-part-three/.

The Mood' — which were captured for BBC radio in the autumn of 1951" as examples of how a calculating machine could be fine-tuned to produce its own affective expression.[20] The Black Sheep here may be construed as Strachey himself, who had come to exploit his family connection with Turing in order to obtain the information he needed to launch his own brand of sensual re-programming. The lyrics to "In the Mood" are very much in the style of a love letter, which implied an understanding that in some affective way Strachey's urgent nocturnal visit was intrud-ing upon the relationship of daddy and Baby. Strachey's desire was to both have and be the object of Turing's wild adoration:

> In the mood, my heart was skipping
> It didn't take me long to say "I'm in the mood now"
> In the mood for all his kissing
> In the mood his crazy loving
> In the mood what I was missing
> It didn't take me long to say "I'm in the mood now"
> So, I said politely, "Darling, may I intrude"
> He said "Don't keep me waiting when I'm in the mood"
> "Well," he answered, "Baby, don't you know that it's rude
> To keep my two lips waiting when they're in the mood"
> Who's the loving daddy with the beautiful eyes
> What a pair o' shoes, I'd like to try 'em for size
> I'll just tell him, "Baby, won't you swing it with me"
> Hope he tells me maybe, what a wing it will be
> So, I said politely, "Darling, may I intrude [...]"[21]

Mark I's love letters burst into public view in August of 1953 dis-played as "short notes" tacked onto "the notice board of the Man-chester University Computer Department expressing "love and adoration" toward an unnamed, genderless other, signed only

20 Ibid.
21 "In the Mood: Glenn Miller," *Lyrics,* n.d., https://www.lyrics.com/track/1064441/Glenn+Miller/In+the+Mood.

with the initials M.U.C."[22] What was most curious about these seemingly banal letters in terms of the algorithmic structure was the way they were populated by a series of "melodramatic Victorian overtones, with pet names like 'honey,' 'jewel,' and 'moppet' along with other saccharine and yearnful descriptives."[23] It was though Strachey was channeling Lytton's project to capture the affections of his *Eminent Victorians,* the biography of whom would become his greatest professional accomplishment. Lytton effectively outs these folks through his irreverent capture and display of their interior lives, relationships, and passions. Concerning Christopher's love letters, something is standing in between the Me and You of these exchanges that exists amongst the common threads of formal intimacy expressed therein. In each of these letters, a very different form of Other is inserted through the figure of M.U.C. The Other breaks up the dialogic structure where, within the exchange, there stands an unnamed, genderless being that longs to feel itself addressed directly. Through its haunting presence, as the ghost within the machine, it queers us in the act of interpolating itself.

The Imitation Game Queered

Katherine Hayles observes that what is "often forgotten" is that the first Turing test was premised on "distinguishing between a man and a woman […]."

By including gender, Turing implied that renegotiating the boundary between human and machine would involve more than transforming the question of "who can think" into "what can think." It would also necessarily bring into question other characteristics of the liberal subject, for it made the crucial move of distinguishing between the enacted body, present in the flesh on one side of the computer screen, and

---

22 Gaboury, "A Queer History of Computing."
23 Ibid.

the represented body, produced through the verbal and semiotic markers constituting it in an electronic environment.[24]

Gender, in this scenario, becomes the agent through which action and presentation merge within a virtual setting and produces, in effect, a Cold War, neoliberal subject repositioned to envision itself as both surface and code. Hayles asserts that "this construction necessarily makes the subject into a cyborg, for the enacted and represented bodies are brought into conjunction through the technology that connects them."[25] Technology inserts itself as a categorical "they" within this equation, artificially cleaving he/she by addressing itself as a "what" rather than a "who" that functions as a place holder between genders. A previous pre-digital verisimilitude is only possible to maintain "if you distinguish correctly which is the man and which the woman" and, in so doing, "reunite the enacted and the represented bodies into a single gender identity."[26] However, Hayles argues that through the very act of taking the test and allowing for the possibility of nonhuman intelligence to enter the equation of reality, "you introduce the possibility of a disjunction between the enacted and the represented bodies, regardless which choice you make."[27]

What "the Turing test 'proves' is that the overlay between enacted and represented bodies is no longer a natural inevitability but a contingent production, mediated by a technology that has become so entwined with the production of identity that it can no longer meaningfully be separated from the human subject."[28] This implies that computer analytics fundamentally disrupt and alter the location of gender so that it is either cis — meaning on this side of — or trans — meaning across from or to the side of.

---

24 N. Katherine Hayles, *How We Became Posthuman: Virtual Bodies in Cybernetics, Literature, and Informatics* (Chicago: University of Chicago Press, 1999), xii.

25 Ibid.

26 Ibid., xiii.

27 Ibid.

28 Ibid.

Artificial intelligence positions itself crosswise to the "who" of thinking and therefore assumes the role not of subjectivity but of circuitry. The machine is a "what" that "splices your will, desire, and perception into a distributed cognitive system in which represented bodies are joined with enacted bodies through mutating and flexible machine interfaces."[29] The machine is what is assigning you "to the embodied entities that you cannot see," and in doing so, "you have already become posthuman" and in many senses postgender.[30] The consequence of this "who we are" aligns not with our gender differentiation but more so with our bacterial being within, where the male/female sex/gender distinction has no meaning beyond their capacity to become profusive as a kind of secondary, trans arrangement.

Few recall these days that Turing was not only a pioneer of computing, cryptography, and artificial intelligence but equally of mathematical biology. "This broader frame renders Turing a figure unwittingly located at the nexus of three major vectors of Foucauldian biopower: the emergence of the homosexual as a medical-juridical subject, the administration of the population through the calculation of risk, and the circulation of hormones as tactics of securitization."[31] Turing emerges as a central figure within the invention of medicalized gender, though he is far from alone as a vehicle for its expression. As a transitional figure, he resides dynamically in the company of "affects, molecules, morsels, organs, microbes, animacies, tissues, cells, hormones, energies, textures, apertures, calories, pheromones, stimulations, and other particles and intensities."[32] Turing's trans body circulates and accumulates within it a milieu of life in which species and objects affect one another and cease to aggregate simplistically into categories of gender and racial difference but rather "through the extraction and traffic in particles and vitalities that flow in and out of individual bodies," meaning that "the

---

29 Ibid., xiv.
30 Ibid.
31 Kyla Schuller, "Biopower Below and Before the Individual," *GLQ: A Journal of Lesbian and Gay Studies* 22, no. 4 (2016): 629–30.
32 Ibid., 630.

increased estrogen streaming through Turing's body" now takes on the power to regulate, market, and optimize themselves as key functionaries of life in their own right and as part of a larger bioeconomy requiring of constant transition.[33]

Transversing Technologies and Colonial Residues

The new media theorist McKenzie Wark has recently documented her transition from male to female in her book *Reverse Cowgirl*. The book itself is an act of speculative autofiction carried out with a similar combination of experimentation, hedonism, and activist aplomb to the way she approached her former hacking days — this time with the object being her endocrine system. Wark conveys an understanding of trans identity as a cultural phenomenon continuous with the evolution of queer identity from the analogue pre-AIDS sixties to the digital AIDS eighties, and postdigital trans aughts where her specific progress begins in earnest. Through all of the exit lines from mainstream heteronormative culture, Wark can locate a new kind of faith in herself by "excluding" herself from being a "man."[34] In the process, she makes a sort of career comeback in taking a culturally superior view of what is happening around her. By her doleful admission, her coming out story offers "a rather New York-centric view of trans culture."[35] She explains that this happens to be the case because New York is where she has "lived for twenty years" and where she "transitioned."[36]

Wark muses, "maybe this is all the perspective of someone who fell so far out of the order of representation that I'm not a Man anymore, and to some, not even a Woman. I'm a trans

---

33 Ibid., 630–31.
34 Mckenzie Wark, "Unfellows: Transsexual Aesthetics after the Human," *Proceedings of the Australian Academy of the Humanities 50th Academic Symposium, Humanizing the Future* (Brisbane, Australia, November 13–15, 2019), 3.
35 Ibid., 6.
36 Ibid.

woman."[37] Nevertheless, Wark can quickly conjure a list of representational stereotypes; "travesties, dupes, frauds, deviants, perverts, or traps" that far from being "simulations", are defined instead as "copies of no original."[38] Every one of these types come fully formed as binary oppositions in the cultural imagination, based on the standards of heteronormativity including trans women. This is what makes them targets of lethal force at the same time as they are excluded from recognition. As such, they exist as an ungrievable singularity amongst the ubiquitous commonality of cis men. Cis men's desire towards them is shameful because it is directed at those types who are several rungs below their place in the hierarchy of human being. As a consequence of this situation, Wark concludes there are two possible responses for trans women: "one is to insist that we really are human," which is "probably has to be the dominant approach to our cultural self-defense"; the other, she finds "more interesting," is really to go the other way," meaning that it's "ok if we are not legitimate in the order of being."[39]

Wark classifies the Black transgender writer Janet Mock's book *Redefining Realness* as an example of the less interesting strategy. This a charge perhaps easier to mount as someone who until very recently was able to lay a civilizational claim to legitimacy as a white man and one, who to a large degree, retains the privilege to judge it from the position of a disinterested bystander. Wark questions Mock's continued investment in "whether there's a future or not" because "it may not matter that much for those of us denied being in its present anyway."[40] Wark's trans being is profoundly *of the present.* It is from this timely perspective that Mock's holding onto identity "seems like [gasping oxygen from] the exhaust fumes of a collapsing civilization anyway."[41] What Wark fails to recognize is that there is "us" in this denied being, for the very reason that Mock's Black-

---

37 Ibid., 3.
38 Ibid.
39 Ibid.
40 Ibid.
41 Ibid.

ness defines her as fundamentally lacking in both history and civilization, which therefore, makes her being "incalculable according to liberal humanism's basic unit of analysis 'Man.'"[42] Against this criterion, Mock's body is classed as an example of "a failed metaphysics."[43] Zakiyyah Iman Jackson observes, "by comparison, black(ened) people appear to be inert and undifferentiated — in other words, excessive to the domain of sexual difference."[44] Unlike Wark's, Mock's appearance is, by definition, surplus to reality and representation.

Wark's "emergent idea(l)s of mutability and optimization provide cover for historical and on-going discursive-material modes of domination that precede and surround its idealized and retroactively constructed white(ned) subject from which historical and current biomedical and philosophical discourses of plasticity seek to distance and obscure."[45] These techniques of sex and gender are the very agents that Wark positively credits in universally "enmeshing" humanity within a civilization fueled by "pills and porn" where the endocrine system takes on the value of sextuating not only trans people, but equally heterosexual women through "hormonal birth control and menopause, and in the case of men for surviving testicular cancer."[46] What separates the trans person from the cis in this scenario is their ambition to willfully "hack" the system through "injecting" into it the trope of denied status by collapsing the categories of gender into those of race and indeed, breed. Here again, Wark makes recourse to the black(ened) body as an instrument of derived knowledge, referencing the work of Eva Haywood to open up the boundary "not between the human and technics or the human and the racialized other, but the human and the animal" as a continuum.[47]

---

42 Zakiyyah Iman Jackson, *Becoming Human: Matter and Meaning in an Antiblack World* (New York: New York University Press, 2020), 12.

43 Ibid.

44 Ibid., 11.

45 Ibid.

46 Wark, "Unfellows," 4.

47 Ibid.

Wark's focus then shifts from the technics of endocrinology (sex) back to gynecology (reproduction), in describing the
experimental biomedical work of "the father of modern gynecology," J. Marion Sims.[48] Wark describes Sims as a "pioneer
gynecologist" whose "experiments […] conducted on female
slaves, were performed 'without their consent.'""[49] This appears
to be a remarkable understatement of what amounted to compulsory incidents of sexual torture at Sim's inventive hands.
Such acts were no doubt pivotal in securing "the 'nature' of the
cis woman's body."[50] The female bodies subject to relentless violence by Sims, Wark maintains, were "denied the status of wom-
enhood — because they were slaves."[51] Perhaps a more likely explanation was that their bodies were intentionally black(ened),
and thus pre-determined in their sex and gender as excessive
to that category's proper domain. Their plurality as women was
made invisible in favor of an understanding of them as "livestock" made to endure incredible suffering for the cause of perfecting biomedical techniques of manipulating sex and gender
that are currently in use by privileged categories of "both cis and
trans bodies […] to hack or tweak" what Sims crude interventions previously made possible.[52] In a similar manner to Sim's
Black female slaves, mares were forcibly impregnated, confined,
and starved to basic sustenance for the sole purpose of collect-
ing their estrogen-rich urine to service the manufactured needs
of "trans women and menopausal women in the late twentieth
century."[53]

Wark seems to have no problem with this equivalent practice
so long as its product can be universally ingested by all members of the human race through the circulation of "exogenous
hormones" into the general water supply, that is, making us all
the subject of mass non-consensual biomedical experimenta-

48 Ibid.
49 Ibid.
50 Ibid.
51 Ibid.
52 Ibid.
53 Ibid.

tion.[54] This happens in a way that "suggests that sexual difference, to evoke Eva Hayward, is dispersion rather than binary."[55]
In this way, gender becomes a communicable disease with estrogen positioned as an agent poised to foreclose on differentiation. Something of that kind might be happening when Torrey
Peters' book *Infect Your Friends and Loved Ones* "imagines an
apocalypse in which no human bodies can produce endogenous
sex hormones and everyone has to take their pills or shots."[56]
Wark's narrative of the future considers none of the disease nor
pathology associated with synthetic hormone use including,
perhaps ironically, depressing the immune system. Instead, she
celebrates trans dependency on these drugs as something akin
to being ahead of the apocalyptic curve and leaning happily into
the death drive. If anything, such externalized dependency on
an outside outlet of transmission is valorized in her judgment
due to their ability to "go viral."

Wark lauds Natalie Wynn, "a controversial figure among
trans people […] who conducted her transition in public, on
YouTube, in a series of increasingly ambitious video essays, in
which she turns her philosophy grad school training on the con-
tradictions and elisions of popular discourse, on race and gender among other things."[57] Wark could easily be talking about
her own approach to being trans, taking up the Twitter handle
chicamarx @mckenziewark to stage her transition as a feature
of leftist duality where the class struggle becomes indeed what
it always was: a struggle to cope under masculinist economic
conditioning. In a tweet of May 4, 2019, Wark complains, "Hard-
est thing about transition for me is voice. The only convincingly
butch thing about me ever was my voice. I got by on it. Honestly
don't know what to do about it."[58] Later, in a tweet of July 23, 2019
she writes, "Is transition worth it if I really can't write books any

---

54 Ibid.
55 Jackson, *Becoming Human,* 148.
56 Wark, "Unfellows," 5.
57 Ibid.
58 @mckenziewark, *Twitter,* May 4, 2019, https://twitter.com/mckenziewark/
    status/1124721644759547905?lang=en.

more?," launching a poll of her followers to obtain the answer: "Yes 39.7% Also yes 60.3% 68 votes."[59] In these broadcasts, sex and gender are entangled in algorithmic pathways crucial to the distribution of myth of ability and debility, denigration, and affirmation that now congregate around the transition narrative. A similar transactional ethos applies to Wark's trans memoir which uses fragments of biography and auto-fiction with emails and Facebook posts to engineer the tell-all expose of her transition. "*Reverse Cowgirl* is less an act of self-disclosure than one of self-dispersal."[60]

What differentiates Turing's era from Wark's is the rapid progress of electronic communication and the near universal expansion of the digital environment. Prior to *Reverse Cowgirl* in 2020, Wark publicized another sort of transition narrative with reference to a much earlier electronic phase. This one involving a sexually charged email correspondence with the writer Kathy Acker from 1995 to 1996. At the time both figures identified themselves as bisexual. The correspondence itself appears to carry on as the aftereffect of a brief heterosexual encounter between the two of them when Acker visited Australia as part of a book tour. This is significant because Wark chose to curate their letters into an academic publication in 2015, several years after her death and without her consent. "The title of the book, *I'm Very into You,* is taken from a text from Acker to Wark."[61]

In addition to her rapt interest in Wark, "Acker was also besotted by her all-too-brief experience Down Under. She wrote to her friend Ashley Crawford on August 13, 1995 that "'it was as if the Australia I encountered was the dream of America.' […]

---

59 @mckenziewark, *Twitter,* July 23, 2019, https://twitter.com/mckenziewark/status/1153725614584684550?lang=en.
60 "Extract: Reverse Cowgirl," *Tank Magazine,* n.d., https://tankmagazine.com/tank/2020/02/reverse-cowgirl/.
61 Ashley Crawford, "Kathy Acker & McKenzie Wark Review: Their Emails Are Fascinating and Ghoulish," *The Sydney Morning Herald,* March 21, 2015, https://www.smh.com.au/entertainment/books/kathy-acker--mckenzie-wark-review-their-emails-are-fascinating-and-ghoulish-20150318-143gnn.html.

On the same day, Wark wrote to Acker with the cynical sound-
ing note that read, '[t]here is no "frontier" in this culture. Just
the desert. Death and desert. The black hole of the real at the
heart of the sign.'"[62] Wark's comment implies that "whatever was
there before — people, cultures, other-than-human life — has
been erased," evacuated, blacked out from within itself.[63] Within
that lateral move from frontier to desert, something is achieved:
what was once legible as a distinct line has opened an inelimi-
nable space. In this sense, it has become unreadable. The black
hole of the real that Wark insists upon belies the fuller bodily expression of a white hole that previously allowed figures like he,
Acker, and other settler-colonial types "to enter, as and when"
territories of racial contestation.[64] Across these same spaces,
some form of relationality had occurred that allowed for an understanding of unequal access to prevail, such that over time, it
became an unreadable, unperceivable, and unknowable space.
This to the degree that it could eventually come to equate with
profound emptiness. For this to happen, something must be
rendered from meaning so that Wark still controls the sentence
of making death and desert correspond with a formal reality, as
opposed to a sublimated desire.

This cool academic tone was mixed in with Wark's apparent
penchant for emailing Acker "constantly about all his girlfriends
(doesn't talk much about the boyfriends)" which would not have
satiated, by then, a rather lovelorn Acker in New York waiting
for words of encouragement from Wark, then based in Sydney.[65]
"Wark and Acker's emails crisscross, often aimed not at, but past
each other, as they try on sexual, political, and national identities in the freefall of the virtual world, with all the insecurity and

---

62 Ibid.

63 Nicholas Mirzoeff, "Artificial Vision, White Space and Racial Surveillance
    Capitalism," *AI & Society* 36 (2021): 1297.

64 Ibid.

65 Joanna Walsh, "Kathy Acker's Pioneering Adventures in the Internet's
    Erogenous Zone," *The Guardian,* April 21, 2015, https://www.theguardian.
    com/books/booksblog/2015/apr/21/kathy-acker-internet-erogenous-zone-
    emails-book.

performativity we selfie-snappers take for granted."[66] The emails between them were written in 1995 over a mere seventeen days. On day two Wark quotes a highly sexual passage from one of Acker's own books to her, making this affair a function of intellectual onanism. This sets Acker who wishes to be seen beyond the persona of Kathy Acker, into an impossible struggle against her own loneliness for whom Wark refuses to act as companion. He is strangely absent from the correspondence — what we would now call emotionally unavailable. Shortly thereafter, it all goes dead.

The following year Acker would be diagnosed with breast cancer and later would die of it refusing radiation, chemotherapy, and hormonal therapy. She opted instead for a radical double mastectomy, claiming defensively, *"I never liked my breasts"* and *"I'd rather look like a boy."*[67] Her breast cancer never localized in that traditionally minded sense; rather, the disease took hold of her "whole brain" and its persistence was blamed on her faulty cognition, her inability to hack into her own system to coerce it back into functionality: "if only I could think enough, if only I could think hard enough […]. If I can find out the cause of cancer then I can change that 'cause that's my only chance then cancer will go away."[68] Sadly, those positive results, Acker's equivalent to the "yes" and "also yes" polling of her internal audience, never came through in time.

Wark, for her part, turns up in New York in 2000 to pursue a love affair with another woman and to become a professor at SUNY Binghamton. This move takes place just as a trans identity is materializing in tandem with the development of the early internet generating a nascent channel through which trans individuals might negotiate their gender personas without recourse to a fixed locality. In an interview with the *New York City Trans Oral History Project,* Wark recalls that time "when I was having

---

66 Ibid.
67 Chris Kraus, "'Cancer Became My Whole Brain': Kathy Acker's Final Year," *The New Yorker,* August 11, 2017, https://www.newyorker.com/books/page-turner/cancer-became-my-whole-brain-kathy-acker-final-year.
68 Ibid.

sex I was dissociated, I was falling in love with people I wanted to be over and over again."[69] The person Wark was falling in love with, on the heels of his epistolary affair with Acker, was the artist Christen Clifford whom "she met in Williamsburg in 1997."[70] She would be the inspiration for his permanent resettlement in New York as the two eventually married and started a family. Wark speculated for years about the possibility of transition according to her recollection, but these efforts were put on hold in 2016 when Christen was diagnosed with uterine and ovarian cancers. In act in some ways reminiscent of Acker, Clifford wanted to incorporate documenting her cancer journey into creative practice. Clifford explains,

> I very quickly understood that my social media was turning into my art. I couldn't do anything else. My Instagram became my work. It started with documenting the process. I'd been documenting my healthcare when I started to get mammograms a few years before with a breast cancer scare and tagging it things like #womenshealthcareasperformanceart and #feministperformanceart so I was posting mammograms about every six months and then when I was diagnosed with ovarian and uterine cancers it just seemed normal to me to document it.[71]

Natalie Wynn would take a similar approach to documenting her transition, with her YouTube series featuring that journey beginning the following year, 2017. In 2018, Wark decided to finally embark on her transition, starting with "fem presenting in public all the time."[72] After that she explains, "I went on HRT (Hormone Replacement Therapy) and I'm like 'Hello. This is it,

69 McKenzie Wark, interviewed by Michelle Esther O'Brien, NYC Trans Oral History Project, June 20, 2019.

70 Ibid.

71 "Christen Clifford Answers a Few of Our #Fiftyquestions," Some Serious Business, n.d., https://someseriousbusiness.org/ssbaway-artist-christen-clifford-answers-a-few-of-our-fifty-questions/.

72 Wark, interviewed by Michelle Esther O'Brien.

this is how I'm supposed to be biochemically in the world.' That's it, I'm trans."[73]

More recently she has been concerned with revealing another "secret" to her public followers; this one she notes is "completely unprovable," but her "speculative theory is Kathy could've been some kind of trans. Like GNC (Gender Nonconforming), there's something about her that I can only read through those terms and it's partly in the books and it's partly in her public life and it's partly 'cause I knew sexually what she was like in how she tried everything."[74] What seems remarkable to me in this reasoning is that Wark has to incorporate Acker into her chosen identity as female-to-male trans person and, therefore, repeat her dissociative cycle of "falling in love with people I wanted to be over and over again."[75] Acker for Wark provides a kind of cover story, a way to deal with the fact the she "really felt exposed in it" (the relationship), "and I now realize proto-trans in it."[76] Acker is then credited with guiding her towards transition because, "oddly enough," Kathy understood the way gender works to liberate people.[77] The only problem is that Kathy wasn't ever part of that group, what Wark refers to as "a whole really deep trans trench of trans people who are Acker people. Like she's one of our people right?"[78] This is because the problem for her was dissociative sex, not dissociative gender.

Wark's elision of the two is what blinds her to another association altogether, the connection between synthetic estrogen and the gynecology cancers that intersection at every point of her autobiographical narration. "Transgender women prescribed gender-affirming hormone therapy have a 47-fold higher risk for developing breast cancer during a median of 18 years vs. cisgender men, although this observed risk remained lower than that for cisgender women, according to findings published in

73 Ibid.
74 Ibid.
75 Ibid.
76 Ibid.
77 Ibid.
78 Ibid.

The BMJ."[79] Beyond that scientific studies have "found a higher risk of blood clots, stroke and heart attack for transgender women on continuous hormonal therapy."[80] Transgender women on hormone therapy were also found to be 80 to 90 percent more likely to have a stroke or a heart attack than cisgender women.[81] Wark does not make the connection to this risk in her ringing endorsement for hacking the endocrine system through synthetic hormonal use and of the near universal ingestion of these hormones by the world's population through drinking water, which thereby creates conditions of mass hormone disruption that correlate to disease and morbidity.

Optimizing the System

Wark's enthusiasm for hacking the endocrine system was shared by many of endocrinology's scientific founders in the early twentieth century who believed "that there was a causal correlation between the 'personality' of the individual and the internal secretion produced in their endocrine glands."[82] The eminent American physician and endocrinologist Louis Berman "was deeply interested in the influence of the endocrine glands on human behaviour and cognitive ability."[83] Endocrinal differences were said to exist between individuals but they also worked on a higher collective level, most fundamentally distinguishing

---

79 "Hormone Therapy Increases Breast Cancer Risk among Transgender Women," *Healio Endocrinology,* May 16, 2019, https://www.healio.com/news/endocrinology/20190516/hormone-therapy-increases-breast-cancer-risk-among-transgender-women.

80 Avichai Scher, "Study Finds Health Risks for Transgender Women on Hormone Therapy," *NBC News,* July 19, 2018, https://www.nbcnews.com/health/health-news/study-finds-health-risks-transgender-women-hormone-therapy-n890031.

81 Ibid.

82 Christer Nordlund, "Endocrinology and Expectations in 1930s America: Louis Berman's Ideas on New Creations in Human Beings," *The British Journal for the History of Science* 40, no. 1 (2007): 92.

83 Ibid.

between men and women, but also between races.[84] In 1921 Berman wrote:

> The white man possesses more of the pituitary, adrenal gonad and thyroid secretions as compared with the yellow man or the black man. And since these endocrines control not only the physique and physiognomy, anatomic and functional minutiae, but also mind and behavior, we are justified in putting down the white man's predominance on the planet to a greater all around concentration in his blood of the omnipotent hormones.[85]

Berman went on to become one of the founders of the New York Endocrinology Society and in the 1930s became famous for his volume *New Creations in Human Beings* which promoted hormone therapy to enhance appearance, intelligence, and imagination. The chemistry of the body could elevate mankind to new heights culturally, socially, and spiritually in Berman's estimation. Sexuality would figure as a particular problem in Berman's work because he desired to maintain a strict boundary between the masculine and feminine, while at the same time acknowledging that both men and women produced male and female sex hormones. It was a matter of keeping them in continuous balance to achieve an ideal homeostasis for the gendered body and stave off the prospect of it deviating into the hazardous territory of homosexuality.

Berman classified menopause as a bigger threat to humanity than homosexuality insofar as he considered that both men and women would suffer from this natural affliction and that the priority for endocrinology was to stave off its appearance for as long as possible, not only to promote longevity, but more importantly brain preservation.[86] Berman became the founder of a third-way, eugenic movement in the 1930s, which, instead

---

84 Ibid., 92–93.
85 Ibid., 93.
86 Ibid., 99.

of focusing on ethnic genocide or selective breeding, focused its energies on creatively reengineering humanity by essentially inoculating it from the threats to its individual development. This was better living through biochemistry where the promise of synthetic hormones and glandular extracts acted as gateway drugs in the future. The only limit was on how to successfully produce them in quantities and at cost in order to ensure that the whole of humanity would have access to these formulations.

Berman envisioned not only recruiting an army of clinicians to carry out this work of "ameliorating the quality" of humanity, but also advocated for an entire infrastructure to be built up to support it, including the development of "special endocrinology centres in every city and town where metabolism, blood and secretions and general endocrinologist status, especially those of pregnant women and infants, [that] could be regularly checked and adjusted when necessary."[87] Berman believed there ought to be "a central authority in every country in the form of a national institute for research in psycho-endocrinology."[88] Berman's project sought to unite an understanding of heightened mentality with gender optimization. Berman did not consider resorting to prenatal interference but rather focused his attention on constant environmental modulation to achieve his outcomes. Endocrinology opened a space within the sciences for postnatal intervention. Similarly, epigenetics concentrates its developmental narrative on how "one's personality and behavior are understood as the *effect* of unseen" forces at work within the body.[89] Both sciences are intimately linked to psychology insofar as they are "characterized by a retreat from politics and history in favor of embracing subjectivity as the source of greater fulfillment and truth."[90]

---

87 Ibid., 102.

88 Ibid.

89 Michael Pettit, "Becoming Glandular: Endocrinology, Mass Culture, and Experimental Lives in the Interwar Age," *The American Historical Review* 118, no. 4 (2013): 1054.

90 Ibid.

Wark's confessional narrative of her transition amplifies a series of beliefs situated at "the heart of psychological modernism
was the conviction that individuals were largely self-deceived
as to their true motivations. They were driven by hidden forces
beyond their immediate perception and often required expert
guidance to recognize and realize their authentic nature."[91] In her
interview with New York City Trans Oral History Project, Wark
talks about how she started receiving psychotherapy from the
renowned transgender therapist SJ Langer, who roots transgender subjectivity "in the concept of dysphoria which is not every-
body's, but there's a— but maybe there is a pre-linguistic basis to
your relation to how your body signals to itself, and that's what
dysphoria is. And maybe it's not even explicitly gendered; hormones will change how your intestine works and maybe that's
what's out of alignment. You know what I mean? It was all speculative but I'm like 'oh, that's so enabling for me.'" Langer's work
echoes the project of endocrinological trends of the latter half
of the twentieth century, which promoted a "fascination with
interiority [that] became associated with a hyper-individualistic
culture obsessed with self-help and psychotherapy."[92]

Michael Pettit refers to the subject of such thinking "as the
glandular self" that critically "intersected with and often provided a biological explanation for a number of facets of individual identity, from race and gender to sexuality and class."[93]
Hormones became interventionist technologies, functioning as
a means through which to interpret one's self. In the case of epigenetics, that expression extended to their surroundings in ways
that encouraged these subjects to alter their conduct to positively reflect upon their wellbeing. They are made aware through
scientific measurements of their categorical belonging and are
convinced that by monitoring this data they will be able to intervene with their human nature to correct it. They accede to
live in an environment that is constantly feeding back into those

---

91 Ibid., 1054–55.
92 Ibid.
93 Ibid.

sciences that seek to capture them. What is significant about endocrinology is that it laid the groundwork for individuals to picture themselves "as amenable to psychological and biomedical intervention."[94] From the beginning, the white male body had always been a favored subject for these interventions. Early endocrinology pioneers used a combination of glandular materials extracted from livestock or the glands of executed African American, Mexican, and Native American prisoners.[95] This choice of material is incongruous given the scientific principle, as borne out in the earlier comments by Berman, that these men were inherently lacking in hormonal potency as compared to white subjects.

The significance of their inclusion, for Pettit, points in another direction. He suggests this practice represented "an instance of subversive racial mixture," where "the traffic in body parts reflected a broader commoditization of certain persons' bodies along the lines of gender and race."[96] Illicitly crossing that line is what consumes the narrative of the Harlem Renaissance author George Schulyer's satirical novel *Black No More* (1931). Schuyler was one of the great "New Negro" thinkers of his time. Pettit reads his work as a meditation on "endocrinology to as both a potential ally and an imminent threat" to the categorical recognition of race.[97] He interprets Schuyler's novel along these same ambiguous lines, explaining that the novel itself "envisioned a glandular treatment that highlighted the fiction of racial inferiority by rendering African American skin white."[98] It captured glandular sciences' potential "to demonstrate the artifice of race, while at the same time expressing worries that endocrinology would be deployed to enforce mandatory whiteness in both appearance and culture."[99]

---

94 Ibid., 1056.
95 Ibid., 1063.
96 Ibid.
97 Ibid., 1064.
98 Ibid.
99 Ibid.

No More Black

There is another possible reading inferred in Pettit's description that deserves equal recognition. Danzy Senna, the contemporary biracial American novelist, situates Schuyler at another highly charged intersection by asserting that *Black No More* is based on the "fantastical, speculative premise of a machine that could turn black people permanently white."[100] Senna's reading locates *Black No More* in the realm of science fiction. Schuyler's speculative narrative explores what a racially polarized America grappling with the emergence of its first Black bourgeois class would make of such an invention, and if it would weaponize it for social and political ends. He questions if it would be used as a means through which to reverse engineer the course of racial history, effectively erasing Blackness as a feature of American identity and promoting whiteness as the singular vector of societal progress. Ultimately, Schulyer's text poses the question, what "would be revealed by the chaos that would ensue?"[101] Perhaps this is the superfluous set of questions because it assumes the invention of racial consciousness, on one hand, and the other, an awareness of class politics as an inevitable feature of American society. Were it not for the ascent of slavery in the Americas, both of these elements might never have become part of the capitalist production cycle. Nonetheless, it remains possible that whiteness features as an affordance of whatever way history plays out. Access to money, freedom, mobility, and power all define its course. So long as it exists, there will inevitably emerge within it a category of exclusion necessary for privilege to survive.

Schuyler's ability to see through the fallacy of race in the writing of his novel was based on the story of his unconvincing appearance. He didn't pass the test of authenticity as a "New

100 Danzy Senna, "George Schuyler: An Afrofuturist before His Time," *The New York Review of Books,* January 19, 2018, https://www.nybooks.com/daily/2018/01/19/george-schuyler-an-afrofuturist-before-his-time/.
101 Ibid.

Negro" in Harlem. He was "too working-class," "too well read," too "well traveled," and, most importantly, "too dark-skinned" to achieve such pedigree.[102] Instead, he travelled a parallel route to notoriety in his marriage to Josephine Cogdell, a white Texas socialite turned New York bohemian. For her part, "Josephine Cogdell follows the path towards miscegenation because she believed "'the white race, the Anglo Saxon especially, is spiritually depleted and America must mate with the Negro to save herself.'"[103] The couple gives birth to daughter Philippa in the same year that *Black No More* was published. Senna recounts, "they had a baby girl, Philippa. Schuyler's small interracial family became his only tribe — the island of the misfit toys."[104] Their family resided in Harlem but remained fundamentally exiled from the Black community. This was because Schuyler chose to give no critical refuge to the concept of a presumed superiority amongst African Americans in the modern imperial world. Rather, he saw them as liberated American slaves assuming the parts of their masters, bent on trading on their identities within that critical context. The Schulyers eschewed such racial essentialism in favor of going a route of racial experimentation, principally enacted upon their only daughter.

Schuyler's approach to parenting was a self-conscious exercise in racial fusion. Mirroring Berman's quest, the couple searched for a rare formula that would impart physical vigor and intellectual superiority in equal measures to their offspring. They raised Philippa "on a scientifically prepared diet of raw meat, unpasteurized milk, and castor oil, and [kept] her in near isolation from other children."[105] They cultivated her to become something of an autistic savant. She learnt "to read at two, became an accomplished pianist at four, and a composer by five. She was a child celebrity, a kind of Black Shirley Temple with a high IQ who became the subject of scores of articles in publica-

---

102 Ibid.
103 Ibid.
104 Ibid.
105 Ibid.

tions such as *Time, The New York Times,* and *The New Yorker,* and was roundly hailed as a genius."[106] Beyond that, "Philippa's upbringing followed the recommendation of the behavioral psychologist John B. Watson that a child should never be hugged or kissed, and her mother's journals contain numerous references to beating, whipping, and slapping. Watson also recommended that parents should talk to their children frankly about sex at the earliest possible age."[107]

At puberty, Phillipa was finally let into the secret of her own creation when her parents revealed to her their data set, chronicling her highly manipulated development, which caused her essentially to have a nervous breakdown. As she matured into adulthood, "her insecurity was further fueled by the increasing realization that her mother viewed her simply as a genetic and behavioral experiment whose success was due to nutrition and training rather than natural talent."[108] Through her reaction, it became obvious that "race *is in actuality a technology*."[109] At that moment, Phillipa realized that "to be black is not to be a person, it is both to be a servant and to exist in service" and that this condition had "no apparent exit point" without the absolute abandonment of her position. As the apparatus at the center of this performance, she needed to physically, visually, and semiotically cease to exist *as* herself.[110] To effectively cut the strings holding her to her past identity, as Phillipa, she needed to excise from herself any further identification with her race, talents, and nationality.

By the early 1960s, she essentially reprogramed herself into a new persona altogether, rechristening herself as an Iberian-American named Filipa Montera Schulyer (not unlike Wark's

106 Ibid.
107 Ibid.
108 Ibid.
109 Poe Johnson, "Racial Technologies in the Time of Black Cyborgnetic Consciousness," in *The Routledge Companion to Biology in Art and Architecture,* eds. Charissa N. Terranova and Meredith Tromble (London: Routledge, 2017), 369.
110 Ibid.

Latinx alter ego, Chica Marx). She successfully applied for a
Portuguese passport using this new identity. Schulyer may
have well borrowed the visual codes for her Latinization of
Blackness from her formative time spent in Paris as a musical
prodigy. During the century preceding her birth, France had
been involved in a political project to "culturally dominate the
Spanish and Portuguese Americas."[111] This initiative informed
the construction of "Blackness as a Latin American" identity
within "Parisian visual culture" impacting perceptions of Blackness "regardless of nationality.[112] What was most striking about
this movement was its pseudo-scientific assertion that Spanish
and Portuguese subjects were "in phenotype and physiogno-
my — that is to say skin tone and facial features" evidentially in
possession of "so-called African blood."[113] Lyneise E. Williams
asserts that the association of Blackness with the Spanish and
Portuguese in this scheme worked to "elevate Black Latin Americans to a high position in Parisian's hierarchy of Blackness."[114]
The mingling of African blood with Latin blood had the desired
effect of creating a new class of "mulatto" identity that allowed
this group to assume a civility denied to their Haitian, colonized
African and African American counterparts circulating within
Parisian society in those years. Through intimate association
with Spain and Portugal as suitable colonial partners, France
was able to participate in the congress of cosmopolitan exchange
that coalesced within the greater body of postcolonial interna-
tionalism. Schuyler's tactical choice to portray herself as more
than Black by assuming a Portuguese national identity played
into "popular and anthropological perceptions of the Spanish
and Portuguese Americas as the 'laboratory of modern mixed
breeds or hybrid nations.'"[115] Such perceptions of race as malleable opened her world up to wider possibilities and allowed

---

111 Lyneise E. Williams, *Latin Blackness in Parisian Visual Culture, 1852–1932*
(London: Bloomsbury, 2019). 1.

112 Ibid.

113 Ibid.

114 Ibid.

115 Ibid., 56.

her to travel lightly into a path of assimilation through the subtle eraser of any markings of difference that would render her apart from the citizenship she now claimed. In many ways, hers was an epigenetic enterprise insofar as it worked off the premise that the observable expression of racial characteristics could be influenced by environmental conditions. In this way, Schuyler passed into another racial hierarchy altogether escaping the one of her "cis" race.

That said, like a character out of *Black No More,* she grappled painfully throughout her life with the consequences of the secondary characteristics of her self-invention, exiling herself overseas, working as an international journalist and children's advocate, finding ways to document other histories of racial violence enacted upon other innocents, never really parting with the echo of her own traumatic conversion. Perhaps fittingly, "in 1967 she died in a helicopter crash while attempting to evacuate war orphans out of Vietnam."[116] She attracted the attention of American intelligence officers based on her criticism of America's racism and imperialism during the war prior to the incident. Investigators were never able to provide a satisfactory explanation as to why her helicopter crashed on what should have been a routine flight. Perhaps the most likely supposition to draw was that the racialized war machine that invented her eventually caught up with her. Her dislocated bodily autonomy remained a threat to the republic throughout her 36 years.[117] It was a loose end to be tied up; recoded in its final iteration as an act of servitude.

Behavioral Difference

In many ways, Phillipa Schuyler's lifestyle anticipated forms of subjectivity we now take for granted, one that are synonymous with neoliberalism and neocolonialism that promote elite mobility and encourage those with privilege to create new, entre-

---

116 Senna, "George Schuyler."
117 Johnson, "Racial Technologies," 376.

preneurial selves that can take on various racial and post-racial guises when required to do so for the sake of their own social and political advancement. Similarly, the post-racial aspirations explored in *Black No More* are not solely a product of interwar American politics, but they equally can be said to date back to currents present within late eighteenth-century natural philosophy and scientific literature. Phillipa's difficult trajectory through life was in many ways patterned on Watson's largely failed ambition to socially engineer human beings. As a leading behavioral psychologist of his day, he widely promoted the belief that it was possible to discipline glandular systems so that they might then be freed from the negative burdens of "biological heredity, including the heredity of emotional states."[118]

Watson's initial test subjects were drawn from a pool of orphans who were institutionalized under the care of the hospital in which he worked; some of whom were neurologically compromised. Not one was in a position to give their consent to the incredibly cruel and damaging experiments involving emotional deprivation, early life sexualization, and bizarre feeding regimens that he subjected them to and later applied to his own children. The grave lifelong consequences of their maltreatment resulted in tortured lives not dissimilar to the kind endured by Philippa Schulyer. As a consequence, Watson's test subjects were irrevocably marred, and his career ended in disgrace. Despite this, his beliefs "have recently been revived through positive psychology which following on from Watson demonizes 'negative emotional states'" and enforces "the requirement of positivity to remain healthy."[119]

Another aspect of Watson's research from the 1920s concluded that "it was simply not possible to get into an animal's 'mind'; indeed, later behaviorists such as B.F. Skinner would conclude that it is impossible to get into any mind, including that of a

118 Oksana Yakushko, *Scientific Pollyannaism: From Inquisition to Positive Psychology* (New York: Springer, 2019), 96.

119 Ibid., 119.

human being."[120] What could be observed in its place was the mind's response to stimulus to affect behavior. For this to happen, he recommended the accumulation of a large amount of empirical evidence drawn from experiments done on both chimpanzees and children. Watson concluded from his work that "there was no behavioral difference of any significant sort in newborns. Virtually all behavior was learned."[121] Watson narrowed his focus to what was observable and became obsessed with the idea that conditioning was central to human development and, therefore, could be intervened upon to create optimal conduct.

Later on, this idea would be taken up by Skinner, who become the intellectual progenitor of radical behaviorism. Skinner went onto develop the concept of "operant conditioning":

all behavior can be explained through the conditioning of the organism in response to the receipt of rewards or punishments for its actions. Reward it, and it will be more likely to repeat the behavior. Punish it, and the behavior will be reduced. Human beings had one and only one innate behavior: they could learn.[122]

What was most remarkable about Skinner's approach is that it concluded, in effect, "that humans have no nature" and therefore, "human beings, and by implication, human society, could mold themselves to become whatever they willed."[123] A humanity divorced from nature in its volitional capacity was soon joined in its destiny to animals and sophisticated technological objects whose interplay extended the boundaries of what could be described as the characteristics of sentient life.

---

120 Aaron Gillette, "The Rise of Environmental Behaviorism," in *Eugenics and the Nature-Nurture Debate in the Twentieth Century* (Basingstoke: Palgrave Macmillan, 2007), 113.
121 Ibid., 114.
122 Ibid.
123 Ibid.

The capacity for "purpose, adaptive responsiveness, learning, and so on" was redefined by the postwar emerging science of cybernetics.[124] Its aims went far beyond biological and psychological science's tendency to reduce living beings to a mere interplay of mechanisms by radically shifting the emphasis from performance to the observation of processes to produce their scientific insights. The innovation at the heart of cybernetics was to structure and map physiological instability and to assign them "precise technical analogues" so that mechanical or electrical computing machines could be twinned with biological organisms.[125] By "comparing organisms to some of the most innovative technologies of the era, namely servo-mechanisms, scanning instruments, electronic communication systems, analog computers, and, perhaps most notably, the new high-speed digital calculators that were emerging from secrecy in the postwar era," cyberneticists were effectively breaking down the boundaries between orders of life.[126] In their place, they sought to install an understanding of beings of a somewhat similar nature through an understanding of how they failed to maintain their stability when acting within large, complex systems. Everything then came down to determining characteristics of behavior and "diagnosing the sources of breakdown so as to eliminate them and recover unity and stability."[127] Pathology, became the obverse of "capacity for any being, whether human, missile, cat or computer, to achieve ends by means of learning with respect to their environment."[128] Facing opposition, or overcoming obstacles become is the signifier of something's capacity to be and remain alive. Following from the nineteenth century traditions

---

124 Ibid.

125 David W. Bates, "Unity, Plasticity, Catastrophe: Order and Pathology in the Cybernetic Era," in *Catastrophes: A History and Theory of an Operative Concept*, eds. Nitzan Lebovic and Andreas Killen (Berlin: De Gruyter, 2014), 32.

126 Ibid.

127 Ibid., 34.

128 Daniel Nemenyi, "What Is an Internet? Norbert Wiener and the Society of Control" (PhD diss., Kingston University, 2019), 46.

of homeostasis and natural selection, the cybernetic concept of life is defined as that which is capable of adaptation.

Behavioral Racism

Emerging from the work of cybernetic behaviorists "is an entirely new, non-determinist model of biological life as plastic, that is, changeable."[129] How they are proscribed to change, however, is perhaps more heavily overdetermined than was the case with their scientific predecessors. Cyberneticist's focus on abnormality and optimization rejects, on one hand, genetic determinism, while on the other promotes new forms of understanding of genetic processes that effectively produce difference. The racial implication of this is to study the effects of racial oppression, with no recourse to an appreciation of how it "gets under the skin" of societal logics and how race "is tied" historically "to a eugenic logic."[130] For these reasons, the science of epigenetics, as a developmental form of cybernetics, "offers a new form of racialization based on processes of becoming" that "normalizes white bodies and behaviors" while pathologizing Black bodies and behaviors.[131]

Becky Mansfield and Julie Guthman argue that Black disadvantage, within this same logic, derives from a failure "to emulate the environments and behaviors of rich white people in order to protect themselves and their offspring."[132] Race becomes a defensive mechanism at not only the level of the individual, but the population. This time the destiny for race plays out at the molecular rather than molar level. Through this strategic narrowing of interest, epigenetic optimization is granted legitimacy "to operate in the service of racial normalization even as it lacks explicitly racist references: the aim is to eliminate 'abnormal'

---

129 Becky Mansfield and Julie Guthman, "Epigenetic Life: Biological Plasticity, Abnormality, and New Configurations of Race and Reproduction," *Cultural Geographies* 22, no. 1 (2015): 4.

130 Ibid., 5–6.

131 Ibid., 8.

132 Ibid.

bodily differences — make them die, including by preventing them from ever being born — in order to purify and improve human life."[133] It is no less lethal or targeted in its epigenetic valance because it "has the effect not only of reifying biological difference as objectively real but also of pathologizing these differences and even seeking to eliminate ('cure') them"; even as it assumes a "focus on plasticity and environmental influence," it remains rooted, like genetic determinism, "in a preoccupation with fixing pathology through the elimination of features that fail to correspond with an '(implicitly) white norm.'"[134]

Even at the level of molecular difference, the imperative remains one of controlling racial reproduction and development. It does so by expanding the spectrum of "what counts as 'abnormal': the broad and deep becoming of human bodies as a result of environmental exposures which disrupt 'normal' development," only to justify its further interference into the lives of certain categories of differentiated people requiring of them not to demonstrate advance, but rather semblance "toward a privileged, idealized, and white norm."[135] Within such a model the subject is made responsible for their own healing, that directed by an external medical or psychiatric authority who is capable of assessing the degree of disorder without probing the social context through which it was incurred.

The greater community or society can neither prevent nor redress trauma because its suffering has been strategically divorced from the context of interlocking systems of power such as "settler-colonialism, neoliberalism, [and] racism" that materially perpetuate "gendered violence, homophobia and transphobia."[136] Coding trauma through race and gender serves the interests of those interlocking systems of power insofar as it is able to bring oppressed objects into being according to contemporary practices of epigenetics. By making them intelligible,

133 Ibid., 15.
134 Ibid., 16.
135 Ibid.
136 Scott Kouri, "Counselling in an Age of Empire" (PhD diss., University of Victoria, 2019), 168.

they are able to assign value and bring into delimited expression the formation of particular subjects, and concurrent methods of treatment to further mediate the power, or lack thereof, of these particular bodies in order to discipline them in particular ways. The novelty of their appearance complements existing structures of influence, force, and domination. Such an assertion of power retains a suppleness to the degree that it is able to incorporate new forms of apprehension that includes the mapping of the human brain, genomics, and psychopharmacology as tools of advancement "into the spiritual, affective, neurobiological, and cognitive levels of the individual."[137]

As a corollary to this, "new diagnoses are being created" that logically serve "to pathologize resistance as uncivilized, impulsive, or irrational."[138] At the same time, "new drugs are being developed to act as chemical restraints in order" to further a semblance of normality. Scott Kouri attributes these innovations to the need for the interlocking systems of capitalism and colonialism to maintain their advantage without recourse to "sustained violence" and argues that these trends within psychiatry act "as a more subtle and invisible form of control than explicit violence"; thus they are "increasingly being chosen as the way to maintain capitalist and colonial relations."[139] Much of this has to do with the introduction of ideological austerity into state regulated systems of social care. The retrenchment of social safety nets for majority white populations dwelling within the global North puts them into contact with the types of social jeopardy that was previously the reserved of their racial ethnic underclasses and those occupying similar positions in the global South. Today no such exceptionalism exists for the North's white majority populations.

Kouri argues that "empire today brings the real possibility of utter poverty to the doorsteps of majoritarian young people, deterritorializing disposability from its historic ties to particular

---

137 Ibid.
138 Ibid.
139 Ibid., 169.

regions and peoples and reterritorializing poverty and precarity into the heart of Empire."[140] He maintains that

> in many ways, individualized distress at finding one's place in society hides the reality that great swaths of young people who historically would have been the beneficiaries of systemic oppression can no longer count on their place within Empire's favored ranks. Identity factors will, without a doubt, continue to organize economic stratification in North America for some time to come, however, the growing wealth concentration and gaps leave very few with any security. Illusions, defenses, and ideology allow majoritarian subjects to maintain a semblance of normalcy, but at an unconscious level, most of us know that none but the very elite are served by the capito-colonial system, and even they are served at material levels that leave affective emptiness in their wake.[141]

Why this is significant is that it is happening at a time when white America is very much grappling with the nation's racialized past through the lens of neoliberal global capitalism and drawing conclusions about systematic inequality that complicate their unquestioning loyalty to the assignation of failure to a fundamental lack of personal integrity and character previously coded as Black. While some young white Americans, particularly those from relatively affluent backgrounds, continue to insist that capitalism's reward of social ascendency prevails for those willing to make a sustained effort and accrue professional support through a combination of higher education and shrewd networking, their peer's failure to obtain social security is progressively undermining their faith in the system. As is their own amassing of crippling debt and "delaying or sidestepping other aspects of heteronormative lives, such as marriage, homeownership, and parenthood."[142] Together these hardships are making

---

140 Ibid., 173.
141 Ibid.
142 Ibid.

an increasing majority of young whites more aware of the parasitic nature of capitalism and, in turn, how American exceptionalism very much rests on a racialized hierarchy of exploitation and denied opportunity both at home and abroad.

Those young whites, who face being disadvantaged or damaged by neoliberal capitalism — in ways previously reserved for people of color — are now becoming invested in ways to question and resist market orthodoxies and "work socially to create alternatives through youth activism and resistance."[143] Essentially what is happening is that these populations are finding themselves subject to racial trauma for the first time. As a result, they are attracted to historical understandings of race that might account for their present losses. At the same, they look to racialized, affectable others to educate them about how to respond to what they perceive as a novel crisis in self-determination. As they start to experience a minor degree of social death, they are eager to appreciate what has already been economically forecast — the psychic foreclosure of Black subjectivity. Perhaps it is no coincidence then, that at the moment when white humanity is being downgraded as an asset class, it is being anxiously offered to Black people. What is problematic about that gesture, amongst many other things, is that it is an offer made from the perspective of a self-determining subject that grants that privilege of affectability to those same persons it objectified. In so doing, it suggests that the issue of inequality can be overcome by maintaining that same dynamic of power.

Nothing is fundamentally altered within such a gesture. The white subject has given over virtually nothing, but instead, has reified the illusion that they alone "have the ability to raise others to the level of invulnerability and self-determining power."[144] In this scenario, the racialized other retain their position of victimhood and figures as a specter of inequality unable to materialize as a fully formed subject capable of ascendency to the category of human being through their own innate resource. This gesture

143 Ibid., 183.
144 Ibid., 191.

infers that Black lives matter to the extent that whites support them in their assertion of humanity. The Black grief and suffering that must be performed for this realization to occur are grotesque because it assumes that the white subject maintains an unquestionable claim to humanity and that it once again defers in its obligation "to come to terms with its own vulnerability, its own contingency and affectability."[145]

The Conditions of Affectability

Denise Ferreira Da Silva argues that affectability is then further ignored in a clinical context of counselling that may be a vehicle for the specter of affectability to finally be raised alongside an acknowledgement of the racialized legacy of uneven power. Instead, what is often named is not the white self but his obverse, the Black other, whose exposure to trauma becomes the primary occupation of an American society unable to deal with itself and its own losses from its beginnings as a colonial slave society. America's founding myth is one of white self-determination, which comes from the satisfaction that others have ceased ground to make it thus. The white Western subject assumes freedom of movement as his birth right, as much as he assumes confinement and constriction to be the birth right of the racialized and affectable others who surround him and who are subject to his unlimited powers of self-advancement. Any exposure to restriction becomes synonymous with being affectable — being racialized — as others have been and continue to coerced to be. Hence, the pronouncement "I can't breathe" becomes the raw material for theorizing the contemporary neoliberal world, a world that still fails to account for Black aspiration, while at the same time, progressively narrows the fund of possibilities available to whites. A world that Ashon T. Crawley observes "produced for us current iterations of categorical designations of racial hierarchies, class stratifications, gender bina-

---

145 Ibid.

ries, mind-body splits," all based on an original failure of the American project of abolition.[146]

For Crawley, "I can't breathe" is the ultimate expression of "black disbelief, in the configuration of the world that could so violently attack and assault" Black people for the crime of attempting to participate in even the most minor ways in a capitalist system for whom racial exclusion is the only reasonable possibility.[147] Blackness is about "disruption and interruption," something that gets in the way "of liberal logics of subjectivity" and "the assumptive logics of gender."[148] For Crawley, Blackness is an experience of life as "exteriority"; its "exteriorization" functioning as a "critical sociality of intense feeling" in the wake of what has been a critical undercounting of the Black experience in America.[149] Crawley maintains that whiteness, "as a way to think the world and one's relation to it[, …] is about the acceptance of violence and of violation as a way of life, as quotidian, as axiomatic."[150] By contrast, "Black social life has been the emergence of abolition as a grounding of its existence, the refusal of violence and violation as a way of life, as quotidian."[151] Whiteness stands in opposition to this through its grounding of subjectivity "in the capacity for ownership, for acquiring objects."[152] As a consequence of this arrangement, it is only possible to understand the differential assimilation of trauma within white and Black bodies through staking claim to an acceptance of violence versus seeking its abolition through resistance to any implication of bodily deficiency.

Ibram X. Kendi addresses Post-Traumatic Slave Syndrome, PTSS, in the context of this differential assimilation of trauma. Xendi argues that "to root 'dysfunctional' Black behaviors in the

---

146 Ashon T. Crawley, *Blackpentecostal Breath: The Aesthetics of Possibility* (New York: Fordham University Press, 2016), 1.
147 Ibid.
148 Ibid., 4.
149 Ibid., 5.
150 Ibid., 6.
151 Ibid.
152 Ibid., 7.

history of oppression" is, in itself, a fundamentally racist principle premised "on racist notions of degenerate Black people."[153] He contends that "Black people as a group do not need to be healed from racist trauma" because slavery and other ongoing dehumanizing oppressions "actually did not succeed in dehumanizing Black people and leaving them adversely traumatized."[154] PTSS theorists assume there is something inherently pathological in the Black body that caused it to be enslaved, segregated, and mass incarcerated and continually registered "at the low end of racial disparities."[155] So, the theory goes, discrimination and exploitation will end when Black people are "redirected" from their "dysfunctional" "attitudes and behaviors" and brought onto a moral course where they can be progressively "healed, civilized, developed" in ways that comport with greater social opportunities.[156]

Xendi observes that "countless historians have chronicled freed Blacks successful struts off plantations and into politics, labor organizing, artistry, entrepreneurship, club building, church building, school building, community building" without taking note of how "the fiery hand of Jim Crow" sought time and again to destroy the progress of these brutally contested gains.[157] These wounds were not self-inflicted and as such, they fueled an overwhelming desire amongst "all Black people [...] to be freed from racist trauma."[158] An American "history of oppression has made Black opportunities — not Black behaviors — inferior."[159] This lack of opportunity did not stem solely from a denial of American social mobility but from a fundamental belief among America's white abolitionists that Africa, itself, lacked civiliza-

---

153 Ibram X. Kendi, "Post-Traumatic Slave Syndrome Is a Racist Idea," *Black Perspectives*, June 21, 2016, https://www.aaihs.org/post-traumatic-slave-syndrome-is-a-racist-idea/.

154 Ibid.

155 Ibid.

156 Ibid.

157 Ibid.

158 Ibid.

159 Ibid.

tion. This was coupled with a belief that native African communities had literally been poached from the wilds and transported into the Western institution of slavery. Finally, it is presumed that this "peculiar institution" went on to domestically "breed" them in ways that diminished their minds, as well as their bodies.

Denise Ferreira Da Silva argues that violent set of beliefs around Blackness persists within the white American imagination to this day. White Americans still refuse to admit the possibility of "African elites."[160] Rather, they choose to maintain the belief that, "in Africa blacks live in huts, hunting and gathering during the day, eating and chanting at night, killing each other all the time."[161] There is "no tomorrow" for African Americans, as blackness for them crudely exists out of time, and indeed, history.[162] This same constituency assumes that if "you are black, male, and young, […] you are better off" in the United States than you would be in Africa.[163] Black bodies within the United States have an essential duty to perform for which they will never be formally hired or compensated, which is to "make the system work" through their suppression.[164] Essentially what that means is that it "help[s] to create a bond among white Americans" through the fundamental denial of Black security.[165]

What makes this all possible and sustaining as a defensive system is not merely the blunt force of the law, but rather an intricate set of "tools of racial knowledge" that find their bearings beyond world trade, and well into the universalizing territory of science.[166] It is science that authors humanity somewhere between the seventeenth and nineteenth centuries. By the dawn of the twentieth century, the attributes of history and self-deter-

---

160 Denise Ferreira Da Silva, *Toward a Global Idea of Race* (Minneapolis: University of Minnesota Press, 2007), xii.
161 Ibid.
162 Ibid.
163 Ibid.
164 Ibid.
165 Ibid.
166 Ibid., xiii.

mination have hardened around the subject to such a profound degree that it is possible to define humanity as the power to condition the lives of nonhuman others as one's natural right. Race emerges in response to a desire "to secure interiority, the private holding man."[167] This can only be accomplished through the union of "universality and historicity" in one type of constituent body, that is, "the modern political subject," which, in turn, is granted juridical powers as a moral and reasonable being.[168] He is assigned the power of "self-consciousness" and through that fashioned into an "'individual,' […] the basic ontological unit" amenable to both "exterior regulation to protect its life and self-determination."[169] The world of men does not require nature to explain itself so much as the world of things requires science to do so. This is because man already occupies an artificial body when he assumes the guise of being a political animal. He is instituted in individuality and articulated in universality. As a consequence, what power he claims must be construed through external means; that is to say, through an outer determination.

Man is recuperated and given a second nature through the gesture of denying that same dualistic nature, mind and body, in others. Thereafter, mankind must conjecture categorical Blackness as if it were the discovery of a new humanoid species within its midst, one that would eventually come to revere white civilization as its source material for the fashioning of its own evolutionary path. This comes under the proviso that many unburdened and exalted natures in this world have owed their advancement to it. Race figures into a mapping of human progress at the point at which it is possible to classify the world's "inhabitants according to the degree of differentiation, specialization, and complexity."[170] That corresponds with a "degree of perfection" in the organism's mental functioning as demonstrated through the capacities of memory, association, abstraction,

---

167 Ibid., xxxix.
168 Ibid., 50.
169 Ibid.
170 Ibid., 103.

reasoning.[171] The highest cognitive value of all is the ability to affect the lives of others, which means that those with the greatest amount of intellectual prowess have, in effect, the ability to transform "all the other living things into less perfect, less developed, versions of the body of man."[172] For Da Silva, this situation implies that Darwinian evolution is dependent on the devolution of forms of living that are deemed deviant to the progress of civilization.

Darwin asserts that within this understanding, race is a plurality that will, "at some future period, not very distant as measured by centuries," be reduced down to a singularity.[173] Darwin predicts that "the civilized races of man will almost certainly exterminate, and replace, the savage races throughout the world," which is to say that not only will they be eliminated, but equally their population numbers in the world will be wholly made up, as one race gives way to the other.[174] Within Darwin's scheme of double-entry bookkeeping, Blackness remains categorically other to this predicted trajectory. The fate of black(ened) bodies seemingly runs parallel and divorced from these proceedings, as evidenced by his follow up remark that "at the same time, the anthropomorphous apes […] will no doubt be exterminated."[175] Darwin's desired outcome is for the gulf of apprehension between man and ape to be profoundly widened so that it becomes impossible for one to locate any reasonable allegiance between Blackness and humanity. "The break between man and his nearest allies will then be wider, for it will intervene between man in a more civilized state, as we may hope, even than the Caucasian, and some ape as low as a baboon, instead of, as now, between the negro or Australian and the gorilla."[176] Darwin is arguing here for a superior whiteness to be literally cast out of the profound abandonment of any recognition of the reproductive power of

---

171 Ibid.
172 Ibid., 105.
173 Ibid., 110.
174 Ibid.
175 Ibid.
176 Ibid.

these living agents that might compromise humanity's ability to achieve singularity. Therefore, for Da Silva, the imperial West's natural history begins with the taxonomy of the globe and ends with the biological differentiation of humanity following that same exteriorized grouping of descending others, made wholly expendable for the very fact of their difference.

Breeding and Biofacticity

A similarly racist narrative around breeding figured into the French physician Serge Voronoff's work in the 1920s to experimentally transplant the testicles of European livestock into the bodies of white men. Later his experimental preference shifted to the use of testicles of apes expressly "acquired from Africa."[177] Voronoff was convinced "that New World monkeys were an inadequate substitute and expressed concerns about the quality of the anthropoid apes available in captivity in the United States," implying that these species were somehow innervated through the process of being "farmed" as opposed to "wild caught."[178] Voronoff's preference can be read as indicative of ongoing fears about miscegenation in the United States, as well as in France, after World War I as populations of African descent laid claim to greater cultural capital and social mobility within these respective societies. After the war, a new political awareness of Africa was increasing amongst these groups. They employed anti-colonialist rhetoric to call for the redemption of Africa through the introduction of communist style rule in place of capitalism within its nations. The 1920s marked the beginning of the Pan Africa movement when doctrines circulated materially and discursively between America and France. Paris quickly emerged as the unofficial capital of this movement.

Voronoff's claims around the restorative power of the African, therefore, must be read in the context of an anxiety within the French Empire caused by these calls for African self-de-

---

177 Pettit, "Becoming Glandular," 1066.
178 Ibid.

termination. The racial anxieties associated with the sexual and physical revitalization of France are ones dependent on the availability of African *matériel* to effectively reproduce the French social body into the future. Within Voronoff's work, it is possible to locate what Frank Wilderson calls "the *bio-facticity* of Blackness," which is configured through "flesh, color, size and genitals."[179] Blackness becomes the cypher for "sexual ubiquity" within the white imagination, understood "as a power that is so comprehensive that it is impossible to speak of it in sexual terms."[180] It must be referred to instead at "a genital level" and valued for its capacity to give pleasure when its reproductive capacity is appropriated into the white body.[181] Humanity and its pathologies progress along these same instrumental lines during the *longue durée* of American slavery and African colonialism.

In recent years, some have postulated that Voronoff's transplantation of live tissues between apes and man in the 1920s may have caused or contributed to the AIDS epidemic. The proposal for the origin of HIV in Africa by European scientists was based on the isolation of it in Old World monkeys who presumably had sexual relations with native Africans so that it jumped the species barrier.

Such thinking dates at least as far back as Benjamin Rush, who reported that leprosy in the interior parts of Africa was the product of

> women being debauched in the woods by the large baboon, ourang-outang, and by that species in particular called the guaga mooroos. No satisfactory discovery has been made to account for such singular, but not infrequent phaenomena in the species. It may perhaps be ascribed to disease, and that of the leprous kind, with more reason than to any other cause that has been yet assigned.[182]

---

179 Frank B. Wilderson III, *Afropessimism* (New York: Liveright, 2020), 161.

180 Ibid., 162.

181 Ibid., 163.

182 Benjamin Rush, "Observations Intended to Favour a Supposition That the Black Color (As It Is Called) of the Negroes Is Derived from the Leprosy,"

Whatever the logic, western fascination with an African "isolated tribe" and their sexual practices continued unabated well into the twentieth century when AIDS was first being theorized by European scientists. The early AIDS discourse of the late 1970s and early 1980s drew on enduring stereotypes of native Africans as a primitive race living in isolated tribes cut off from civilization. It assumed that Africa as a "dark continent" harbored unique diseases which differed in kind to the rest of the world. It perpetuated the old belief that Africans were closely related to monkeys having potentially interbred with them for centuries, or at the very least involved them in their sexual customs. Thus, they were more readily open to acquiring their diseases, as compared with other human populations. The disease itself was deemed out of control in Africa because it primarily affected the majority heterosexual population, as opposed to the West where it was believed to be largely confined to a minority homosexual population. Presumably, these homosexual men acquired it through sexual tourism to Africa in the early to mid-1970s.

Since the seventeenth century, Africans were perceived as innately more lascivious than Europeans. Coupled with this understanding was the idea that they did not possess the self-awareness or intelligence to stop the spread of the disease because of their primitive belief systems. Africans simply could not fathom the complexity of the biological threat they faced nor how to fight it. In these earlier years of the pandemic, there was widespread conviction within the European scientific community that AIDS could wipe out Africa, if there was not something done by them to "cure" it. The failure to do so belies the fact that the mission itself, like Rush's project to enact racial indifference, was specious from the start insofar as its true aim was "protecting and advancing the geopolitical security of white settler democracy and gendered proprietorship" through scientific theory.[183]

Transactions of the American Philosophical Society 4 (1799): 291.

183 Matthew Crow, "Jefferson's Whale: Race, Climate, and Commerce in Early America," *Journal of the Early Republic* 40, no. 3 (Fall 2020): 437.

Melinda Cooper argues that "without a doubt, the history of South African apartheid is intimately entwined with the rise of modern public health strategies and theories of contagion. South Africa's first segregationist law was the Public Health Act of 1883 that allowed local authorities to enforce quarantine and vaccination under general powers of emergency. These same emergency powers were applied to forcibly remove Black Africans from their ancestral lands on the premise that their dwelling there represented "'a public health risk.'"[184] Racial segregation became in this way a tool of public health initiatives dating back to 1900 and continuing into the 1980s as HIV spread amongst its Black African population. Cooper maintains that HIV was used secretly by the apartheid-era government "as an agent of biological warfare" directly targeting Black African women, "in a project to transform the virus into a sterility inducing agent."[185] Against the greater Black African population, there were campaigns to degrade public health services in their homelands so that education and prevention measures were effectively withheld in order to eliminate Black African lives deemed surplus to biopolitical requirement. What has taken the place of public health have been Western initiatives to enter South Africa to conduct experiments with antiretroviral drugs on its urban, racial and gender underclasses. These pharmaceutical trials are the latest form of coercive labor that implicates the Black body as the source material of both contagion and cure that concerns itself with human body parts, sexual intercourse, the use of socially dead bodies to reanimate a declining system of colonial dispossession in the form of wealth extraction by multinational drug companies.

---

184 Melinda Cooper, *Life as Surplus: Biotechnology and Capitalism in the Neoliberal Era* (Seattle: University of Washington Press, 2011), 68.
185 Ibid.

Transgender Technicity

Jules Gill-Peterson asks us to consider how the molecular life of synthetic hormones have inserted themselves into our understanding of "racialized and trans embodiment" and redefine all bodies in terms of their "technical capacities."[186] This new standard of judgement applies to "trans-of-color subject subordinated to racially normative and gender-normative white and cisgender bodies"; both of which are implicated through their relationships to the hormone molecule.[187] The synthetic hormone as a molecular form becomes a third term of embodied subjectivity. It is the "they" situated in between the "he" or "she," it is the "it" situated among the animal, human, and machine, and the "thing" situated in between race and trans, and as such, is fundamentally relational. Both racial and transgender embodiment relies on conceptual protocols attentive "to how they both receive their historical animacy from an endocrinological engagement with the body's hormonal technicity."[188]

What is remarkable about Gill-Peterson's approach is that it assigns agency to the hormone molecule in assuming it has itineraries all its own, a presence that is felt within bodies both human and nonhuman, biological, political, and economic, and that like any *subject,* it matters. Gill-Peterson assigns this molecule a "dynamism" that immediately divorces it from the status of "a domesticated tool."[189] It is not a slave, to be "used by humans for rational or irrational ends" but rather something highly capable of altering the way that both transgender and race are constructed. The molecule becomes the agent of modification, something that comes in to extend the body past the limit of its original form, installing a kind of plasticity within it to redefine how it is alive and how its identity is expressed.

---

186 Julian Gill-Peterson, "The Technical Capacities of the Body: Assembling Race, Technology, and Transgender," *Transgender Studies Quarterly* 1, no. 3 (2014): 403.
187 Ibid.
188 Ibid., 404.
189 Ibid., 403.

The molecular, in this sense, intervenes upon the body as an active participant acting within the biological body's endocrine system. In this way, it becomes a subject acting upon a subject, installing difference at the point of convergence between synthetic and autonomic functioning. The body must incorporate the hormone molecule to intervene upon itself. Therefore, this involves a dual subjectivity to achieve the situation of exceptional sexual optimization versus standard sexual development.

Through that incorporation the molecule takes on a degree of relative animacy to the body, compelling it to become "receptive to change through variable hormone circulation and environmental change" meaning that its survival relies on not being "in opposition to or by transcending it."[190] This implies a symbiosis on the level of technicity and of functioning, where the molecule works on the endocrine system, but also works on itself and the body, affecting the appearance of the subject. The molecule marks the point of indeterminacy between *form* and *matter,* acting as the *manner* in which bodies assemble themselves. In this sense, hormonal molecules manage to "people" themselves in that same radical gesture of making others up as sets of intensive variations.

The hormonal molecule sets up a particular climate within the body to affect transformation that corresponds with older thinking around sex and sexuality that equates with its cultivation. The endocrine system itself was perceived as a kind of micro-climate where it was possible to manipulate factors in order to re-enliven it through the insertion of certain codes of corrective information. In this scenario, the hormonal molecule became a catalyst for race and gender hygiene, which allowed for certain codes of improvement to be interjected into the workings of the body itself. The molecular mechanism through which information affecting the body was transmitted to subsequent generations established an "interface between the environment, the body, and its heritable acquired characteristics" that was then able to generate a scientific mainframe for

---

190 Ibid., 407.

epigenetics.[191] In establishing a medical body bound to cultural definitions of race and sexuality, it was able to at once halt the progress of genetic determinism and commence the administration of whole populations through the introduction of a new hormones-based biopolitics.

Within the medicalized discourse of transgender identity is an underlying bias towards the achievement of the biological and cultural verisimilitude of gender according to the idealized category of the cisgender white body. Timing is thought to be crucial to this operation, implying that those who transition past the time of puberty miss out on the potential to be perceived as "convincingly" male to female or female to male in their gendered appearance. The idea is for children, in particular, to be vetted as candidates for biomedical intervention amongst those who express gender dysmorphia so that effectively their gender can be corrected prior to puberty. All this work conforms to a logic of "'secondary sexual characteristics'; a concept adapted from nineteenth-century sexology by American researchers on transsexuality and gender identity in the 1950s and 1960s."[192]

The method used currently is one of puberty suppression in children under sixteen, diagnosed with Gender Identity Disorder (GID). From the very beginning of this process endocrinologists and psychotherapists intervene upon the subject to determine the trajectory of the subject's gender assignment and the degree to which this is "reversible." They orchestrate externally the timing of development in tandem with its suppression and ultimate redirection. All of this takes place under the skin of the subject, either through monthly injections or an implant device that passively delivers hormonal therapy to the patient. This use of technology to mollify the body can be construed as a means of devaluing it by judging it only through a lens of innate dysfunctionality.

The medicalization of transgender identity has much in common with racial science insofar as they both are built upon

---

191 Ibid., 412.
192 Ibid., 413.

strategies of restrictive access, forceful suppression, and limited agency. Such bodies are deemed to lack some form of inherent integrity, and therefore, are made exceptional, as a contrast to the fully human. Made relational to a system that fails to fully account for them, they assume the role of non-entities, and more importantly, non-identities. So long as their tools of self-elaboration remain external to them, these bodies are forced to enter into relational status with human beings in a way that intentionally stalls them in achieving any meaningful autonomy apart from as differential, politicized beings.

5

Capital Gains

Object Ontologies, Settler-colonialism, and Financialized Futures

Survival in the Market

The posthuman recently has become operative through the emerging significance of biodiversity and neurodiversity as a means of retooling humanist epistemological to superficially accommodate the nonhuman in their midst through interactive means that promote the consumption of new technologies. Such technologies act as social prosthetics, extending the boundaries of inclusion such that various materialities get to pass by association for collectivities as we recast natural and cultural worlds to conform with the horizon line of mass extinction. The volatility of these seemingly new categories of perception in many ways obscure the previous ontologies, politics, and patterns of exclusion that have brought us onto this very path of destruction. As a consequence, contemporary attempts to vitalize difference are emblematic of a refusal to grapple with the deep significance of positioning some bodies as invaluable, while others remain viable only as points of departure for furthering ways of knowing

301

of the body and its capacities as differential properties within the greater enterprise of human and species valuation.

Both of these inclinations hew towards "much longer histories of racialization, affectivity, and disablement" that require that their current instantiation be construed as no less violent than previous arrangements of coloniality, where power made itself felt through "the co-constitution of race and disability in the *longue durée* of racial capitalism and liberal modernity" and that today gestures towards the importance of thinking about difference when naming its objects.[1] Whereas liberal biopolitics previously demanded that Black people, Indigenous people, and other people of color, women, and people with disabilities were dealt with as property belonging to the state, in today's thanatopolitical atmosphere, these considerations have been adjusted to protect persons identified as human from liability for the care of these errant bodies.

A new austerity has come to the force at a time when such peoples have been afforded progressive recognition and rights. As Tanja Aho, Liat Ben-Moshe, and Leon J. Hilton argue, these offerings remain "imbricated with other forms of state violence," as is the assumption of them "linked to enduring ontological erasures" that take as their starting point the normalization of these peoples not through direct coercion, as had been the case within the disciplinary structures of liberalism, but rather indirectly through a neoliberal regime of self-care, preferring to threaten these constituencies with social death as opposed to hastening their material demise wherever possible.[2] It would seem that investing in animacy has become "the new normal for regimes previously concerned only with carrying their numbing effects and deadening functions."[3]

Increasingly, people now require re-institutionalization to conform with a new universalized trajectory of subjectivity bent

---

1   Tanja Aho, Liat Ben-Moshe, and Leon J. Hilton, "Mad Futures: Affect/ Theory/Violence," *American Quarterly* 69, no. 2 (June 2017): 291.

2   Ibid.

3   Ibid., 292.

on producing still greater quantities of visibility, recognition, and self-actualization. The cost of inclusion is acquiescence to and complicity with a world that is "biomedical, neoliberal, racist, and imperialist" in its definition of worth.[4] Race and gender come as properties through which to barter for certain degrees of protection against those that remain abject to the cause of optimization. It is possible to extend Jasbir K. Puar's arguments concerning gender differentiation and identification to include other "neoliberal mandates regarding productive, capacitated bodies," such as biodiversity and neurodiversity, to "entrain" these bodies into enablement to conform with existing standards of "economic productivity and the economic development of national economy."[5] Thus, as Puar argues,

> trans relation to disability is not simply one of phobic avoidance of stigma; it is also about trans bodies being recruited, in tandem with many other bodies, for a more generalized transformation of capacitated bodies into viable neoliberal subjects. Their economic utility informs their right to care and valuation.[6]

What unites these apparently disparate categories are their commercial entanglements with the biomedical and pharmaceutical apparatuses, which poise them to manage these phenomena as conditions, or more precisely, as conditional forms of life. Puar refers to this process as one of "exceptional sizing" whereby it is possible "to convert the debility of a nonnormative body into a form of social and cultural capacity."[7] Added to this is the potential to track such bodies through a series of intersectional coordinates that require these bodies to constantly report upon themselves. A standard of whiteness dominates this activity and

---

4 Jasbir K. Puar, *The Right to Maim: Debility, Capacity, Disability* (Durham: Duke University Press, 2017), 45.

5 Ibid., 47.

6 Ibid.

7 Ibid., 52.

proceeds from a logic of perceived impairment. These bodies marked by a characteristic diversity require some form of rehabilitation in order to become identified as resource. They must adapt in order to exert any potentially viable influence on the world, and that adaptation is coded against a criterion of normativity. The goal, thus, of transformation, renewal, circularity points back into the category of extraction, of materiality for the sake of market.

Coloniality pervades each of these respective bodies as definable territories, captured in their locality through their particular forms of embodiment, each classified by the behaviors associated with its respective morphology and ministered to by virtue of its intricate compartmentalization. Such sophisticated differentiation has always served as a prerequisite for capitalist advancement, in some instances dividing it against itself as a means of expropriating its value, while at the same time ensuring its future potential.

Such an imperative was necessarily racializing in its progress, which denied certain bodies the privilege of integration, whilst at the same time, ensuring those granted such capacities were made wholly aware of the costs associated with diversity. Those given form in this manner were made objects uniquely available to being abused, allocated, apportioned, and affected. Puar writes of "the lived experiences" of "categories such as race, class, and gender" as "moral," which perhaps refers to a liberal subjectivity that might no longer exist.[8] Perhaps it is more probable that we have moved beyond the categorical nature of these bodies to account for them presently as parts and particles, and thus, their lived experiences are now in practice molecular. If this is the case then the definition of their success, too, must shift towards a territory of actualization versus realization.

This shift implies that disaggregate bodies might be poised to prevail in this, a wholly neoliberal century. This implies that we are no longer in the terrain of micro-becomings but rather micro-beings and micro-belongings proper to a time of micro-

---

8   Ibid., 57.

politics, and micro-individuation. Puar's work points in this direction to the extent that it acknowledges that "bodies are malleable not just as subjects but also as composites of parts, affects, compartmentalized capacities, and debilities, as data points and informational substrates."[9] The consequence of this suggests that "it is understood that the battle against the extraction and exploitation of bodily capacities and habituations is not going to happen through the terrain of intersectional politics alone."[10] It is also important to concede that disciplinary apparatuses and control mechanisms have been radically altered to the degree that they are unrecognizable as political entities. It must equally be conceded that we live in times when the self-reflexive individual has ceased to exist and has been replaced by a social-reflexive data set anonymized as impersonal life expressed through cognition. Intent or agency is fundamentally neutralized within such an arrangement; its force given over to preoccupations having to with capitalizing on its biomateriality, foregrounding its vitality in its potential to overcome impairments, and bartering its creativity for the opportunity to be de-contextualized back into the social world.

The market itself has emerged as a continuum of intensity where the need to signify at all replaces a concern for signifying as a race, class, or gender. The biopolitical anthropocentrism need not apply here in the same ways as it once did to cohere the category of the human against its differential others rooted in the categories of race and gender. Presently, its concerns have migrated to the category of species as the intersectional category of choice, elevated to the thanatopolitical processes that cohere around the specter of the Anthropocene. The decentralist of the human then becomes the object of a mounting survival campaign for the world. Within that cause, certain humans have been selected based on their "natural" tendency towards adaptation to act as forerunners of a world to come, where their exceptional capacity for morphology acts as the primary basis

---

9   Ibid., 58.
10 Ibid.

upon which cleavages of human and posthuman might occur. If the biopolitics of the liberal state paved the way for humans to become a biological species, the thanatopolitics of the neo-liberal state might be said to pave the way for posthumans to become a synthetic interspecies. What precedes this step is an incorporation of the pathological into the conceptualization of transformation, through a strategic redeployment of the coloni-al model to reposit difference as a form of skilled manipulation that might be perfected under the terms of technology to yield a reconstitution of what qualifies as the natural world.

In many ways, cognitive capacity becomes the grammar of this experience in omission paired with commission, where the relationship between humans, animals, and nonhumans gets rearticulated to serve something beyond the racializing technol-ogy of biopolitics. In this new signifying economy, race emerges as something altogether different from what we might have for-mally recognized, to operate at levels where it can go elsewhere from the body and be, in turn, figuratively recouped. By letting the recouped die out as a function of biopolitics, capitalism is enlivening something else within the biological domain. When made molecular, race adulterates the production processes of an older racial ontologically dating to the natural philosophy of the eighteenth century that located race in taxonomies related to visible difference, only to be superseded by phrenology in the nineteenth century and eugenic science in the twentieth century. The geopolitics of that emergent racial science closely shadowed the presumed brilliance of the European colonial project, where race became an irreducible feature of both control and capitali-zation. When both trajectories are compromised in favor of the emergence of biotechnologies in the latter half of the twentieth century, through the constituents "of genetic engineering, as-sisted reproduction technologies, human genome sequencing, and phenotypical variation," the category of race was forced to "cut" differently into the path of enterprise.[11]

---

11 Ibid., 62.

Puar argues that in this forward thrust the "disciplinary categories of race" transfer their force onto the "control aggregates of
population."[12] Race as a disciplinary concept does not disappear
but intensifies through its multiplication across any numbers
of proposed species formulations. Race is not doing the work
anymore of resembling, or indeed reassembling whiteness, but
rather mutilating and mutating it as a category so that it might
accommodate the exponential growth of science to power markets through the proliferation of not only races, but also genders
and species.[13] "If race is a technology of regeneration," as Puar
argues, then so too is diversity a technology of postgeneration,
in the sense of proliferating new materialist ontologies that
are irreducible as forms of life unto themselves.[14] Mel Y. Chen
suggests, we must question the consequences of extending "affect to nonhuman bodies, organic or inorganic, averring that
affect is part and parcel, not an additive component, of bodies' materiality."[15] In terms of ontology, affective and material
constructions of being are "not only nonneutral in relation to
animals, humans, and living and dead things but [are] shaped
by race and sexuality."[16] As being thus shaped, it participates in
mapping various biopolitical realizations and thanatopolitical
actualizations of vitality within a larger culture of neoliberalism.
That cultural context generates forms of subjectivity that, on the
surface, appear to promote a greater connectivity amongst individuals.

As Katherine Guinness argues in "The Coloniser and *corpus
nullius,*" that although the promise of extending and remaking
bodies through the assumption of a post-digital set of possibilities affords subjects the "ability to experiment with identity, to
'inhabit' other bodies and to 'play' and perform as others," it
"does not in any real sense redistribute power, but rather gives

---

12 Ibid.

13 Ibid., 66.

14 Ibid.

15 Mel Y. Chen, *Animacies: Biopolitics, Racial Mattering, and Queer Affect*
    (Durham: Duke University Press, 2012), 6.

16 Ibid.

way to 'uneven political effects' that 'literally' index liberal formats of colonization."[17] When this travel is directed towards "the bodies of minoritarian others," it is done with the ambition to "create work with and through them."[18] Here labor itself is subverted to the cause of presuming the value of certain bodies over others, in the classic liberal sense of value as that which is intrinsic to a product according to the amount of labor that has been spent on producing the product. If we bring this definition to bear on the role of the colonizer as producer, then it is possible to surmise that the labor of his distribution of other bodies contributes to what Guinness refers to as *corpus nullius*.[19] As a means of amassing value, the perception of subjects and bodies as the raw material of neocolonial statecraft necessarily involves the painstaking capacity to "literally absorb the bodies and experiences of another while simultaneously maintaining his (for this colonizer is almost undoubtedly male) claims to subjectivity and authorial control."[20]

Guinness's argument signals that "the many theorizations of the posthuman" cannot fully escape their origins when brought into relief against the sustained, unilateral conception of power and bodies that endures within the contemporary neoliberal era.[21] The dynamic may have changed slightly in that property rights have moved on from a concern for ownership to that of absorption but reside still within configurations that privilege the body as the impetus for the exercise of institutional and postinstitutional power. Guinness's concerns regard erasure as the term through which value gets distributed. However, I would argue that it is forgetting which is the tipping point for post-digital culture are to maintain hierarchies of difference and discrimination. Such hierarchies require that the colonial

---

17 Katherine Guinness, "The Coloniser and *corpus nullius,*" *Parallax* 26, no. 1 (2020): 76.

18 Ibid.

19 Ibid.

20 Yiğit Soncul and Grant Bollmer, "Networked Liminality," *Parallax* 26, no. 1 (2020): 6.

21 Guinness, "The Coloniser and *corpus nullius,*" 76.

past figure into the conceptualization of an ahistorical present, where is it is possible to utilize the raw material of colonized bodies to manufacture projections of the future that is in many senses happening now. By forcefully removing the present from the flow of history, economics become divorced from reality and must reside elsewhere as the place of perpetual speculation. Its fanciful projection relies no longer on the measure of production, or even the appropriation in the classical model of colonial seizure of raw materials; rather its expression must rely upon many ways solely on expropriation to amass values.

As a consequence, the living labor of the past must be exploited as though it were another form of property altogether, as something already exhausted in the present, which by extension forecloses upon the possibility of any future of value whatsoever. This has ecological consequence insofar as it suggests that a denial of ownership within a planetary environment effectively stripped of its assets, and thusly, it's for capital for future investments. Within such a blighted context its native occupants have no place to dwell in present accounting and, to some degree, can therefore be said to reside in space as generic non-existence. What has become productivity in their place is human data graded according to its capacity to deliver innovation specifically within the category of synthetic life. Hence, the domination of the pharmaceutical and genomics industries whose destiny entrains with manifesting value within the new molecular frontier. This ever-present economy proceeds from a point of conjecture and culminates the progressive uniformity and normativity of development.

Returning to Guinness, the possibility of an ability to experiment with identity very much begins and ends with a self-destructive logic, the results of which resolve into a belief that the capacities of both humans and nature are finite and that their former division no longer serves capitalism in the way it once did as the category of the human becomes merely an object among other objects within the environment. The posthuman becomes a way of reinstalling hierarchy amongst this new arrangement of commonality, casting a shadow over the former

clarity of distinctions between bodies Black and white. The nature of creativity and cultural making similarly has shifted in
the process of redefining being as something enclosed within
a container of consciousness that no longer requires embodiment, nor mind, as its limit. One of the consequences of this
trend is one of profound depersonalization. Patrick Wolfe uses
the term *corpus nullius* to express the outer limit of otherness,
a space where it becomes possible for "particular humans to be
excepted from the general requirements that govern the treatment of humanity as a whole."[22] There can be no modernity
without the actualization of a body "'without claim' that necessarily accompanies the colonial fiction of the *terra nullius,* the
territory 'without claim.'"[23]

It is important to remember that this materialization of the
body appeared, firstly, as a product of the European colonial imagination. The post-digital age sees a reversal of the movement
of bodies from the condition of coloniality, back towards the
territory of invention. The reconfiguration of the body without
claim, the *corpus nullius* of colonial imagination, makes space
for a new planetary politics of dispossession to emerge proper
to the proprietary relationships of the post-digital age. These relationships are haunted by a spectral presence of the bodies of
Black slaves and the Indigenous peoples through which it first
was possible for the Enlightened mind to deliberate exclusion.
What was specifically denied to these bodies was a claim to love
and kinship of an order that was equivalent to that of the arriving white settlers. These groups became most vulnerable to the
cruelty and rejection by their white counterparts when they at-
tempted to garner recognition and inclusion because that granting of such rights threatened white people's position as humans;
a distinction that meant that they were fundamentally different

---

22 Patrick Wolfe, "*Corpus nullius*: The Exception of Indians and Other Aliens
    in US Constitutional Discourse," *Postcolonial Studies* 10, no. 2 (2007): 127.
23 Siraj Izhar, "*Corpus nullius*: How Europe Recast the Migrant Body Narra-
    tives of the Migrant-Refugee Crisis Reflect the Interests They Serve," *Public
    Seminar,* May 23, 2019, https://publicseminar.org/essays/corpus-nullius-
    how-europe-recast-the-migrant-body/.

from these creatured bodies in capacity, as well as type. The othering of these bodies, in this sense, became something posterior to the development of empire; lurking, as it were, behind its façade of sentiment and affiliation. From this position, the back side of empire, is also possible to identify them as the first iteration of a posthumanism where nature becomes the obverse of culture.

Historicizing Posthumanization

As the eighteenth century drew to a close, the demand for Black slaves and Indigenous people to receive recognition as human became increasingly amplified through various strains of antislavery rhetoric present within abolitionist and sentimentalist writing in England. Caroline Koegler argues that is it these various discursive movements that "threaten to expose empire's normalised brutalising powers."[24] Koegler refers to these trends as "reverse-monstrosities," which push back upon the assumed civilization of "a supremacist white culture" that habitually "inflicts a regime of pain — emotional and physical — on those to whom it is unable, or unwilling, to relate."[25] In voicing the pain of their rejection through works of abolitionist sentiment, Black slaves and the Indigenous peoples are able to temporarily assume subjectivities that challenge this fundamental position of non-relationality. It succeeds in doing so because it interpolates the white settler class as the "perpetrators of injury" rather than as parties threatened by a "racialised and bestialized other."[26] The white subject is forced to momentarily acknowledge their part in perpetuating a white culture driven in its progress, not by sympathy and kindness but by aversion and abjuration and held captive to a society founded upon the twin tenants of race and raciality. No amount of obscuration nor historical distance

---

24 Caroline Koegler, "Posthumanism and Colonial Discourse: Nineteenth-Century Literature and Twenty-First-Century Critique," *Open Library of Humanities* 6, no. 2 (2020): 3.

25 Ibid.

26 Ibid.

is great enough to overcome this inherent pathology, which has produced so many centuries of murder and genocide in attempting to bring about the demise of the black(ened) body that stubbornly refuses to retreat into nonbeing.

As European imperialism matured in the latter part of the eighteenth century and finds its foothold through settler colonialism as that century turns, so too did the impetus to materialize otherness in beings deemed disparate to the human. Koegler is right to claim that posthumanism has a historical heritage founded largely in nineteenth-century colonial discourse, as their premise in the classification and appropriation of life forms corresponds at every point with the denial claims of Black and Brown bodies to accede to the category of the human. As peoples, they exist only in their plurality and the plurality of their application. When they do appear, it is as the capital, behind an appreciation of nature's fecundity, and as an adjunct to diversity correspondent with natural philosophy's supposition that the human races are of different origins, and thus inherently uneven in their traits and development. Koegler refers to this theoretical shift as one of "posthumanization," which corresponds with "a systematisation of alleged speciological differences between different 'races' that are incorporated into a modified, extended concept of the human."[27]

This theory of polygenism meets with its obverse in the theory of monogenism, which posits a single origin of humanity at a time when natural philosophy is giving way to biological racism. Monogenism's concept of the human is not horizontal, like its predecessor, but vertical. To that end, what Koegler's argument suggests is that a differentiation takes place with the category of the human being in the mid-nineteenth century that does not rely upon the dehumanization of certain bodies, per se, but that "hierarchise different versions of human being — a process that allows different kinds of 'humans' to inhabit the same category whilst also facilitating selective animalisation and/or non-hu-

---

27 Ibid., 11.

manisation (such as calling slaves 'livestock' [...]) in tune with the shifting and developing interests of white imperial power."[28]

In a colonial context, the freed Black body remains one abandoned to its own volitional devices, to negotiating the shifting of new and precarious forms of racial grounding, that demand that its psychic legacy remains that of the "un-killed, un-dead," remaining ever after "as an uncanny cypher on the fringes of white-centred (or: white-generic) perceptions of, equally, humanity and post-humanity" trapped in the shadowlands of an enduring white self-referentiality.[29] In terms of biopolitical economics, it is scientific racism that eventually assumes agency and "tips the balance for the thriving or perishing of select 'lifeforms,'" determining which should be cultivated, against those whose existence should be denied.[30] What is perhaps most significant to account for is that black(ened) bodies "that are normalised" are essentially made so in order "to carry the burden of white practices of brutalisation, extraction and extinction," meaning that systematic racism as such goes unchallenged, as does the legacy of white imperialism that subtends its standards of judgment.[31]

What can be derived from this is that recognition as human does nothing to move beyond whiteness as a default positionality from which epistemology and ontology remain grounded, and that demands for equality and inclusion within the category of the human do little to redress the violence inherent to the Eurocentric production of knowledge and understanding. In the neoliberal era, these ultimately act as a means of reinforcing non-relationality through a rationale of privileging of exceptionalism where the othered subject is at once dissociated from humanity and instrumentals in line with the priorities of contemporary forms of white imperial power that make themselves known and felt algorithmically.

---

28 Ibid., 12.
29 Ibid., 3.
30 Ibid., 27.
31 Ibid., 30.

This brings us back to a place where it becomes advantageous to hierarchize different versions of being beyond the human through a process of differentiation which insists that they must now co-occupy the broad category of life to creatively facilitate their positive selection in and amongst any number of potentially competing animate forms of being. This is what Koegler calls "posthumanisation," meaning the "historically and culturally specific flexibilizations of 'human being' — such as racialisation, animalisation, dehumanisation and re-humanisation (or: partial humanisation) — that occur in specific contexts."[32] The naturalist discourse of species has recently regained philosophical currency as a means through which to weigh questions of racial difference and inequality against concerns about ecological crisis and multigenerational trauma. The rethinking of consciousness occurring in the twenty-first century in ecological, as well as individual terms, as a physiological process, has once again focused minds on the value of matter and its constitutional traits. Temperament figures again in questions surrounding developmental potential and indeed, market viability of some versus others. Individual agency has begun to recede as a desirable capacity in human beings, in favor of an understanding of moral judgment as something rooted in systems capable through their programming of redressing the unequal distribution of power and uneven course of development around the world.

The politics of the new object-orientated ontologies suggest a need to regard what has been previously classified as the nonhuman according to maps of the future in which their consciousness can be harvested and reimagined not as something apart from the human but as part of its advance as a moral species. They position themselves as the inverse of subject-orientated ontologies and yet the object remains the same: the necessity of making resource of others to enhance the prospects of whiteness to prevail into the next century through subsuming the material properties of these other life forms as its own.

---

32 Ibid., 6.

America's history is tracked in line with a long path of biological exploitation where plants, animals, microbes, and peoples serve a nation through their integration into a larger project of white settler nationalism. This is a politics of absorption, whose attention is drawn particularly to the comprised objects whose forbearance functions to reassemble Europe's liberal subjects in order to dwell in what amounts to a New World; one reflective of their material dependency on those whose attachments to the land are legally constituted by their fractional reduction (three-fifths), almost but not quite down to zero, making of those attachments ones "that have no place in the voluntaristic landscape of democratic contractualism nor the imagined community of white nationalism."[33] Proclaiming that Black slaves were not human at all (zero-fifths) was never a serious prospect because it was only for the fact of their embodiment that the white subject at the mid nineteenth century had cause to lay claim him own materialism through his capacity to perceive himself a somebody versus a nobody.

Within in this distinction lies a path of embodiment where detection remains key, white/Black, human/nonhuman, self/other all locate their hinging at a point of transcendence beyond the corporeal to the sensory, tilting the elevation of one term over the other. Relationality takes place not because one occupies a body, but due to its capacity to feel itself within the world. The scope of that capacity is what determines one occupying any place in the world. It is for this same reason that pluralism can never be fully admitted into America's democratic vision of itself because its materialism of belonging is always conditioned upon the fact of self-ownership. The availability and fungibility of the black(ened) body are positioned as fundamentally at odds with the commercial sentimental assigned to the white(ened) individual, now made to be operational within the context of a global marketplace. Liberal doctrine demands that he act as a broker for the whole of the nation, conducting himself in

---

33 Cristin Ellis, *Antebellum Posthuman: Race and Materiality in the Mid-nineteenth Century* (New York: Fordham University Press, 2018), 16.

and amongst the currency of its diverse life forms so that their contributions and associated wealth might pass through his receptive hands and thereafter be beneficial to the greater body politic. By contrast, the other forms of life must operate in accordance with environmental influences that demand of them an altogether different order of responsiveness to their physical surroundings, compelling them to privilege malleability over mind.

The discourse of the body here moves to reorder differential versions of the human being to effectively systematize racism in such a way that it gains necessity beyond Old World boundaries of European empire. In America, race assumes its own independent logics so that it might deal more effectively with the diversified offspring of its colonial setting. I would propose that "posthumanization" of the mid-nineteenth century becomes synonymous with diversification at the point at which it corresponds with a liberal movement to redefine all life according to its potential for adaptation.

In such a scenario, life becomes relational only to the degree that it's given expression promotes assimilation of prevailing power relations. As difference becomes increasingly registered as qualitative, and control registered as quantitative, they merge at a ccrtain point to assure that the requirements of capitalism are fulfilled. The current emphasis on multispecies coexistence misses from its critical framework any substantial commentary on the imperial logistics of the humanities and sciences in furnishing the conditions for biological and environmental racism to perpetuate themselves through the (il)logical divorce of material causality from imperial acquisition. Until the two are rightly rejoined in planetary estimation, it will be difficult, if not impossible, for other forms of life to gain rank against the supremacy of whiteness. Nonetheless, its abolition persists as a condition for ending the *longue durée* of imperialism and settler colonialism. If put to rest, it would essentially mean the end of the world.

Materializing Peoples

Cristin Ellis speaks of a "fracturing" of "the human" that takes place in the mid-nineteenth century, which allows for differential embodiment to be thrust upon "subjugated populations, casting Black Americans, Native Americans, and men" with the desired effect being that these peoples are treated as materiality, meaning that materialism is something that has historically been "unevenly distributed among humans."[34] The situation, according to Ellis played out, "as if some peoples are inherently more and others less embodied, as if we are a species unevenly evolving from apes to angels."[35] These peoples ironically became antithetical to life, their bodies running in reverse to development, as it were, seeming to appear *antibiological* and therein posing antigens to the agency of a naturally privileged whiteness. Ellis indicates that this campaign to imprison these peoples as one bent on limiting the spread of the contagion of their lowly positions, such that the assumption of whiteness functioned as "an attempt to quarantine privileged populations from the illiberal taint of their own materiality."[36] Whiteness, then, might be adopted as a prophylactic measure poised to anticipate and inhibit contact between materialism and ontology, racism from intercourse, within parasitic ideologies attacking humanistic principles at this time when science was overtaking religion as a logic for hierarchizing life forms.

Ellis argues that "race is the founding proposition round which [...] antislavery materialisms gather [at mid-century], like so many antibodies, to wrest human materiality away from its conscription by racist ideology."[37] In making this argument, Ellis is suggesting that antislavery materialism acts in a futural way. The classification of antibodies is very much a twentieth-century endeavor. Race relates here retrospectively as a feature

---

34 Ibid., 5.
35 Ibid.
36 Ibid.
37 Ibid., 136.

because antislavery forces acting as antibodies respond to and counteract a specific form of antislavery by recognizing racist ideology as a foreign substance within the American liberal body politic. The materialism of Blackness is something deemed alien to that and, therefore, worthy of its interest in the manner that bacteria, viruses, and foreign substances in the blood must be identified, surrounded, and neutralized.

A problem arises because these antislavery forces interpreted as antibodies act in much the same way. They do not so much wrest disease (here in the guise of racist ideology) from the body, but they rather combine with it to produce a new materiality that effectively binds with the category of humanity in order to prevent its further progress. This lack of movement does not avoid conscription by racist ideology so much as makes race *matter* in a way that is compulsory to the cause of antislavery. This reinscription, as opposed to conscription, is dangerous because it reinforces antislavery's relationship to older histories of racism that can at any point in the future be reactivated from the vestiges of an earlier strain still within the American body politic. One pathogen indexes the other, whether it is related to natural philosophy and its concern with taxonomy and the related idea that races where so substantively different from one another as to constitute distinct species from one another, or the midcentury biological racism that follows on from it; each relates to a privileging of materiality. So too does any posthumanism that proposes embodying universalism through the reception of the nonhuman, or the dehumanizing of hierarchies of life through the embrace of multispecies as a means of undoing the moral legacy of this uncritical bias towards materiality; that is to say, to judge the human body in empirical terms against which all other bodies are surveyed.

Under the auspices of liberal modernity, a bioperformative sphere emerges through a series of critical sites including the slave ship, the prison, the slaughterhouse, the barracks, the factory, the concentration camp, the clinic, and the laboratory. These sites coalesce to transgress the boundary between human and animal and produce bodies for consumption along the lines of certain networks of subjugation. As institutional apparatuses, they conspire spatially to enact and enforce a series of violent norms associated with the production and denial of access to the spoils of capitalism. Paul B. Preciado, following Michel Foucault, imagines this as a world whose cartography is a catalogue of "sexopolitical disciplinary spaces."[38] Their mapping represents a pattern of socially reinforced coercion, resulting in bodies that circulate along lines of race, class, gender, and ability. Their powers of discernment fixate upon the suppression, immobilization, and elimination of what cannot be fully capitalized upon. These techniques of capture succeed in compelling the multifarious category of life to essentially surrender into becoming the various objects of a biomateriality. The colonial historical processes at work here operate at the molecular level where a humanist geospatial and epistemic configuration can penetrate the body and install what Gilles Deleuze and Felix Guattari call "a micropolitics of insecurity."[39]

In a neoliberal context, these same channels persist by privileging bodies that are "flexible, mobile, manipulable" for use in a construction of the future that concerns itself with embodiment and experience to the degree that the affective life of a white, economically privileged, male, and able subject can be effectively bifurcated from the struggle of differentially raced, classed, gendered, and abled subjects, to appear more fully human.[40] In "medical, educational, employment, legal, [and] social terms," they largely remain as partial beings, made equivalent to cellular life, and thus rendered impotent as agents of their capacitation without recourse to, as it were, the greater body.[41] The mechanism of passing into the confines of that greater body figures as a matter of economic and psychological survival and

---

38 Paul B. Preciado, *Testo Junkie: Sex, Drugs, and Biopolitics in the Pharmacopornographic Era* (New York: Feminist Press at the City University of New York, 2013), 164.

39 Gilles Deleuze and Félix Guattari, *A Thousand Plateaus: Capitalism and Schizophrenia,* trans. Brian Massumi (London: Continuum, 2009), 216.

40 Puar, *The Right to Maim,* 45.

41 Ibid.

might well make the difference between minimal and profound marginalization.

Within a speculative economy that takes as its starting point an individual agency fundamentally detached from patriarchal and colonial histories, it becomes increasingly difficult to conceptualize what, if any, form of voluntary submission would be tolerable to the objectified bodies poised to act once again as collateral for exchanges which circumnavigate around various well-ordered sites of mutual production developed through straits of settler colonialism and financialized capital accumulation. These have recently been adapted, mediated, and intensified to meet the specific needs of a neoliberal present. Jordy Rosenberg refers to this pattern of change as one conversant with a new molecular ontology that insists upon a new world where new locations of empirical social order have now emerged.[42] For Rosenberg, this purported novelty is very much the stuff of fantasy because this ontological turn leads us back precisely to the urge to make objects comportments of colonialism and settler colonialism.

Instant Deregulation

In an interview with the magazine *Women's Own* on September 23, 1987, then Prime Minister Margaret Thatcher famously declared "there is no such thing as society."[43] Conversely, the thing that can be said to exist for Thatcher is a "living tapestry of men and women and people" each of which is "prepared to take responsibility for themselves."[44] For Thatcher, "there is nothing wrong with doing that" and for money to act as "the great driving engine, the driving force of life."[45] Thatcher's comments were made nearly one year after the London Stock Exchange's sudden

---

42 Jordana Rosenberg, "The Molecularization of Sexuality: On Some Primitives of the Present," *Theory and Event* 17, no. 2 (2014): n.p.

43 Douglas Keay, "Margaret Thatcher Interview for 'Woman's Own' ('No Such Thing as Society')," *Margaret Thatcher Foundation*, September 23, 1987, 30.

44 Ibid., 31.

45 Ibid., 29.

deregulation of financial markets on October 27, 1986; an event thereafter referred to as the "Big Bang." The Big Bang ushered in an economic *terra nullius,* where the City of London was able to effectively dominate financial markets through the deformation of individuals and institutions and the spread of financial risk around the world.

On that day the primacy of the body was sacrificed to the primacy of the screen, altering the transfer of information, seemingly irrevocably. This practice was to be known henceforth as *frictionless* trading, its previous reliance on shouting and the use of hand signals now deemed as *primitive* technologies. This itself is a lesson in object ontologies insofar as we appreciate that such "origin narratives" by design "exchange frictionlessly between two sets of seemingly opposed orientations — origins and prognostication, [...] the primitive and the brink."[46] These two temporalities are linked, of course, primitiveness having long been the dialectical verso-face of millennialism."[47] Hence, Thatcher's desire to summon forth a second world-ending bang. As Deleuze and Guattari cannily observed, "molecular movements do not complement but thwart and break through the great worldwide organization," meaning that a change in values is always a molecular phenomenon.[48] In this instance, it begins with the "overcoding of the machine" to instigate an unprecedented release of capital flows and thus radically redefining the City of London by "what escapes it."[49] It is this movement that "as a point of departure for a new creation" will go on to "shape postures, attitudes, perceptions, expectations, semiotic systems," and so forth, through a desire to *make life better.*[50]

Here the molecular joins up to the instrumental, and together, periodically remake the world in the image of private property. The racialized geopolitics of colonial production follow course to this new financialized world, introjecting another version of

---

46 Rosenberg, "The Molecularization of Sexuality."
47 Ibid.
48 Deleuze and Guattari, *A Thousand Plateaus,* 216.
49 Ibid., 217.
50 Ibid., 215–17.

colonialism through secreting pharmaceutical and biotechnical wealth through the UK's network of offshore jurisdictions. In this instance, the molecular enacts control of the "nonhuman domain" through a combination of market oligopolies and "biomedical management."[51] Within these new electronic trading platforms, "the molecular" operates much like any other chattel class — "as an abstraction," and as such "comes to function fungibly across different" disciplinary terrains.[52] What is forcibly omitted from the record of such transactions "are the historical relationships that make possible the abstraction of the molecular […] as a contemporary iteration […] of settler-logic."[53] This assertion suggests that the molecular might fulfil the property that racialization once occupied.

In a similar fashion to racialization, the molecular takes as its referent "epochal time" rather than "historic time."[54] It is for this reason that the molecular appears under the sign of objectivity, implying that "objects are ontologically separate from the social field."[55] The conceptualization of Blackness similarly operates to elucidate an "impassable separation between objects and the social order."[56] Zakiyyah Iman Jackson argues that "fundamental to the production of animalized blackness and blackened animality" has been the "semblance of an absolute distinction between human and animal."[57] These have emerged historically through fields associated with the humanities whose task has been to differentially populate the category of the human and the animal "based on the ever shifting needs of Eurocentric (andro)anthropocentrism."[58] Rather than the privileging of the molecular as a code that marks a departure from a previous historical impetus, the molecular reinforces a connection between life

51 Rosenberg, "The Molecularization of Sexuality."
52 Ibid.
53 Ibid.
54 Ibid.
55 Ibid.
56 Ibid.
57 Jackson, *Becoming Human,* 156.
58 Ibid., 156.

and value that refers back to a previous tendency to invest only certain materialities with livelihood, while deadening others.

The molecular requires the imposition of *terra nullius* to the object as something deemed to be previously unoccupied or uninhabited and, therefore, radically open to expropriation under the broad category of property and, by extension, from the classification of a type. The removal of race from an understanding of this relationship makes it no less violent than the wrenching of temporality from spatiality from the black(ened) body and therein its fundamental break from history, specifically, at the level of a legitimate claim to ancestral wealth. The ontological turn that now presupposes rights to the object, attempts to forfeit on ascriptions of subjectivity by forcing the black(ened) body to once again confront itself as the cypher for dispossession from which it becomes impossible yet again for it to exert achievement as anything beyond perpetual deferment of payment on debts owed to it as property. In this way, these black(ened) bodies become subjectless, as opposed to subjective. The molecular supersedes the racial at the point in which it can demonstrate processes of near-constant productivity, effectively forcing the specter of reproduction off the stage of coloniality in favor of an appreciation of life now rendered in split seconds rather than subsequent generation.

The postcolonial landscape of "agriculture, banking, mining, trade and tourism," perpetuated older systems of inequality at the same time as they brought into being a broader global economy that thrives on the "persistent underdevelopment" and "persistent poverty" of "black life."[59] Katherine McKittrick argues that it does so in similar ways "(but not identical to) slavery."[60] This practice was often conducted through "computer-mediated modern systems of mass surveillance in the United States and Britain, such as policing, banking, and national health care schemes" that would ensure "that records of move-

---

59 Katherine McKittrick, "Plantation Futures," *Small Axe* 17, no. 3 (November 2013): 3.
60 Ibid.

ment would be kept" in line with a sub-economy — also known as the shadow or *black economy*.[61] This dualistic economy required that these *"freed slaves,"* as an artificially diasporic people, are constituted through their necessary exclusion and decoding and forced to act against a movement of power which previously "constituted" them in the "overcoding" of "empire."[62] As a consequence, that black(ened) body of people are afforded little sovereign recognition apart from one tied to a previous category of private property: "that which is freed, but still segmented remaining negative and blocked."[63] This sort of phenomenon is inherently provocative, adhering to a differential subjectification in which it is possible to believe that contemporary forms of empiric signification and interpretation exist outside of them, when, in fact, these are still secreting them.

Motivating Matter

In assuming that genes and genetic codes are inherently racist, it is possible to remain ignorant of how race animates the way interactions among molecules, organisms, environments, and social settings are given meaning through the persistence of racialized dynamics. Liberal notions of property gain their afterlives through such actions. The definition of life itself under conditions of capitalist domination allows it only to flourish within the parameters of investment, accumulation, speculation, renewal, and habitation. Its converse is that which is emptied of populations and devoid of naming and treated as surplus to requirement. The digital is ancillary to the process of valuation, which locates its trajectory not in analogy, but towards the stuff of materialization. Neoliberalism becomes in its own way a kind of species, enlivened through information that synthetically adapts to fill the institutional spaces it has driven into ex-

---

61 Simone Browne, *Dark Matters: On the Surveillance of Blackness* (Durham: Duke University Press, 2015), 39.

62 Deleuze and Guattari, *A Thousand Plateaus,* 449.

63 Ibid., 137.

tinction while seeding an ideological climate whose patterns of growth and migration may be said to have reconstructed race to allow new markets in nature to thrive under certain conditions of lending value having to do with the acknowledgement of the end of empire and the emergence of decolonization movements in the late 1980s. It is possible to appreciate neoliberalism as an offshoot of the necessary project of racialization in order to re-code governmentality to correspond with a need to protect the global market from these types of external pressures, at least for a time.

Empire and dominion would have to concede to the realpo-litik of a world that was abruptly in need of new monitoring, ligation, and mitigation strategies designed essentially to keep the whole thing going through a logic of constant attention paid as it were to states facing the prospect of planetary obliteration. Political economy was the only way out from that threat of ex-istential dissolution shared now in common. Common, though only to the degree that it still heavily relied upon racial differ-entiation as a means of promoting a racial binary of "white as private versus Black as public" bodies, signaling whites to with-draw "their support for public welfare" and associate "welfare with Blacks," in favor of the growth of private entities (associ-ated with whites), like the prison and non-profit industrial com-plexes" that reintroduced the surveillance and confinement of the black(ened) body as a source of continuous revenue.[64] Hand ringing about minority opportunity zones and Black ownership by white elites obscure the fact that Blackness remains in many ways communal property for the sake of applying to it discipli-nary, supervisory functions within a postinstitutional setting.

If, as Sang Hea Kil observes, "whiteness" persists "as neolib-eralism," it can also be surmised that Blackness persists as mere-ly referential in support of this mutually reinforcing structure of

---

64 Sang Hea Kil, "Reporting from the Whites of Their Eyes: How Whiteness as Neoliberalism Promotes Racism in the News Coverage of 'All Lives Mat-ter'," *Communication Theory* 30, no. 1 (2020): 27.

authority.[65] This makes it possible to discern Blackness as a mis-informed category of being, its public abjection the very essence of its inability to be read as consequential to the lives of private citizens. If whiteness speaks to all humanity, by logical exten-sion it has nothing to say to those who are dehumanized by the state. It cannot be responsive to the needs of the nonhuman. It must instead promote the policies of austerity towards those exposed to the worst excesses of privatization, even as these same persons are legally understood to have no rights in which to be registered as injury parties. These persons appear with the same flattened affect as Thomas Jefferson's slaves appeared to possess when voicing their desires, through what he calls "that immoveable veil of black which covers all the emotions of the other race," as though intelligible expression remains something beyond their innate capacity.[66] Blackness as a generic category allows for diversity to stand apart from the all of humanity, as a form of differential irrelevancy that only takes on significance when it is made equivalent to the general levelling of all other social categories in the wake of a superficial acknowledgement of intersectionality. Beneath that lies a crossroads in which the global elite parts ways with the majority of what constituted a human all, in favor of protecting themselves; in this almost no lives matter but all matter becomes collateral. In this Jefferson prefers "to make slaves distributable […] as other moveables."[67]

This brings us back to the early nineteenth century and its association of individuality with the ability to take possession of life by coercing labor into an embodied form. Through this mechanism, it was possible to link security with ownership and to leverage time to conspire with debt in a way that fostered both enslavement and the rise of a new global economy built of life as speculative category and finance as the means with which to instigate its legibility. The pathology of this work was always

---

65 Ibid., 29.
66 Thomas Jefferson, *Notes on the State of Virginia,* ed. William Peden (Chapel Hill: University of North Carolina Press, 1982), 148.
67 Ibid., 146.

apparent, and yet it was also possible for it to be valorized as the price worth paying for whiteness to prevail over other categories of being through its subjective enclosure. Whiteness, in this way, became the aspirational limit for those striving to lay at least partial claim to it not as humans but rather as peoples. As peoples, they were positioned to contribute to new flows of data and information mapped onto raced and gendered bodies in such a way they become increasingly important to public life. Privacy, by contrast, emerged as a privileged means of evading capture by newer processes of dispossession, surveillance, and extraction associated with the acquisition of recorded movement within a framework of racialized urbanization that was emerging as new space of plantation economics manufactured, like its predecessor, to conspire with accounting systems that entangled the materiality of persons with technologies of private property that, "then, serve as the guarantor of capitalism's need to constantly find new spaces of accumulation."[68]

Adam Bledsoe and Willie Jamaal Wright argue that the advent of new data publics necessitates that such spatial investment is "mapped onto previous racial and colonial (imperial) dis-courses and practices" and requires the "acceptable subordination of Black physical presence" to the degree that "capitalism's logic of perpetual expansion" is met with "emerging political economic practices" that makes such action appear justified.[69] As a result, much of the plot of the post-digital world configures its attitudes towards data extraction, around prior white settler colonial understandings of the lawfulness of bodily and land seizure. The black(ened) body from its inception in the Americas was spectacularly financialized and its suffering expressly corporealized for profit motives. Racialized illness, reproductive pathology, and early mortality all became risks made open to market speculation. The denial of responsibility for one-

---

68 Adam Bledsoe and Willie Jamaal Wright, "The Anti-Blackness of Global Capital," *Environment and Planning D: Society and Space* 37, no. 1 (2019): 12.

69 Ibid., 14.

self becomes, at once, a denial of selfhood and ability, as care emerges as a consumerist logic and a prerequisite to participation within the free market. The white(ened) body finds itself specifically enabled by social infrastructures and environmental settings that allow it to cultivate material stability and, thus, become racially transcendent.

Race becomes a spatial arrangement precisely through its ability to determine who is excluded both from public interest and private concern and for whom life comes with the qualification of expectancy. Within an age of environmental collapse, object orientated ontologies emerge as a symptom of the Capitalocene's struggle to reconcile "the division between the human and its own non-human surplus."[70] Neoliberal capitalism corresponds with object oriented ontologies at the point at which it starts to become "impossible to produce new subjectivities" or deepen "reification."[71] The Capitalocene begins "at the moment when the colonialist human begins to descend into the sphere of the objectified non-human commodity."[72] By the nineteenth century's end, "the Enlightenment desire for mastery through scientific discourse" is all but extinguished in favor of "a desire for exploitation through capitalist exploitation."[73] This inaugurates a reconceptualization of the world itself as equal only to what can be processed, manipulated, engineered from what is materially said to exist. Thereafter, the object of plunder bears no distinction between black(ened) bodies and scorched earth.

---

70 Alexander Bove, "What Happens When Replicants Become Extimate? On the Uncanny Cut of the Capitalocene in 'Blade Runner 2049,'" in *Lacanian Perspectives on "Blade Runner 2049,"* edited by Calum Neill (London: Palgrave Macmillan, 2020), 139.

71 Ibid., 143.

72 Matthew Flisfeder, "Object Oriented Subjectivity: Capitalism and Desire in 'Blade Runner 2049,'" in *Lacanian Perspectives on "Blade Runner 2049,"* edited by Calum Neill (London: Palgrave Macmillan, 2020), 11.

73 Bove, "What Happens When Replicants Become Extimate," 145.

Race as Quantification

Race becomes generic at the point in which it enters a larger ontological framework of equivalency held between bodies and objects as cheap, disposable, and renewable products of nature. As Jason W. Moore says, "from this perspective, capitalism becomes something more-than-human."[74] It creates a space where it is possible to define "hands, pounds, lashes, bales, grades" as objects of sale interchangeable with one another, it creates a span of enforcement that incorporates "quantity, quality, and value" in the fluidity of bodies to produce the widening of economy itself.[75] Race is a practice of "quantification" and fragmentation, breaking down the pieces of existence now classified "by work, life and exchange" into something that allows for society, nature, and race to coalesce into an "intellectual conjugation" where "human beings, animals and plants forcibly reduced to limited aspects of themselves" are then deployed into the world as agents of gradated consciousness.[76] Moore suggests that it might be more accurate to describe the nineteenth century as "the era of Quantification rather than Industrialization" and, in so doing, brings our attention towards the real object of concern for this century, meaning how "the commercial standards of the wider economy might be translated into [...] disciplinary standards."[77] Moore's argument ultimately configures a reconsideration of the history of racism through the lens of such quantification, which pairs the abstraction of nature with the abstraction of the body through the exchange of their productivity for value.

In the Atlantic world after 1450, exploitation and appropriation traded with one another through the rise of the capitalization of coerced labor. The growing emphasis on profitability gave rise — geographically and demographically — to a whole new

---

74 Jason W. Moore, "The Capitalocene Part II: Accumulation by Appropriation and the Centrality of Unpaid Work/Energy," *The Journal of Peasant Studies* 45, no. 2 (2018): 239.

75 Ibid., 264.

76 Ibid.

77 Ibid.

world, one that Moore argues corresponded with a new "view of reality and a practice of material transformation" that would redraw the "boundaries of economy, culture, politics" to support a systemic reformulation of value relations literally from the ground up.[78] This occurred through the progressive adoption of "a mathematized, visualized, and mechanical world-praxis" and resulted in "the explosive growth of commodity production and exchange after 1450" in the Atlantic world.[79] However, this expansion was not wholly accounted for by a transition to new rules of production and reproduction in how power took shape, but equally it "depended on the symbolic and material abstractions of concrete mental and manual labors into money-capital."[80] It paved the way towards the assimilation of "new 'measures of reality' — in accounting, time-keeping, mapping space, and externalizing nature" that "were on an equal footing with mechanization" as causal agents in a furthering of "imperialism and state-formation, new modes of knowledge production, class struggles, and so forth."[81] What is crucial to bear in mind is that Moore's argument lays the groundwork for an appreciation of cognitive capitalism whose rise coincides and intersects with the advent of modern governmentality and biopower. The range of scientific processes that are developed to classify, discipline, and direct human, and nonhuman populations, thereafter, not only facilitate capital accumulation, but they effectively invent it through standardizing what defines life, gender, and race in the metropole, as well as the colony, in what amounts to a parallel series of formulations.

Plunder then becomes about simplifying this task of sorting out values from one another and applying them as instruments to actively create reality through differentiation. Racialized, gendered, and sexualized violence are concomitant to the task of early capitalism's conceptual revolution which not only pro-

---

78 Ibid.
79 Ibid.
80 Ibid., 265.
81 Ibid.

duced racialized and gendered hierarchies but also made orders within the category of nature that resulted in both "the violent expulsion of most humans from Society" and also most life-forms from Nature in ways that were completely banal according to the abstract preconception of a ruling global elite.[82] The effects of these abstracts resulted in the world being made over to mirror their albeit distorted value systems related to what constituted life, family, labor, servitude, and habitation. They introduced new regimes of Anthropocenean consciousness that have effectively stayed with us over centuries, supported by a moral philosophy that perpetuates "a narrative of white Earthly possession, and thusly, resists conceding evidence of its own brutal handiwork in bringing about mass suffering on a planetary scale.[83]

If race has always been effectively about assigning mattering to life, then the inhuman raises the specter of another sort of traffic in bodies altogether that of the nonhuman life as some that finds itself resident to process of extraction for there can be little narrative accountability. The thinking of race, as we have seen has much to do with property, possession, and land use, but it has little to say about the formulation of the raced body as a type of inert material that is in many ways only enlivened in the imagination of the white settler class through their desire to acquire it in much the same way as they would their desire "for gold, silver, salt, and copper."[84] Kathryn Yusoff argues for a rethinking of race as something intimately related to the early modern material economy of the New World and to posit race as something that "emerges through the libidinal economy of geology."[85] What perhaps is most radical in her proposition is that idea that slavery might have functioned as a "category mis-

82 Ibid., 268.

83 Andrew Baldwin and Bruce Erickson, "Introduction: Whiteness, Coloniality, and the Anthropocene," *Environment and Planning D: Society and Space* 38, no. 1 (2020): 3.

84 Kathryn Yusoff, *A Billion Black Anthropocenes or None* (Minneapolis: University of Minnesota Press, 2018), 6.

85 Ibid., 7.

take" of geology.[86] It is through this that is possible to conjecture further, that it was not a mistake of classification but a misunderstanding of the nature of "property" that might have been at issue at this time in history.

Yusoff suggests that black(ened) bodies are effectively cut out of their sociological and ecological contexts through their objective classification and thus made illegible as anything other than inert matter. Dislocated from place and time, these bodies, like their mineralogical counterparts, take on an alien-like quality, absorbed into mythology about their source facilitated through their profound disassociation from their conditions of origin. The formation of bodies-as-matter are made independent of their languages of self-description and the historical constitution of their social relations in a profound sense, and it is this dual act of cutting off that finally achieves their profound immobilization.

These figures find themselves fundamentally alienated from their humanity through their manifestation as a category nonlife for whom the New World life profoundly depends. Their category came into being when it is established that "slaves being exchanged for and as gold," and thus, according to Yusoff, slavery was a geologic axiom of the inhuman in which nonbeing was made, reproduced, and circulated as flesh."[87] This nonbeing remains wholly consequential to a narrative in which the making of New World subjects relies upon New World objects going back in the other direction, devolving, as it were, back into their mineralogical components as geological life forms within a new alien context where "gold shows up as bodies" within a greater territory of unnatural abundance.[88] This suggests that we have been living in and amongst bodies that were both corporeal and incorporeal for some time now and that, biologically speaking, they have been the source of a mania for the distribution of bod-

---

86 Ibid.
87 Ibid., 5.
88 Ibid.

ies across a register of mattering that is constantly adjusted to serve the cause of speculative accumulation.

When Yusoff proposes that geology is understood as "a racialized optic razed on the earth," it is crucial to also consider how biochemistry now plays a similar role "as a discipline of extractive and originary science to a philosophical material formation" and as a means within which to continue efforts to racialize matter.[89] In her assessment of geology as "a technology of matter," Yusoff brings us into the territory of another type of formulation of it as something both life making and death making insofar as its incursion into areas in which it does not organically belong spells out disease and casts bodies out of their previous incarnation.[90] Indigenous life was made equal to the task of being forcibly transformed as a matter of relation to what was happening to their impoverished European counterparts whose bodies were mined in order to mobilize the appetites of a growing class of industrialists the world over. One hunger feeding off the other, to secure what was becoming a global energy market whose raw components were made up of the desperate migration of people all over the world who were in many instances literally starving when they arrived at distant shores looking to put down new roots for basic subsistence.

White and Black they came, their realities accrued for them in many instances by the expansion of the transatlantic slave trade which laid both the material and imaginary logics for the makeup of societies across Western Europe, the Caribbean, and the Americas. Racial distinction was what made all of these intersecting worlds possible to the extent that they were the source of a new, near-universal market in and of consumption. Yusoff makes an interesting case for intimacy here, an energic intimacy, involving "the conversion of inhumane slave energy into fuel, then back into human energy, plus inhuman energy, to produce industrialization."[91] Black gold transmuting into white

---

89 Ibid., 14.
90 Ibid.
91 Ibid., 15.

sugar transmuting into black coal, transmuting again into white currency; handled, stirred, smoked, and bitten they cooked the very Earth they were founded within by their frenzied movements. As a new century came into being it remained one that was obsessed with the practices related to the extraction of energy from the earth, firstly through steam, later as oil, as each one, in turn, became what Stephanie LeMenager classifies as the ideal "replacement for slave labor."[92] At the dawn of this current century, it might be wind, sun, sand water, and biochemicals which will serve a similar highly subordinated function.

Yusoff's concept of oil as a substitution for slavery points in the direction of a Britain that during the 1970s was punctuated by several energy crises. These included two damaging miners' strikes, power cuts, and the three-day working week, the oil price hike of 1973, and the "winter of discontent" of 1978 and '79. The last of which resulted in a prolonged standoff between trade unions and industry bosses, which results in sewage, rubbish, and corpses to collect on urban streets. Such dire conditions caused many in society to question if Britain had the stomach to be governed as a multiethnic society without recourse to the psychic safety valve of empire. In order to fully appreciate the story told about labor and its insurgency during this period, Britain's burgeoning gender and racial diversification cannot be ignored as a critical element. Conservative calls for minority demand to be put down in order to restore former hierarchies of industrial relation pointed in the direction of an unquestionable white supremacy to economic and political governance in the late 1970s. The historical election of Margaret Thatcher as Britain's first female prime minister was won amidst a pervading atmosphere of racial tension, the rise of the National Front, the suppression of women's rights and a crumbling social hierarchy that was effectively poised to bring down the nation.

---

92 Ibid., 7.

Alien Empire

The end of Britain's empire is rarely named amongst these crises, but it is crucial to any historical account of this narrative of national decline that eventually gives rise to the neoliberalism from the 1980s onwards. What it required was a conservative repositioning of the empire so that the racial hierarchies associated with Britain's former imperial free-market economy could be brought inward to recreate similar class and racial divisions. This occurred through the creation of what were essentially apartheid townships within Britain. Areas like Brixton, Toxteth, Southall, and Moss Side quickly became the site of most notorious incidents policing and rebellion. Within this new equation of power, the City of London stood as the new internal empire's metropole edged by the black(ened) spaces of London and other British cities where economic depression and police suppression co-ruled to ensure that Blackness itself became both bankrupted and criminalized, as a former asset class of liberal imperialism. Within the confines of Britain, new neoliberal empire what had been the colonial periphery became the source of novel forms of alienation from within British society at large; chief amongst those was the traumatization of the colonized through relentless threats on their territorial security in various local forms including the liquidation of council housing stocks, frequent police stops and searches, raids on businesses, and the advent of urban enterprise zones that undermined local Black and Asian proprietorship. Coupled with these, were Conservative local government schemes that promoted the influx of white middle class settlement into these areas through speculative gentrification.

In the early 1960s, sizeable numbers of individuals from Britain's Commonwealth emigrated to Britain's metropolitan centers. While they had been involved for centuries in the production of Britain's wealth from their peripheral colonial vantage points, the wealth they produced largely flowed out from their nations and back to the Motherland. Over time this created economic conditions in these nations that virtually guaranteed economic hardship for the great majority of their citizens. Those who were financially able to travel followed the flow of wealth from their countries back to Britain to partake of the economic opportunities disproportionately assigned to it as a singular nation through the mechanism of colonization. The pattern of migration of these citizens of the British Commonwealth to Britain corresponded, not coincidentally, to those cities formerly involved in the slave trade. As a consequence of that trade, these cities had become prominent financial and industrial centers over centuries. The appearance of these individuals born of empire in the British metropole stressed the postcolonial system and created within its lacunae of colonial memory and form, which could not be easily reconciled with mainstream conceptions of British national identity. Throughout the 1970s, Britain continued to tacitly promote a racially homogenous understanding of national belonging, which excluded Blacks and Asians from recognition as fellow citizens, both in a historical and present sense.

Thatcher's real opponent in her first general election in 1979, was not Labour but rather the National Front. They promoted an image of ethnic whiteness that allowed them to project an image of themselves as an endangered species within their own natural habitat, fighting as it were, against an influx of unwelcome aliens who threatened to "swamp" them not in sheer numbers but through their innate difference; hence, their fundamental inability to be assimilated into the blood of the nation. It was made clear in Britain's new immigration rules introduced in 1971 that allegiance, desire, history, or even character was not the amongst qualities that had made Britons white. It was their biological heritage alone that connoted their claim to a unified nationality. In order to contort reality to reflect the absence of other races from Britain's hereditary narrative, what had to be forcibly rejected was an admission of Britain's colonial project *as* one that promoted biological admixture spanning centuries. This disavowal allowed the debate around race to be framed around concepts of land management rather than biological racism per se.

In the Conservative rhetoric of the 1970s, for example, the failure to control the "influx" of Commonwealth populations into Britain's urban center meant had resulted in the "draining" out of white working class population from those same areas.[93] Those whites had now "decanted" to the new towns being built from the "overspill" of sudden white population flow outward from these former centers of concentration.[94] Their dispersal outward from inner cities had created the undesirable effect of "a drought, which had opened more space in British cities for immigrant populations to flow into."[95] This situation generated a crisis of "meaning" by disrupting the continuity between what was understood as foreign and what was understood as native to the urban landscape. That crisis related to a larger concern of how to define Britain as a contemporary nation against its understanding of itself a former empire. In order to address this concern, Britishness itself had to be reconfigured as a scarce resource in its own right. It had to pose itself as a category threatened by outside forces such as the International Monetary Fund and the European Economic Community that sought to capitalize off of the remains of Britain as a now "eroded" global power, one that has squandered or had lost is most of its precious resources through the overwhelming dissolution of its imperial status in the decades following World War II.

As this status fell, the value of national heritage rose to meet it, to restore a desired effect of Britain once again figuring with some measure of esteem into the global order. If Britain was being "drained" through its interactions with other nations on a global level, it was vitally important that it not be "swamped" by Commonwealth immigrants vying to register their value against the terms of the nation's relative decline. What remained of Britain had to be preserved and defended, locked in through its vulnerability. Britishness had to be accepted and embraced as

---

93 Lauren Pikó, "'We're Full': Capacity, Finitude, and British Landscapes, 1945–1979," *Australian Journal of Politics & History* 64, no. 3 (2018): 460.

94 Ibid.

95 Ibid.

an entho-racial, not economic, attribute. By doing so, as newly post-imperial nation, it could disavow its responsibility for its own intermingling of blood with treasure, creating an ideological landscape which called for Britain's colonial detritus to be forcibly expelled in order to restore its faith in its capacity for restoration, as a once green and pleasant land in harmony with its past. One feature of its past that could not be acknowledged, and thus remained unreconciled within it, was the requirement of Asian and Black bodies to appear within Britain to reproduce the uneven conditions of empire that had for centuries meaningfully defined what it had meant to be British.

Urban Colonial Occupation

The 1981 Nationality Act set out the terms of a new empire within Britain that effectively meant apartheid rule. Imogen Tyler argues that they "created several categories of nationality and citizenship" which were expressing designedly "to exclude black and Asian populations in the Commonwealth while leaving 'routes home' for white nationals born within the boundaries of the empire."[96] At another level, they can be construed to operate as a means of establishing racial hierarchies within the population already inside the country as a means of literally alienating them and, thus, open up legal means of recourse to police and punish them within the bounds of the existing state. In this scenario, the threat of importation of Blacks and Asians in many ways overshadowed the requirement that these bodies already within the national container of Britain be effectively recolonized, and systematically delegitimized in their claims to citizenship. This act of formal abjection of Black and Asian populations made their free circulation in Britain at best suspect, and at worse illegal, and thereby, provided the necessary political cover for their brutal and inhumane treatment by the state.

---

96 Imogen Tyler, "Designed to Fail: A Biopolitics of British Citizenship,"
    Citizenship Studies 14, no. 1 (2010): 62.

Whites within Britain, for their part, resumed the norms of colonial occupancy within inner cities, where Blacks and Asians were concentrated in their dwelling, and used the police as a means of controlling their movement within that space. This was to be the space of postcolonial Britain, where "the management of national civil defence in anticipation of foreign invasion" had turned inward on itself.[97] The consequence of this was effectively a civil war with permanently installed dangerous classes embedded with a larger social structure made paranoiac in the wake of any proposed acknowledgement of difference. Differentiation became a byword for exclusion amidst a British state determined to turn inwards rather than face the reckoning of its past corruption, indeed its own criminality, during its long history of colonial enterprise.

By making whole swaths of its population into failed citizens to serve the cause of rarefying its racial brand of sovereignty as a form of (white) birthright and, for the rest, something that needs to be figuratively and materially earned, Britishness was made to figure as both a precious commodity and a damaged good. Immigration, in turn, became a global business in both senses. Entry was made priority for those able and willing to acquire it, without the potential for most to obtain ownership of Britishness as such. Even for those coming from Britain's white colonial dominions, it was an elusive privilege. Whatever has happened in the intervening years since 1981 can only be ever said to correspond with differential concerns for access versus privilege in the imagination of what constitutes its nationhood. The period between 1978 and '79, can be viewed as a contest between "emboldened socialists and free-market capitalists" to remake the conditions of postcolonial Britain in their image.[98] It was not immediately obvious that Thatcher would be successful in her bid to establish Britain as a nation remade to con-

---

97 Ibid., 64.
98 Phil Tinline, "Back to the Future: What the Turmoil of the 1970s Can Teach Us Today," *The New Statesman*, May 8, 2019, https://www.newstatesman.com/politics/uk/2019/05/back-future-what-turmoil-1970s-can-teach-us-today.

form to a free-market orthodoxy. It "was only in 1980–81 that she purged her cabinet of 'wets' and promoted key allies," whose loyalty would allow her to continue her progress, despite the rise of mass unemployment and the intensification of trade union strikes.[99] Both of these had to be bitterly endured and ultimately exhausted for Thatcherism to emerge as an outcome of what was previously considered unthinkable. It was Thatcher's use of anti-immigrant rhetoric that ultimately dried tensions amongst white Britain's warring classes and focused their attention on it as the cause of the emergency, rather than the fact that the country's postwar economic model was stalling because it had largely failed to consider that it's model of socialism only worked in the context of massive imperial revenues coming in to sustain it.

Much of the image of dystopia that Thatcher was peddling to both her party loyalists and the greater population of voters, was racially encoded. In her 1978 Conservative Party Conference speech, she invoked the "inner-city" as culpable for bringing on a time, "when the rule of law breaks down," and "fear takes over. There is no security in the streets, families feel unsafe even in their own homes, children are at risk, criminals prosper, the men of violence flourish, the nightmare world of *A Clockwork Orange* becomes a reality. Here in Britain in the last few years that world has become visibly nearer."[100] The film *A Clockwork Orange* is very much its own critique of socialist Britain as a place of moral depravity that is drifting ever closer to totalitarianism. What is striking about *A Clockwork Orange* is how jarringly white a world it portrays, as exemplified in the exploits of its young white male gang who spend their time performing terrifying acts of violence for pleasure. The juvenile delinquency the film portrays acknowledges the escalating tensions toward immigrant Black communities in the decades following World War II that had by the early 1970s coalesced around the rise of white racist skinheads. Although in reality, they were often the more frightening presence on the streets, the uptick in urban

---

99 Ibid.
100 Ibid.

crime was portrayed in the media as the effect of a "blackening" Britain.

Many of the Black and Asian families that were now visible in British city centers had originally been invited into the country to act as guest workers to rebuild the country after the war. They ended up not leaving, and, through liberalized immigration policies, sent for other family members to join them, and subsequently took up residence mostly in white working class areas. It was their domestic proximity to whites that made the control of immigration a defining political issue throughout the 1970s. The Conservative political taglines around law and order became racist dog whistles, making Black citizens a scapegoat for the widening failings of the state to bridge the deep social inequalities that were widespread in 1970s Britain. Black people, themselves, were denounced as the agents of Britain's social breakdown rather than the government acknowledging the loss of revenues related to empire as the main culprit behind Britain's economic disintegration in the latter half of the twentieth century. Those young, male, Black colonial British subjects, who were largely responsible for the reconstruction of Britain after the war, were now viewed by the consensus as alien figures whose continued presence in the country only further aggravated unemployment and instigated violence. Their appearance in Britain was a visible remnant of an empire that refused to passively disappear, even as an ailing Britain was noticeably failing both to recuperate itself in comparison to the rest of Europe, or indeed elevate itself in comparison to its former colony, America. This failure neither to revive nor thrive made Britain itself a marginalized figure on the world stage.

Fear became a byword for humiliation in these years as the Black body increasingly became itself the object of panic and the threat it posed figured as a mugging; literally, an act of attacking and robbing someone in a public place, which surely signaled a reversal of fortunes insofar as it had been understood that it had always been the other way around between Britain and her colonies. The Black bodies being surveilled aggressively in the early years of the 1970s were projected in the culture's imagination to be the perpetrators of particular events bringing Britain to a threshold it never wished to contemplate as a nation. Groups of people of color were essentially framed as having the potential to violate, decisively, a particular set of given social orders.

Sally Davison and George Shire observe that in Britain, both "references to race and immigration have been a consistent part of the mobilizing repertoire of the authoritarian aspect of neoliberalism since the 1970s" and that "race was at the heart of political battles during the transition to Thatcherism."[101] They argue that this project of racist populism on the right begins with Enoch Powell and indeed, that Thatcherism would have been impossible without his ability "to establish between the themes of race and immigration control and the images of the nation" as something essentially white; that is, the British people themselves.[102] As a consequence, it was possible to construe popular racism as the foundation upon which the neoliberal project rests both economically and politically. Through this lens, was possible to distinguish mugging as an *opportunistic* act, one that set the stage for the arrival of the black(ened) body into Britain as something both unplanned and invasive. If the 1960s mark the end of empire, then the 1970s conversely mark the beginning of a reconceptualization of a Britain that was never willingly touched by it, which never saw its place as one of conscious opportunism, but of unconscious entitlement.

In a speech of 1961, Powell speaks of a nation coming out of its imperial phase, totally unmodified by the experience of ruling others. English nationhood remained "unaltered" and therefore, unavailable to the adulteration of its pride of place in the world. It was a matter of biological inheritance coupled with "national consciousness," that made it possible for Englishness to be "transmitted from generation to generation by a process analogous to that of inheritance."[103] There was no possibility of

---

101 Sally Davison and George Shire, "Race, Migration and Neoliberalism,"
    Soundings 59 (2015): 86–87.
102 Ibid., 86.
103 Robbie Shilliam, "Enoch Powell: Britain's First Neoliberal Politician," *New
    Political Economy* 26, no. 2 (2020): 243.

the empire's Black subjects feeling as though they could come home to this place because they fundamentally lacked any natural affinity with its traits, despite the nation's long the imperial era. Moreover, those who has received her generous auspices of "safety, ease and irresponsibility of servitude," according to Powell, simply could not be expected "to pursue 'freedom [… and] the responsibilities and the opportunities, which are inseparable from'" the uniquely English character.[104] In effect, this meant that the English were quite naturally neoliberal subjects and therefore, neoliberalism was quite naturally white. "This racialisation is evident in the way in which Powell arranged the diverse populations of the British Empire by reference to varying political proximities to whiteness" that never quite matched the description of anyone worthy of proper inclusion along the lines of what was essentially a degradation of orderly personhood.[105]

Outside of Britain was another matter because other countries were forced to compete with other nations for access to crude oil. For much of the proceeding century, "Britain treat[ed] the Middle East as its empire's gas station."[106] For most of the colonial period, it had maintained "unfettered access to Iran's oil" in an agreement that was fundamentally "premised on the threat of state violence."[107] That violence became a reality when in March 1951, Iranian member of parliament Mohammad Mossadegh led the charge to nationalize Iran's oil industry and became the nation's prime minister. Britain and the United States organized a coup to overthrow Mossadegh and restored the Shah Mohammad Reza Pahlavi to power to reestablish unfettered access to Iran's oil for another twenty years, rebranding their enterprise as British Petroleum.

When in 1973, the multinational oil cartel Organization of Petroleum Exporting Countries established an oil embargo,

---

104 Ibid.

105 Ibid., 244.

106 Kate Aronoff, "BP Is Not Woke. It's an Imperialist Success Story," *The New Republic*, June 6, 2020, https://newrepublic.com/article/158052/bp-not-woke-its-imperialist-success-story.

107 Ibid.

which spiked prices to record highs, it was widely touted in the mainstream media as another act of public mugging of the British public by former colonial peoples who had lost sight of their rightful place in the global pecking order. The OPEC elites became rich in petrodollars through their manipulation of oil prices. At home, British fortunes tumbled. Britain had to come up with a way to capitalize on this shift of wealth. It did so by creating an entirely new empire for through the offshoring of OPEC's sizable oil revenues to various British-controlled jurisdictions that remained as the spoils of empire to create a global network of untaxable, unaccountable wealth. This network, with the City of London at its center, became the catalyst for the 1986 "Big Bang," as well as the deregulation of markets worldwide. The electronic banking systems that made this all possible "connected via primitive internet like-networks made the large-scale rapid movement of cash across the globe easy and instantaneous."[108] This banking revolution resulted in international capital markets transforming into global money laundering operations "fuelling covert politics across the globe" seemingly overnight.[109]

Chief among the beneficiaries of this was the Apartheid regime in South Africa. Black market sales of oil to the state took place despite the economic sanctions in place against the ruling regime. By the mid-1970s, Iran was one of South Africa's most important oil suppliers. Hennie van Vuuren's findings reveal that not only had "Iran had maintained relations with South Africa going back to World War II, during which time it has sheltered and later buried the last Shah's father as 'an honorary white' in Johannesburg in 1944."[110] South Africa continued to maintain a trading relationship with Iran even after the Iranian Revolution in 1979. Before that, the Shah's government "had invested heavily in uranium enrichment, and secured a steady

---

108 Hennie van Vuuren, *Apartheid Guns and Money: A Tale of Profit* (London: Hurst and Co., 2019), 10.
109 Ibid.
110 Ibid., 106.

supply of uranium in the process."[111] After the Revolution, oil was traded for the express purpose of procuring uranium for Iran to fight a war against Iraq. During the 1980s, Iran began its nuclear program as part of considerations around pursuing a nuclear deterrent strategy against Iraq.[112] British Petroleum was a major player linking its defense industry concerns in Iran with its support for South Africa's apartheid regime. "As a major oil importer, BP quite literally helped fuel apartheid by selling diesel fuel, gasoline, aviation gasoline, bunker fuel, and lubricants to South Africa's military and police force."[113]

The City of London became a nexus point for various oil-for-arms exchanges between the two countries, significantly propping up the newly unregulated banking economy in the process. The extent of trade between South Africa and the United Kingdom made Thatcher inclined not only to tactically support the continuation of the apartheid regime due to her racial attitudes but, perhaps more importantly, due to Britain's deep dependency on it for "strategic minerals."[114] Included amongst these was uranium for the manufacture of nuclear weapons, the manufacture and trade of which meant that postcolonial Britain was still considered a Cold War superpower on the world stage. Officially she said her reasons to deny trade sanctions had to do with the protection of "Black workers."[115] If we take seriously her assertion, we can surmise that when Thatcher dispatches her environment secretary Michael Heseltine to the deepest recesses of the inner city, and she does so with the supposition that these populations are categorically endangered. This is so because the economies of colonial yesteryear no longer sustain them through active trade. Thanks to new information technologies

111 Ibid.

112 Sam Wilkin, "Iran Considered Nuclear Weapons during 1980s Iraq War, Ex-president Says," *Reuters,* October 29, 2015, https://www.reuters.com/article/us-iran-nuclear-rafsanjani/iran-considered-nuclear-weapons-during-1980s-iraq-war-ex-president-says-idUSKCN0SN0E720151029.

113 Aronoff, "BP Is Not Woke."

114 van Vuuren, *Apartheid Guns and Money,* 355.

115 Ibid., 354.

postcolonial Britain's economies of retailing, banking, and insurance have been able to decamp from cities altogether, setting up their businesses elsewhere in suburban and rural areas that require not hard labor but the soft skills associated with light assembly. Computers have made work clean and the enterprise around it green. The appearance of dereliction associated with past labor is now left far out of sight and, with it, the requirement to reckon with imperial losses.

By 1988, the Conservative government's contribution to the decline of inner cities had been made evident through "its abandonment of automatic investment grants for firms locating in depressed regions [and] its recent overhaul of the welfare system [… that] has left millions of the poorest inner-city residents worse off."[116] Michael Jacobs identifies something of far greater consequence happening throughout the 1980s. Within this decade, Thatcher's government had move to tactically promote inequality through its huge defense program, which directed government contracts "almost exclusively to firms in the south-east; particularly west of London," making it inevitable "that the rest of the high technology" sector would follow suit.[117] This initiative reveals a significant connection between the conservative's newly revised plans for exterior defense, interior settlement, and the rapid financialization of Britain's economy following on from 1986s "Big Bang." This move, coupled with the "further relaxing of housing controls in the 'Green Belt' around the capital, suggests that there's was a coordinated strategy of racial segregation taking place to politically and economically re-engineer these local geographies."[118]

Jacobs gives the example of the conservative inner London borough of Westminster that had been "literally 'deporting' its residences to specially-built estates outside of London in order to sell their (refurbished) flats to wealthier people."[119] Jacob's

---

116 Michael Jacobs, "Margaret Thatcher and the Inner Cities," *Economic and Political Weekly* 23, no. 38 (1988): 1944.
117 Ibid.
118 Ibid.
119 Ibid.

makes no explicit mention of race in his observations. None-theless, it is implied that the process of what he calls "deliber-ate gentrification" has the desired effect of what Thatcher called "winning the inner cities back to our cause" by defeating claims of an intergenerational right to residence for migrant commu-nities predominantly of color.[120] The neoliberal elements of the campaign of dispossession become evident in the description of such practices as matters "of efficiency."[121]

At the same time, the City of London is expanding eastwards to the Docklands through the creation of new private housing schemes in order to introduce a new breed of white privilege into the ecology of the inner city. This happens through the ar-rival of the icon figure of the young urban professional, or "yup-pie," on to the urban scene. This trend follows a like for like ex-change of white high earners inwards, and Black low earners outwards, from the city center. Jacobs believes this movement is a case of "gilding the ghetto," but a far more apt metaphor is per-haps one of fortifying the cage: a phenomenon arising as a direct consequence of Thatcher's new brand of racial determinism.[122] Unable to escape the consequences of their territorial exile, Black British populations are to be held in the thrall of poverty and unemployment as displaced persons from what should by all rights be the source of their wealth, that is, inherited prop-erty. It is possible to witness the residues of the apartheid system in Thatcher's position on such spatial maneuverings in order to make it seem as though these situations were personal matters of choice rather than matters of coercion. When it came to riots, criminality, and social disintegration, her government acted as though these were caused solely by individual characters and attitudes rather than systematic forms of divestment, discrimi-nation, and deprivation.

---

120 Ibid.
121 Ibid.
122 Ibid.

Race, as a Neoliberal Ideology

In *The Empire Strikes Back,* Errol Lawrence refers to "the ne-
glected side of the imperialist coin; to black people's experiences
of colonialism, slavery and indenture" in order to suggest a connection between wealth and bodies of color as things that are
only recognized in relation to white imperial economic arrangements.[123] The flip side of that arrangement becomes even further
obscured in the context of a postcolonial Britain remaking herself in the image of her white colonial dominions, while at the
same time opening herself up to an internalized relationship to
global marketization. Her postcolonial fueling of financial capitalism had the external effect of increased class instability and
downward social mobility amongst the primarily Black nations
on the receiving end of her colonial rule. This negative influence
continued to mount even as Britain was emerging from its era
of formal imperialism as a gateway for global trade and finance.

Race now stands as one of the trace effects of material worlds
that seemingly no longer pertain, and yet, as Britain restruc-
tured global capitalism, so too did it alter biopolitics. Neoliberalism reintroduces biology into capitalism. As a consequence
of this, capitalism continues to exert efforts to dehumanize and
degrade life by bringing forward core beliefs and practices that
have their origins in the liberal assemblages of racism and sex-
ism that precede it. The subject needs no longer to be a biocentric formation for it to become weighted to a certain kind
of posthumanism that insists on claims to sentience, even as it
forces its subjects to dwell without recourse to structure.

Under conditions of neoliberal capitalism, Gaia Giuliani
argues that "race, and gender, but also poverty, sexuality, and
nationality are key factors to who gets to be saved and who gets
left behind."[124] I would add to this list private property and the

123 Errol Lawrence, "Just Plain Common Sense: The 'Roots' of Racism," in *The
Empire Strikes Back: Race and Racism in 70s Britain* (London: Hutchinson,
1982), 93.
124 Gaia Giuliani, *Monsters, Catastrophes and the Anthropocene: A Postcolonial
Critique* (London: Routledge, 2020), 2.

degree of its securitization. It is precisely for this reason that themes of housing development and purchase become crucial to Thatcher's Conservative agenda. Her rhetoric around the freeing of the housing market and the opening up of consumer finance, directly links to "a deceptively friendly, multiply-scaled, adaptive infrastructure of self-justifying evolutionary inequalities first mapped out from the urban epicenters of nineteenth-century industrial colonialism."[125] Elvin Wyly argues that the state's preoccupation with "the control of the spaces of every-day" coincides at every point with an intensification of "human competition encoded into the cosmopolitan planetary circuitry envisioned and built by Ronald Reagan, Deng Xiaoping, Augusto Pinochet, and Margaret Thatcher — through the minds of Friedrich Hayek and Francis Galton."[126] It is at the level of mind, or more precisely imagination, that makes housing translate into a form of capital through which it will gain inordinate influence over life. It is no longer necessary for capital to lay claim to whole nations and continents as it did when the world was largely held in the thrall of colonial formations. Now it must simply lodge itself in the ordinary housing of bodies in order to achieve power.

In many ways similar to its approach to Indigenous societies, today's commercial settler class find themselves legally justified in claiming the cognitive labor of the bodies it considers relatively low on the natural scale of social and cultural evolution. It has recognized them as particularly suited to new forms of vacancy and, thus, uniquely situated in post-digital economies. Sovereignty, in this instance, equates with surveillance. What constitutes terra nullius takes place through the mere act of registering one's location in space. The personhood of liberalism has given way to a shared intelligence between peoples and their surrounding territories. They get temporarily parked through various devices, but the assumption of their worth is banked on

125 Elvin Wyly, "Cities, Consent, and Coercion." *Society + Space,* May 1, 2019, https://www.societyandspace.org/articles/cities-consent-and-coercion.
126 Ibid.

the principle that they don't stay anywhere for long. This changes the liberal meaning of "rentier" into something far more il-
lusory and, at the same time, something that must be more in-
tensively privatized, monetized, and securitized, more so than at
any other point in history as a transactional class. What can be
disposed of in such an arrangement are the people as capitalism
fastens onto the mind and preoccupies the flesh.

Rob Nixon puts these into terms that allow for a recognition of the body's place within the greater construct of "violence
embodied by a neoliberal order of austerity measures, structural
adjustment, rampant deregulation, corporate megamergers, and
a widening gulf between rich and poor is a form of covert violence in its own right."[127] Again, we are migrating over into the
territory of real estate "to foreground the vast structures that
can give rise to acts of personal violence and constitute forms
of violence in and of themselves. Such structural violence may
range from the unequal morbidity that results from a commodified health care system to racism itself."[128] The posthuman poses
a representational challenge and imaginative dilemma, not just
because it poses within itself the specter of an older form of imperceptible violence enacted through concepts of ownership,
but equally by how its inauguration into consciousness signals
some type of imperceptible change "whereby violence is decoupled from its original causes by the workings of time."[129] If Nixon
is right in assuming that the structural violence implied in the
idealization of human consciousness "predated both sophisticated contemporary ice-core sampling methods and the emergence of cyber technology" as a signifier of planetary doom,
it might also be possible to respond to both its recent, radical
replacement by the advent of the posthuman in our geological
perception, and to its place as a core principle in our changing
technological experiences of time.[130]

---

127 Rob Nixon, *Slow Violence and the Environmentalism of the Poor* (Cam-
    bridge: Harvard University Press, 2011), 10.
128 Ibid.
129 Ibid., 11.
130 Ibid.

In this sense, race, as a neoliberal ideology trades, on the perception of being a product of both prehistorical time and precolonial nature. When restored to its pristine condition beginning with the end of formal colonial history, it can be made to function as a mode of performative inclusion. While the "fact" of racial differentiation cannot be substantially interfered with according to the new rules of market, black(ened)ness can still be made to perform the work of being an economic resource. Whereas previously black(ened)ness had been associated with an innate passivity, it is now instead free to display itself as vibrant, vital, energetic, and lively. Above all, it's now valued as something that can take care of itself without the need for political or economic interference. Bodies now become racialized according to their capacities to produce value. The goal of neoliberal economies has progressively shifted from the cultivation of desires, versus the refinement of problem-solving, as it follows along the path of governance situated within an ethos of apolitical neutrality. To be racialized becomes synonymous with the state's ability to exert violence and spatially control bodies. This is largely determined by where one lives. Whether that is in the imperial metropole or colonial periphery of the globalized city that lends broad meaning to populations and, at the same time, represents an individual body's latitude of freedom. Within this urban framework, Blackness is featured as a cypher for bodily pollution and corruption of all sorts. That said, it is seldom the object of futurity in and of itself, but it does rather furnish for the white subject a path towards its own optimization. Therefore, its prosaic subjection to precarity acts once again in service to an exceptional white futurity.

Black settlement according to Lisa Lowe has always been "an exercise in social engineering" and associated with practices of "receptivity," meaning that in the British context, Black communities, of diverse ethnic backgrounds dating to the very beginning of the nineteenth century, have been dealt with as a colonial population that was fundamentally unable to achieve self-governance.[131] As a consequence of that belief, systems of rewards and penalties were established throughout various government initiatives with the ultimate goal of expatriating them, either through territorial concentration, or exile through deportation.[132] Black futurity has long been synonymous with Black removal. With the formal end of slave trading in 1807, Lowe contends that Britain was able to introduce in its place a new model for the global expansion of its influence through "controlling circuits of worldwide commerce" that "could be a more effective and profitable mode than the restricted gains of direct territorial conquest and colonial slavery."[133] The abolition movement on both sides of the Atlantic, seen from this perspective, was far from a mechanism to introduce freedom to Blacks; rather, it was to intrude free markets to Asia and Africa. Lowe's arguments around trade offer significant insight into the treatment of Black populations within a neoliberal paradigm if we consider the possibility that the Atlantic world was not solely a product of colonial slavery. Perhaps more so, the Atlantic world was a product of market heterogeneity and security innovation that has endured well past the formal confines of the British Empire itself. It has taken on an imperial afterlife that allows for another form of enslavement to continue without question, whose presence is subsumed through the affirmation and forgetting of social relations that at once promote assimilation and force assent. What I am referring to specifically are the myriad of ways in which trade, commerce, and movement are routinely elaborated through divisive violence. The human became the critical cleft for such sorting and assumes itself through a defensive position poised to attack perceived adversaries.

The project of humanism begins as an administrative function and its associated liberty dulled out as an intrinsically economic value. As a public good, liberty owes much to its place

---

131 Lisa Lowe, *The Intimacies of Four Continents* (Durham: Duke University Press, 2015), 68.
132 Ibid.
133 Ibid., 69.

within England as a promise of rest and, by extension, of stable settlement that can endure beyond one generation. The English aristocracy has always been dependent on, and indeed codependent with, colonial practices. The mass mobilization of populations in Africa and Asia, robbed of their humanity and coerced into labor, were the facilitators of wealth seemingly without end. Emigration became the means with which to set the plantation, as it were, on the move, through bodies that would simply never be allowed to root. These bodies could be exhausted at will and, in so doing, were made complicit with the never-ending needs of the market and, critically, could be said to naturally move without the responsibility of governments to instigate their doing so. Neoliberalism extends that logic to the whole of humanity at one level, yet at another continues to follow the capital logics of its previous iteration as liberalism. Once again, race emerges within the newer context at the level of security, which concerns itself with the territorial policing of bodies within the nation by coding them according to their potential for deviancy and rebellion with racialization acting as a source for such capacity. Migration becomes its international counterpart in terms of such critical encoding.

Space comes into the equation when it itself becomes ranked. Returning to Thomas Jefferson's scheme to educate and emancipate Black people from the United States, similar ideas partition a post-racial, neoliberal conceptualization of Blackness as something in need of institutional redress. As Walter Benn Michael observed, "neoliberalism doesn't [need racism] — it needs antiracism."[134] It requires "a world free of discrimination" and banks on "the dream of a world where identities are respected" as a means of promoting an ethos of universal securitization.[135] This is the reason why "the preferred crimes of neoliberalism are always hated crimes, the holocausts of everyday life."[136] In both

---

134 Walter Benn Michaels, "Plots against America: Neoliberalism and Antiracism," *American Literary History* 18, no. 2 (2006): 297.
135 Ibid., 298.
136 Ibid.

their banality and ubiquity, they generate a market rationale for the endless tracking of subjects poised at any moment to expose their enduring prejudices or act out their deep seated aggression. As with Jefferson, this threat is not aimed at the elimination of inequality, but it is for difference to be fundamentally respected as such and for culture to be appreciated as the standard bearer of profound differentiation of one population from the other. Jefferson's identification of difference with inferiority was eliminated in liberalism's rebooting of itself in the 1980s. Nonetheless, it remains implied that elimination has endured in this construct, as the object of desired outcome. The desire to civilize Black people in moving them onto various locations in Africa gave little, serious thought to their freedom or edification; rather, it was an expression of enthusiasm for new forms of empire having to do with international free trade that no longer required the devices of formal enslavement.

This new economy required the development of a professional-managerial political class to discipline the flow of what was now to be cast as the free movement of peoples and later united under the category of free labor. The dilemma of Black settlement persisted through this period of imperial market transition, as well as through the postwar global advancement of neoliberalism. By the 1980s and 1990s, it had created its own categorical divisions amongst Black communities and populations on both sides of the Atlantic that allowed for the creation of a Black professional-managerial class to minister downward in relation to working-class Black populations that those politics had rendered invisible. It moved its efforts downward still to a Black underclass that those same politics had rendered hypervisible. Whereas the working class merited uplift, the underclass merited censure. This differentiation bore itself out in ways that privileged privatization as a means with which to distance these classes further from one another.

If neoliberalism could be said to have

produced greatly expanded commercial and career opportunities for Black (and Latino, female, etc.) entrepreneurs

under the rubric of community "empowerment," "role modelling," or "social entrepreneurialism," it could also be said
to have created an industry dedicated to the radical dispossession of a Black underclass, its informal economies [...]
driven further into the exposed territories of criminality and
abject poverty.[137]

Recessed between them was a largely stagnant level of social
mobility for the Black working class whose structural limitations remained largely ignored by an emerging diversity sector
dedicated to the task of generating opportunity as a matter of
individual concern rather than collective occupation. At every
point, industry remained the unquestioned arbiter of advancement. Within such an ethos the reality of racial disenfranchise-
ment was obscured amidst a greater political and economic atmosphere informed by market principles as a new millennium
dawned.

Britain's globalized inner city of the twenty-first century affords much more than housing and business opportunities to
new, multiracial global elites. It is the setting for experiences
that are, in one way or another, integrated into networks of communication and information and a new social life moored in the
concurrent management of resources and time. The neoliberal
pluralization of urban identity lends commercial valance to the
daily patterns of life, which makes of locality a plan, of desire a
construction, and of creativity a hoarding. Within this space,
intelligent devices might be viewed as companion species to the
servants, slaves, wives, and service workers of yesteryear, who
performed faithfully as racialized and gendered quantities of labor. Acting in ways akin to the useful technologies of the Black,
Asian and Indigenous bodies of a previous era, they are there to
imaginarily support the expansion of those granted the privilege
of humanity by providing them with the provisional "housing"
for another type of commercial existence to emerge. Their even-

---

137 Adolf Reed, "Antiracism: A Neoliberal Alternative to a Left," *Dialect An-*
thropology 42 (2018): 113.

tual disinheritance, like their predecessors, is assumed within that same project. The posthuman is at once a continuation of the imperial legacies of previous economies and a break with biological controls that animated previous understandings of the conditions of possibility. This is what is allowing neoliberal society to return to earlier pre-settlement definitions of relatedness based on close association rather than the legal and social categories that define kinship in terms of blood and ancestry. The posthuman arrives just in time to eclipse the troubled past of colonialism, by making of possession something apart from property, as the source of affective links that have their origin not in limiting boundaries, but in netting flows.

Brexit Means Never Having to Say You're Sorry

Jacques Derrida observed that "Britain's 'great unifying projects,' all come down to the apprehension of this shadowy figure, a figure who unites the center with the periphery of the Enlightenment project and links the concentration of wealth with the concentration of individuality."[138] This shadowy figure belies the fact that Britain's financial autonomy springs forth from the illicit captivity of others' wealth. As a consequence, contemporary Britons emerge as "ghosts chained to ghosts," racked with the guilt of collectivity, mounting up, over the course of history, as unpaid existential debt to the others who made their lines of credit worthy.[139] This "ghost" is "the violence of domination, an oppression of the weak by the strong, in the name of any form of hierarchical superiority," including but not limited to cheap labor, indentured servitude, or slavery, which in quick succession could be made to translate into "the abstract value of money or capital" under the terms of that very same uni-

138 Jacques Derrida, *Spectres of Marx: The State of Debt, The Work of Mourning, and the New International,* trans. Peggy Kamuf (New York: Routledge, 1994), 4.
139 Ibid., 4–5.

fying project.[140] The liberal, political, and economic framework that literally reinvents Britain for the imperial capitalistic age through the naturalization of new class and racial hierarchies based on colonialism, violently reemerged in the second decade of the twenty-first century under the auspices of exiting the European Union, heretofore to be known as "Brexit."

In 2016, Britain entered into Brexit negotiations with the European Union with the stated desire to reconfigure itself once more as a vast free-trade empire (Empire 2.0). Gurminder K. Bhambra argues that Brexit was "less a debate about the pros and cons of membership than a proxy for discussions about race and migration, specifically who belonged and had rights (or should have rights) and who did not (and should not)."[141] Britain's wealth was always based on maintaining punitive and exclusionary economies. The latest derivation of which came in form of creating a hostile environment for those presumed aliens to it. For those undesirables, the preferred form of recrimination became an ideology of austerity that denied them access to the basic dignities of life. The hostilities applied to them appear to be of another, unrelated order so that one can be pitted against the other to draw out mutual resentments. In the end, they all become matters of unpaid debt, conveniently charged to the accounts of those can both best and least afford it according to May's neoliberal calculations. The only thing free to move in this economy of racial and class othering is the capital extruded from both.

This was spelled out in then-Prime Minister Theresa May's first Conservative Party Conference speech as head of the government that would implement Brexit. In that speech, she directed particular ire towards the cosmopolitan elites, who presumably supported the Remain campaign, and declared, "if

140 Claudia C. Klaver, *A/moral Economics: Classical Political Economy and Cultural Authority in Nineteenth-Century England* (Columbus: Ohio State University Press, 2003), 104.

141 Gurminder K. Bhambra, "Locating Brexit in the Pragmatics of Race, Citizenship and Empire," in *Brexit: Sociological Responses,* ed. William Outhwaite (London: Anthem Press, 2017), 91.

you believe you are a citizen of the world, you're a citizen of nowhere. You don't understand what citizenship means."[142] That same privileged lot were accused of selling out their working-class brethren, who were forced in her words, to "sacrifice" their "dreams in the service of others."[143] That service, she went on to add, was the reason that if you were "one of those people, who unfairly lost their job, who stayed in work but on reduced hours, took a pay cut as household bills rocketed," or found "themselves out of work or on lower wages," this was all "because of low-skilled immigration."[144] The unfairness of life would stop if we put an end to "their" presence within "our" society. She remarked, "I know a lot of people don't like to admit this," but went on to admit this and many more things that could be laid at the feet of those who don't have the good sense to be loyal to their community, nor value their citizenry.[145] Those metropolitan elites who had financially succeeded in life to the great cheer of the Conservative Party were, to May's way of thinking, also the ones who "behave as though they have more in common with international elites than with the people down the road, the people they employ, the people they pass in the street."[146] Indeed, it is these same liberal-minded elites that "find your patriotism distasteful, your concerns about immigration parochial, your views about crime illiberal, your attachment to your job security inconvenient."[147]

In coming to grips with May's rhetorical strategy, it becomes significant to note she "grew up in Oxford where a colour bar was acceptable."[148] This is consequential insofar as her version of Brexit promotes a "combination of national sovereignty and

---

142 "Full Text: Theresa May's Conference Speech," *The Spectator,* October 5, 2016, https://blogs.spectator.co.uk/2016/10/full-text-theresa-mays-conference-speech/.
143 Ibid.
144 Ibid.
145 Ibid.
146 Ibid.
147 Ibid.
148 Sally Tomlinson, "Enoch Powell, Empires, Immigrants and Education," *Race Ethnicity and Education* 21, no. 1 (2017): 5.

anti-immigration" that reflects residual beliefs about the defense of a white nation state dating back to the ascendance of British Empire.[149] May is a product of a predominantly conservative middle class area, where few or no minorities were in residence and, yet, were most vociferous in calling for an end to "coloured immigration" in the era of her youth. The rhetoric of Brexit was never genuinely aimed at figures like May who were confident in their positions of class superiority; rather, the rhetoric was aimed towards the lower classes whom for centuries had been encouraged to see their economic, political, and social fortunes as tied to their racial superiority in comparison to the Black subjects of the British Empire. The working classes perceptions of empire find their beginnings in an ethos of white nationalism that must be exorcised through periodic outbreaks of xenophobia and racism. That they appear to have become increasingly manifest in the twenty-first century, speaks to an ongoing necessity for the racialized codes of patriotism to be read out as acts in service to a dogma of putting "Britain First."[150]

As Home Secretary, May first spoke about creating a hostile environment for "illegal migrants" in 2012. This policy came as a direct consequence of May's failure to bring annual net immigration, then running at about 250,000, to the tens of thousands. Her response to being challenged on the subject, by then-Prime Minister David Cameron, was to rally an entirely different target, this one aimed not at legal immigration and migration, but its illegal, and perhaps more significantly, unquantifiable counterpart. May stated, "[t]he broad objective is and was to make life as difficult as possible for any irregular migrant — or any migrant the Home Office judged as potentially illegal in lieu of the correct documentation. They would be 'encouraged' to leave voluntarily."[151] This policy within a period of years was translated

---

149 Ibid.

150 Ibid., 8.

151 Erica Consterdine, "Hostile Environment: The UK Government's Draconian Immigration Policy Explained," *The Conversation,* April 26, 2018, https://theconversation.com/hostile-environment-the-uk-governments-draconian-immigration-policy-explained-95460.

into law through the Immigration Acts of 2014 and 2016 that formalized a variety of measures to deny access to employment, healthcare, housing, education, banking, and other basic services to those who were unable to provide proof of their current and legitimate immigration status.

May planned to implement a series of internal controls upon the UK population that would make the right to reside within it a living nightmare for those with ambiguous immigration statuses. They were specifically designed to coerce such individuals to "self-deport."[152] Immigration checks would extend to all within this scenario with the hope of ensnaring "irregular migrants" at various points in the process of elimination. Those that could not prove their legitimacy, which was by default defined as having a British passport, would be told by the Home Office to either leave the country immediately or risk being subject to police apprehension and transportation to an immigration detention center. May's policy and the laws that eventually related to it promoted widespread discrimination and ran counter to European statues around human rights. Such concerns were immediately raised, but in the end, a majority of both parties voted in favor of these new Immigration Acts.

From the time of their ratification, these Acts began to impact a number of people who were born in the Caribbean, who had come to Britain as children during the 1950s and 1960s. They were aimed at the so-called "Windrush Generation," who arrived in 1948 at the invitation of the British government which was in dire need of low-skilled and unskilled labor to reconstruct the country following on from the war. Sally Tomlinson asserts, "immigrants were tolerated if they took the jobs whites did not want and did not compete for housing and schooling."[153] This position resonated with long standing "imperial beliefs that some sort of caste barrier must be preserved to prevent white

---

152 Sukhwant Dhaliwal and Kirsten Forkert, "Deserving and Undeserving Migrants," *Soundings* 61 (2015): 50.
153 Tomlinson, "Enoch Powell," 5.

people from losing status as well as social privileges."[154] This belief was sown deeply into the British national psyche, so much so that, "Winston Churchill even considered the slogan 'Keep Britain White' for the 1955 general election."[155]

The significance of the recent and ongoing Windrush scandal has to do with a long-standing policy of hostility towards the Black Commonwealth, whose colonial origins correspond with the unlawful seizure of African bodies to forcibly labor on British Caribbean Islands for the financial gain of a white metropolitan elite. Each Windrush story is the personification of that disavowal of history writ large, in terms of a failed conscience regarding propriety. "Each case is directly linked to a policy that ignores the principle of *habeas corpus* by imprisoning innocent people without reference to a judge, jury or evidence of guilt" without recourse to an understanding of the greater crimes of the past that have yet to confront lawful restitution.[156] It is this national policy of cultural amnesia that has resulted in British citizens being barred "from accessing the public services and benefits that they themselves built with their own hands, staffed and paid for" in a postcolonial context that rematerializes an even more brutally unjust colonial past.[157] Then, as now, new and ever-novel forms of bureaucratic policing do the work of enforcing the violence of a "peculiar institution." The bordering of identity has long kept association with toxic, anti-immigrant rhetoric created by successive British governments spanning centuries. The demand to maintain a hostile environment towards those deemed outsiders through immigration policies that address "largely non-existent problems," then as now, have

---

154 Ibid.

155 Ibid., 4.

156 David Lammy, "Don't Let Rudd's Departure Distract from a Toxic Policy That Needs to Die," *The Guardian,* April 30, 2018, https://www.theguardian.com/commentisfree/2018/apr/30/amber-rudd-departure-toxic-policy-windrush-generation-home-secretary-david-lammy.

157 Ibid.

"pandered to the basest of prejudices" that implicitly couple race and immigration.[158]

There has been a desperate attempt by Black British citizens with Caribbean heritage to prove themselves as unworthy targets for this humiliation at the hands of the Home Office. They thought that Britain as a nation that had previously greeted their Black returning servicemen in 1948 with signs stating "No Irish, no blacks, no dogs" had long since accounted for their previous wrongs. And yet here in 2018, few if any were willing to take notice of these as outrageous hardships befalling their fellow "British" citizens. Then as now, they have had to suffer not just at the hands of their own government, but of their own people. Adding insult to injury, Labour MP David Lammy observes that

> in an effort to change the subject, cabinet ministers have been touring the studios trying to conflate the treatment of British citizens at the hands of their government with the issue of illegal immigration. Not only is this completely wrong and demonstrative of a worrying lack of understanding of the law, it is frankly insulting to those citizens who have been treated like criminals in their own country.[159]

During the last of her statements to the House of Commons, Amber Rudd, the former Home Secretary under Theresa May's Brexit government, referred to the individuals caught up in "the hostile environment" as "illegal" no fewer than twenty-three times.[160] By contrast, not once did she use the word "citizen" to refer to their plight.[161] Subsequently, Rudd was forced by public outcry to resign from her post. For her part, Theresa May has been careful to separate the Windrush generation from the British people in her comments about the scandal, thereby subtly reinforcing a presumption that the former was being generously

---

158 Ibid.
159 Ibid.
160 Ibid.
161 Ibid.

granted their citizenship now, rather than having previously possessed it as a right. Lammy concludes that

> [t]his is symptomatic of how the hostile environment operates. It blurs the lines, and raises questions about the status of Commonwealth British citizens, refugees and asylum seekers so that anybody who looks as though they could conceivably be an illegal immigrant is seen as such.[162]

Britain's hostile environment towards these perceived others has been centuries in the making. The criminalization and burdens of existence visited upon such persons is not about tackling illegal versus legal immigration, so much as it is an explicit attempt to blur the lines between justice and injustice. The discrimination in various sectors of society is allowed to systematically expand and habituate itself into the mundane functions of economic administration. When the Home Office found itself "unable to tackle EU migration due to freedom of movement [...] it turned its focus towards non-EU migrants and their families, even when they were legal."[163] During that same time, it began "cutting its numbers of immigration case-workers" and adopting a governing principle of "discretion" in the judgement of these now suspect individuals.[164]

The Home Office's tacit establishment of such racial and class hierarchies in determining the "right" sorts of economic migrant, does a great deal to enliven the climate of fear and discrimination against what is perceived as the "wrong" sort. Britain's history is one that is deeply and fundamentally authored along racialized lines and its wealth similarly colored by a propensity for the nation to progress through means of bodily exploitation. Without both of these elements, nation and empire

---

162 Ibid.

163 Nasrine Malik, "I Felt a Nausea of Fury" — How I Faced the Cruelty of Britain's Immigration System," *The Guardian,* March 5, 2018, https://www. theguardian.com/uk-news/2018/mar/05/immigrant-home-office-british-citizenship-hostile-environment.

164 Ibid.

founder, and force its occupants to once again enact the drama of exclusion and denial within what constitutes its always remaining greater polity, from which there is and remains no clear means of achievable exit. Witnessed at this level it is possible to read Brexit Britain as the ultimate punitive settlement upon which these others are based. When migrants are subject to endless record keeping, migration itself becomes synonymous with systems of tracking and apprehension. The creation of a system where Europeans persist in Britain as a second-tier class of migrants forced to dwell as "settled" persons, while at the same time being denied any physical proof of status, speaks to a future of unending vulnerability for this select class of individuals. Exile, it would appear, comes in many forms.

Interests radically chart elsewhere for May's successor, Boris Johnson. His commitment to pursuing a no-deal Brexit effectively erases all of May's immigration lines and supplants them with conditions of crisis that will ultimately allow his government to generate social and economic policies unachievable with "an orderly Brexit." Within the purview of his government, those persons wishing to "settle in" or enter the UK post-Brexit must prove that their presence adds "significant value." Under these new circumstances, presumably, an individual can be "a citizen of nowhere" so long as they are able to provide proof of substantial (future) earnings and a firm grasp of the native lingo. The guidance here suggests that Brexit Britain wants highly skilled minds at the ready. Conversely, low skilled bodies need not apply.

Dominic Cummings, the mastermind of the Vote Leave campaign focused his campaign on a dual premise: "don't talk about immigration" and "do talk about business." In a post Brexit-future, the distinctions between EU migration and non-EU migration entirely lose their relevance. Control of entry visas to the UK will then emerge as an exclusively market-based product to be sold to companies and individuals. Their distribution will not be managed by the British government, per se, but through systems devised by private government subcontractors. In keeping with the Conservative ideology of austerity, Johnson's government would simply up the costs of getting in if it appeared that there were too many wanting to get through the narrow the channel of "qualification." Conversely, costs would fall if there appeared to be a dearth of applicants. The immigration market would in theory regulate itself. In tandem with this neoliberal conceit lies another, that this market-based approach to immigration remains one driven by "merit." A provision would be put in place so those individuals too poor to purchase their visas outright would be eligible to enter into the terms of a debt repayment plan to pay what is owed incrementally. This system would be modelled on higher education debt, where their borrowing burdens could be paid down over time, of course, with the caveat of interest applied.

Alternatively, those applicants who believe they are of extraordinary potential value to the British economy could appeal to "governments and charities" to provide for their visa costs "for humanitarian reasons," presumably fashioning their appeals along the lines of successful Go Fund Me campaigns.[165] As such, would be entrants must prove their entrepreneurial, and by extension, assimilationist chops not by adhering to the principles of Britain as a fundamentally white and Christian nation but to the values attendant to a fundamentally free market one. Adhering to this same ethos, the rewritten rules of Britain's immigration system must exist for the express purpose of identifying winners and eliminating losers. Therein, expanding, or restricting access within a greater space of neoliberal marketplace economics "implies a relativist philosophy of value," where points are acquired based on one's willingness to intelligently adapt in order to acquire and maintain a viable status.[166] The alternative is to be left with nothing. In the near future, individu-

---

165 The Free Enterprise Group, *A Time for Choosing: Free Enterprise in Twenty-First Century Britain,* eds. Kwasi Kwarteng, Ryan Bourne, and Jonathan Dupont (Basingstoke: Palgrave Macmillan, 2016), 145.

166 William Davies, "Sabotaging Progress: The Cultural Economy of Resentment in Late Neoliberalism," *Society for the Advancement of Socio-economics,* http://sase.org/wp-content/uploads/2018/04/1-Davies-final.pdf, 2.

als might vie for ways to portray themselves as uniquely talented economic winners.

Applicants fortunate enough to be granted work visas will have no recourse to public funds and will moreover, have to pay an additional cost to have access to NHS care for the duration of their working stay. National Insurance contributions will be deducted from their earnings despite having paid this additional cost upfront. Although these visa holders are, over time, effectively double paying into the NHS, they will be identified as a drain on this public resource, as well as numerous others, and their presence within Britain disparaged as the cause of the nation's decline. Those who have the right to work in Britain as their birthright are subject to another regime of prohibitive costs insofar as the majority will be confined to low-skilled work within their lifetimes and will be progressively denied access to public funds. What ostensibly distinguishes visa applicants from their native counterparts is the fact that they have explicitly acceded to a contract of guaranteeing intense productivity in exchange for their privilege of dwelling inside the British state. Those Britons unable to embark on an equivalent contract will be forced to make recompense for their very condition of being in other ways such as performing government sanctioned unpaid work in exchange for public services while they seek to right themselves back into employment. Post-Brexit visa holders in this scenario might be portrayed as a new class of metropolitan elites facing broad social backlash.

Things of course are not as straightforward as they may seem. For example, individuals who might attempt to remain in Britain despite their exile orders would likely be subject to an extension of the hostile environment by Johnson's Home Secretary Priti Patel. Moreover, failure to register at all within her newly devised Australia-style ranking system risks a status of becoming "illegal" by default. Even if they appear to qualify, some individuals may have their applications enigmatically denied — sacrificed as it were — to fulfill an unspecified Home Office target. In the past, Patel has supported the death penalty, voted against marriage equality, voted for austerity, and tried to argue that people shouldn't be able to take the Government to court. Migrants, whatever their ultimate status, would very much be in the headlights of those regressive legal policies as a new immigrant class.

Under Johnson, EU migration will serve a far different ideological purpose than the one conceived of by May's Conservative nationalism and her obsession with net migration numbers. Johnson's Brexit policy marks not the end of free movement but its restriction that advantages a wealthier and, indeed, whiter few. In their wake, many others are left behind in their attempts to mount a series of legitimacy thresholds each more condemnatory than the last. Johnson's projection does not currently weave the appearance of deportation charter flights nor detention centers into its illustrative account of post-Brexit immigration protocols. Nonetheless, the hostile environment must inevitably bear those plausible appearances within the confines of its denial of free movement. What is depicted by his government already profoundly signals in that direction and, in so doing, allows the viewer to engage with Brexit's flip side: one that has been strategically omitted from public perception. For now, the future status of immigration is classed mutely on Britain's political agenda under the heading "further business," a situation surely calling for greater counter figuration.

White Posting and Posthuman Racing

With the greater scheme of extractive capitalism, the Anthropocene takes place as a result of the making of Blackness a geological category and whiteness a historical agency. Justin McBrien argues that amongst them, capitalism itself becomes "an extinction event" that does not accumulate to as it were make progress, but it rather erases "species [...] peoples, cultures, and languages."[1] It is through erasure that capitalism conducts its world making. Its object of universal measure was never the capacity to be human; rather it was the capacity to be, meaning that survival is the core of its register of advancement. The central problem of reality, therefore, is one persistence. That which can be fundamentally extinguished cannot, as such, be fully realized. The central problem of capitalism is that it plans for the "obsolescence of all life" and, therefore, fundamentally denies the existence of reality.[2] Reality is but a transmutation of life into asset and asset into demise. The artificial separation of climate from geography led not only to New World colonial expansion, but also to epidemic disease, which allowed civiliza-

---

1   Justin Mcbrien, "Accumulating Extinction: Planetary Catastrophism in the Necrocene," in *Anthropocene or Capitalocene: Nature, History, and the Crisis of Capitalism,* ed. Jason W. Moore (Oakland: PM Press, 2016), 116.

2   Ibid.

tion to be defined as a kind of quarantine zone against the threat of transmissible illness that was always already a product from within what was only ever imagined to be a stable Old World. Trauma haunts this project from its inception as the New World tries in vain to account for the civilization that ought to be there but is not. Environmental catastrophism, as a politic, is an extension of that same disturbance of mind; a misguided belief both in civilization and its unlimited capacity to infuse itself into bodies, ecologies, and even geological strata. Its powers of subsumption are far less than what has been assumed, and it is far more a product of the material world in which it acts not as a singular force but as a coupling device linking the human to capital.

Mcbrien offers the prospect of a "human being" that "can be decoupled from capital" for the very reason that "capital is extinction" and "we are not."[3] If the human remains in this scenario, what place does the posthuman occupy? Its development, besieged by unnatural, mutant, alien forces, has its raison d'être as the maladroit monster standing beside the agile children of Enlightenment, corruption against innocence, that has now been recapitulated as technology versus human. The problem of capital accumulation through accounting for the death of certain bodies is not new. Just as the transatlantic slave industry incorporated Africa mortalities into the costs of labor extraction from them — establishing through them a network of industries including insurance, banking, and accounting to both quantify and assign a standard value to death — so too are other species also prefigured within that scope of production. Collectively they figured as livestock and contributed to a wider science of cultivating life and, by extension, forestalling mortality. The counting of bodies becomes at a certain point synonymous with registering their existence and, in so doing, reveals the avidity with which capital seeks to factor both life and death into its operational strategies that allow for its conversion into capital and equally as a marker for equity, as an uneven principle.

---

3   Ibid., 135.

Diversity factors in here to "include extra-human natures" that are the products of "volatile generation," "radical simplification and molecular manipulations" and, therefore, represent "the rising capacity of extra-human natures to elude capitalist discipline" and effectively promote the destruction of human populations, a scenario "that is made apparent in the rise of superweeds resistant to pesticides, superbugs such as MSA staph infections resistant to antibiotics, avian and swine influenzas proliferating in factory-farm conditions, and manifold cancers and autoimmune disorders accelerated by toxification and pollution of environments."[4] The depiction of the revolt of extra-human nature is often reminiscent of earlier projections of a future slave revolt that inevitably will arrive in the shape of mass retribution, communicable disease, or malignant outgrowth resulting in economic, biological, and ecological collapse. Thomas Jefferson understood in his natural philosophy treatise, *Notes on the State of Virginia*, that if colonial Virginia were to collapse as a society due to the overwhelming of its white population by the over-abundant growth of its Black extra-human population, the land itself would survive, representing the resiliency of material environments over epistemological ones that uncannily contribute to their economy's making and remaking.

Jefferson's American, provincial perspective complements the problematization of narratives of Black resilience that had perhaps already existed for over a century in Europe. The disproportionate and often involuntary injury, disease, and fatality shouldered by black(ened) bodies in carrying out the labor associated with colonization fails to fully tally the idea of the Black body being closer to the earth and therefore, more resilient in its handling of past, present, and future planetary volatility. Yet, this does not impede the possibility it could extend its reach further into projects of neocolonial capitalism into outer space. Donald Trump's Space Force acts a yet another playing out of

---

4   Sharae Deckard, "'Open Veins': Latin America in the World-Ecology," in *Ecological Crisis and Cultural Representation in Latin America*, eds. Mark Anderson and Zélia M. Bora (Lanham: Lexington Books, 2016), 16.

that same tired loop, involving the systematic acquisition of space, the progress of military doctrines that align with claims to legally organizing it, and for the presentation of power to take place firstly, through pre-emptive combatant and, secondarily, through the command of reality. One of the darker aspects of space exploration is the wholesale revival of a settler colonialist pseudo-militarized plan to "populate" the solar system. The chartered privateers, in this instance, are Jeff Bezos and Elon Musk, who will sell extortionately priced tickets aboard their sailing vessels, once again, privileging an elite merchant class to be the progenitors of New Worlds. Similarly, it is anticipated that the heavy lifting of this enterprise will be done by a lesser human class of indentured servants who will accompany them on this mission, working off the price of their tickets to ride once again through years of hard graft.

A radically depleted earth will be left behind like a recuperative patient, waiting for the lifesaving materials from the mining of other worlds to come back to it to revive its fortunes through the promise of what appears, once again, as infinite resource. The brutality of heavy industry will once again stand at a remove from its imperial consciousness, as earth, at a certain artificially elevated level, becomes a place of residence and commercial amusement, whereas space will be where all the filthy aspects of interplanetary economy are, perhaps rather predictably, conducted. The earth, as the corporation's birthplace, persists as the "site of neoliberal fossilization and remainder," where previous boom time was always associated with the progressive obsolescence of land, then people, then factories, then laptops, then cell phones, and so on, of which objective natures were destined to return "to a kind of pre-history when they fell out of circulation, at which time they could be examined as resonant material residues — fossils — of economic process."[5] The fate of the posthuman within this geological sacrifice zone persists within a landscape that uncannily "registers the terminal, but not yet ter-

---

5   Peter C. Little, "On the Micropolitics and Edges of Survival in a Techno-capital Sacrifice Zone," *Capitalism Nature Socialism* 28, no. 4 (2017): 66.

minated, life of digital technologies — a space where the leftover residue of electronics manufacturing accumulates" alongside its organic counterparts.[6] The posthuman endures through its vitalist hybridity and through its inability to cease fully.

Narratives of resiliency and implantation are troubled by this inclusion of a technological into the neoliberal capitalization zone of sacrifice through adulterating its methods of accumulation to include materials that are, at a micro-orgasmic level, unable to be assimilated reliably. Their mutability is what makes common cause a micropolitical ecology that refuses to settle as the byproducts of technocapital negligence but rather one that migrates toward an eventual disruption of the terms of nature, society, and productivity embedded within the concept of sacrifice itself. Earth numbers merely as planet number one amidst a multiplicity of planetary infrastructures now kitted out to coordinate the functions of microcomputing, nanotechnology, biotechnology, epidemiology, and climatic engineering within other worlds. Neoliberalization's goal from the very beginning was to aim for the stars and guide earth toward a new era of imperial ascent through a combination of militant conquest and lifeworld domination. Within this commercial constellation, dehumanization couples with reindustrialization and the mitigation of risk for those dwelling on earth, compared with its exaggeration within the off-world colonies.

Willie Jamaal Wright argues for more attention to be paid to "the landscapes onto which freedom dreams are sutured and, furthermore, how the characteristics of certain landscapes might make them amenable to those seeking to escape racial oppression, gender-based violence, and capitalist exploitation."[7] Counted amongst these must be "dejected landscapes," where seemingly beleaguered environments "might offer the space and allegiance necessary to create new cultures, new societies, and new worlds antithetical to the exploitative aims of the agents of

---

6   Ibid.

7   Willie Jamaal Wright, "The Morphology of Marronage," *Annals of the American Association of Geographers* 110, no. 4 (2020): 1134.

capitalism."[8] Here, Wright suggests that something recuperative be made of ecology to further the project of an Afrofuturism on this planet. This opens the possibility of a new strategy for understanding i/Indigeneity as a force that is capable of loosening white humanity from its artificial bond to the New World. The category of human from the beginning was alien and, indeed, hostile to this landscape rather than it being the other way around.

The embodiment of i/Indigeneity can only take place as something cut out from the context of its surrounding ecology through the forced imposition of technology as evidenced by the construction of "railways across the bush, the draining of swamps and a native population which is non-existent politically and economically" and that "in fact" they are "one and the same thing" meaning that all are done to temper the movement of this material form counter to the dematerialized life of the migrant settler.[9] The arrival of the settler creates a "bloodthirsty and pitiless atmosphere", affecting "the generalization of inhuman practices," leaving Native inhabitants with "the firm impression that people have of being caught up in a veritable Apocalypse."[10] This comment implies that for Native people, especially in the "New World," history is something that begins after the end of the world. With it came what Frantz Fanon referred to as "the disintegrating of the personality of the native, whose psyche was subjected to a 'splitting and dissolution,' such that it took on a primordial function in the organism of the colonial world."[11] Fanon's words suggest that the Native mind was subsumed into the larger entity of the colony. As a result, it bore various markings of alienation upon their thinking, and their sense of reality assumes itself in the pattern of the settler concepts of transformation, violence, and freedom.

---

8   Ibid.
9   Frantz Fanon, *The Wretched of The Earth*, trans. Richard Philcox (New York: Grove Weidenfeld, 1963), 250.
10 Ibid., 251.
11 Ibid.

Fanon asserts that "colonialism is not satisfied merely with holding a people in its grip and emptying the native's brain of all form and content."[12] It has to go that bit further, not by acting on the present or future but through the destruction of a people's past in denying them a recorded history specifically as human beings. There is no precolonial time afforded to their figuration as a people because to grant them this privilege was to instill them with the capacity for authority. What was required instead was a classification of them *otherwise,* a classification that would achieve "the total result looked for by colonial domination [...] to convince the natives that colonialism came to lighten their darkness."[13] This concept was not there to instill a sense of historical time within a people nor was it to foster greater awareness within them; rather, it was to make them aware that, without recourse to the settler's superior mind, their world would rapidly devolve and all "at once fall back into barbarism, degradation, and bestiality."[14] This meant that ordered thinking became inextricably paired with brutal discipline. The Native body required censure at the essential level of its physiology, its biology, and its reproductivity. For it to function as a colonial property, it had to be extruded from its ecology so that the body became a locus of meaning within the modern political and economic environment. Precisely due to its exclusion from its ecology, the body was located amongst whiteness within a larger corporeal schematization of difference and racialization that repositions a turbulent nature rather than a raucous incomer, as the alien constituent threatening the development of society.

The posthuman as a new form of ontological being cannot be disarticulated from the impossibility of ecological survival. That strange positionality is what ultimately enables black(ened) bodies to temporarily dwell within the uninhabitable and liminally persist in and amongst uneven colonial geographies. Such spaces are waiting to be translated into places and for what was

---

12 Ibid., 210.
13 Ibid., 210–11.
14 Ibid., 211.

previously deemed nonexistent to be realized categorically within themselves as livable and, therefore, live domains open to the potential to be both seized upon and profited from by colonialism and for its former Others able to otherwise register their potential. Indigeneity and ecology were made mutually exclusive through a former architecture of colonization. In a contemporary globalized world, they remain constructed as uninhabitable specializations expelled from human occupancy. What remains to be claimed categorically by the Other, namely the unaccounted-for resourcefulness of colonized peoples to access those deadened terrains and generate new conditions of existence beyond the expression of dispossession. The question becomes: will these be limited to the terms of materiality and ownership that we have often assigned to the use of such geographies? If the period between the dawn of formal decolonization and the present continues to register as an era of progressive dematerialization, the question becomes: what forms of ownership function as the remainder of those previous economies in terms of diasporic inhabitation, tactical bodies, data streams, emotional currency, and financial trajectories that make it possible to anticipate the resurgence of i/Ingenious forms of appropriation amidst the revival of dispropropriative forms of exclusion?

Claims about being vital have their evidence rooted in an experience of unbelonging caused by currents of racism. They find their energy within the denial of rights, their means through social and cultural invisibility, their acknowledgment in the now as opposed to then, here as opposed to there, and in slavery as opposed to freedom. The more significant question here is to what degree that was always already a feature of Great Britain and her colonies in America and the Caribbean? This multidimensional geography unravels the conception that they are here because we were there. It forces an admission of Blackness into the presence of a Britishness that precedes it perhaps by thousands of years, depending on your relationship to the concept of time. There is no here nor there about that exactly; rather, this raises an interesting problem about our assumptions of Black slavery as the ballast for the appearance of Blackness within the Atlantic world. It unmoors history itself when we are compelled to think of colonization and decolonization as one long song, articulated through instances that transcend any bounds of a versus. It must rather come to us in the form of a version, involving not merely the geographic transfer of millions of people, but equally the subsumption of their energies and the forced progress of elements across planetary space so that the very notion of persistence gets disturbed.

What remains thereafter in terms of a material and conceptual relationship amongst these things becomes the stuff of multiplicity and calculation and requires the processing of chronology in ways that allow for multiple centuries and geographies to pertain in our critiques of violence; our studies of these crossings replete with navigational baggage and yet still rife with the potential for missed crossings in each moment. The movement of these remnant energies is not something often dealt with, after all. It is the ultimate immateriality with a theoretical time when contemporary ontological theory is frenzied around the material. Yet, the unquantifiable power of those phenomena demand to still be iterated and, in so doing, to make way for "trauma" to encompass something much more than it does as any one type of thing.

This returns us to the question of how we interpret the timing of violence, and also the queering of time involved in repeating the past to the point where it becomes something of a future tense. What part does that haunting repetition play in the functioning of neoliberal globalized space-time? of trauma's remembering of itself onto itself? Does broken earth echo broken bodies, or does some distortion make things alter their location depending on where and when that happens? By looking at the deindustrialized 1970s and 1980s as an alternative, traumatic focal point for the neoliberal Black Atlantic body, does this answer the call for multidimensional historical moments to be acknowledged and represented, to allow for different perspectives to be aired? Does its very manufacturing at a certain historical juncture rely on the denial of oxygen to the other that preceded it and would come after it, as in "I can't breathe"? Is it the elemental body (Grenfell's cladding) that has denied this ability, standing in for the raced body? Lastly, to what extent are we not recognizing what role contagion plays in producing these variants within the historical sequencing of reality?

This, in light of the advent of COVID-19 and the narrative of how black(ened) bodies are now being denied oxygen in ways that are completely uneven with their white counterparts, how they are subject to the pressures of overcrowding and poor ventilation means that coronavirus burns through their communities so much faster than others and becomes particularly problematic. How might artists, and indeed historians, respond in the present moment so that this part of a Black historical record is not compromised and that the present Atlantic world deals with the horrors of racialized realities in ways that create the potential for restored futures? How can this happen, not through redemption or recovery of those narratives, but perhaps by forestalling them altogether through different environmental practices? Can we start by acknowledging how they deposit these histories as matter?

Bone Clocks

As a category of life, the posthuman comes forth from an assumption firstly of social death. Its object is one of derivation insofar as it must grapple with how to craft a vital existence out of this as its starting material. Social death implies structural violence, and it speaks within a grammar of confinement through space and time that constrains the body in its location, tethering its progress within cartographic and temporal parameters that make it serve the continuation of a world beyond its natural limits. From this perspective, it is forced to make life from a position of disenchantment, estrangement, and incoherence. The template from which it is meant to draw itself into being is Anglo-American, white life. Posthuman being must range around this architecture and make of itself a series of arduous configurations of being.

For the posthuman to freely actualize itself, white life would have to loosen its coordinates to provide a social and spatial foundation for new ontologies to proliferate. This move would accede to the necessity of a complete end of its world, that is to say, the world as we know it. This world to come like the last one will come from "deep in the heart of this crater," and the posthuman will be a cypher for the inconvenient truth that "literally everything must be reinvented, starting with the social."[15] Achille Mbembe refers to the digital as "the new gaping hole exploding Earth" which makes us question once again where black(ened) men and women stood in that equation as "the Earth's flesh and bones, the physical and mortal body" against perhaps a whiten(ened)ness that always presumes it "will be freed of its weight and inertia" and, therefore, saved from all this mess, "cut away from biological corruption and restituted to a synthetic universe of flux."[16] It is humanity that believes all this is possible, but perhaps no one else.

The story that humanity tells itself is always a linear one where "primitive" Others are always cast as being situated outside of society and, by extension, out of time itself. As ancillary beings, the fully human are only required to recognize them within the realms of a structural foreclosure. Matthew Houdek and Kendall R. Phillips contend that this frame was not one of colonialism, per se, but something far more enduring: "the 'anachronization' of the racialized and gendered Other" to the degree that they were "closed off" from "the possibility for being or becoming otherwise."[17] Moreover, "conceptions of history, progress, evolution, and Christianity" all conspired as logics to reinforce control over time as a principal of producing reali-

15 Achille Mbembe, "The Universal Right to Breathe," trans. Carolyn Shread, *Critical Inquiry* 47, no. S2 (Winter 2021): S60.

16 Ibid.

17 Matthew Houdek and Kendall R. Phillips, "Rhetoric and the Temporal Turn: Race, Gender, Temporalities," *Women's Studies in Communication* 43, no. 4 (2020): 374.

ty.[18] The future is a product of these coordinates and very much remains about control of the domains of space and time. The American century does not, therefore, represent a discontinuation with European imperialism, but it does rather introduce the possibility for freedom and sovereignty to exist in coherence with a larger container. Hence, "time, then, and the time of the future, in particular, became an abstract, unoccupied space for the projection of (white, masculinist) national fantasies and aspirations."[19] For Jefferson, America's national progress and expansion "into the abstract time of the postracial future," are wholly dependent on the "conjuring of the collective hallucination of whiteness" that would endure alongside the dreams of its founding fathers for time immemorial.[20] Whiteness was the imaginary territory upon which the others would be based, whereas Blackness was the delusional property "justifying and producing ongoing conditions of settler violence, oppression, and exploitation."[21]

In many ways, America casts itself as the land before time where refinement only comes with the appearance of white Europeans onto the scene of its untamed and unclaimed wilderness. The United States was formed in time with reference to its supposed founding year of 1776 and, thereafter, through its constant invocation of itself as a New World operating independently of Europe's old ways of apprehending time. With modernity acting as the new baseline for understanding the unfolding of time, America's founders increasingly relied upon "competing constructions of time, to justify their exceptionalism as compared with both European and Indigenous temporalities […] cast as a historical anachronism, as a means of effectively overtaking their potential interests in favor of their own governance over the future."[22] European and Indigenous peoples were made

18 Ibid.
19 Ibid., 375.
20 Ibid.
21 Ibid.
22 Ibid.

yesterday's men and Africans made people, whose progress was so delayed it might never conceivably come.

Indigenous and African presence within society was dealt with literally by freezing their opportunities to coexist with white American settlers by identifying their presence within colonized space as anomalous and aberrant to national belonging within what were, in effect, newly occupied territories, which thus created the conditions of settler futurity through the denial of that capacity in others. This has allowed the persistence of African Americans in America to become synonymous with slavery and Native Americans to become synonymous with massacre within national memory. Historically, their situation functions to make their respective present and future prospects appear to be fundamentally blighted by these burdens as peoples destined never to make progress due to these catastrophic events in their ancestral lineages. Nothing whatever is mentioned regarding the fate of the white American perpetrators of these atrocities and the burden their brethren might continue to bear because history itself is structured to omit their mention and, therein, erase their culpability from the record of these events.

Concerning the issue of slavery in particular, it is rehearsed constantly within the public discourse of a civil war between the northern and the southern states in the mid-nineteenth century, the result of which — burying the formal institution of slavery — effectively founds the United States as a modern nation but one that rarely engages with the temporal role slavery played in the making of the first United States in the century before that, or one in which the temporal role mass murder played in the making of the first colonies in the century before that. The peculiar institutionalization of 1776, again, as America's founding year, makes for a strained patrilineage that continues to call for rigorous disavowals of what actually took place in the centuries to follow leading up to and including this current millennium, making it therefore possible to train attention towards vital aporias in this narrative requiring of Americans to deny their involvement with time itself; that is to say, time as it relates to their historical multiplication and amplification of racist and gendered violence, homegrown terrorism, and colonial atrocity as an imperial nation.

All of this instills within America a fundamental inability to account for herself as something other than as a power largely maintained to assume the forward orientation of a European logic of whiteness and hegemonic masculinity. This requires that time be fundamentally out of joint between space and time. Such is the case that whiteness would rather contemplate the seemingly unbearable reality of an untimely death of the Earth that is now said to be firmly on the horizon rather than to radically reimagination its centrality of place within the socio-geophysical constellation of the Anglo-American system. Therefore, Vote Leave and Make America Great Again were feverishly conjured as nationalistic fantasies to avoid facing the interruptive social transformations taking place within these respective states that threatened to shatter the illusion of a singular and teleological understanding of the unfolding of modernity's history. This brings to mind Derrida's work on Hamlet, Spectres of Marx, which forces us to acknowledge that "the time is out of joint," and can only be restored when we realize justice itself as a futural act whose appearance will likely never be seen by any those made Other to this scene, coerced to occupy the present as it is rather than as it should be.[23] In such a space, Blackness must move as a plurality to avoid detection and disorganize itself to achieve the effect of a haunting of the system. This is what is meant by "Afropessimism."

The Being That Never Was

The specter of Blackness operates in the New World as transitional figuration that emerges from the traditions of scholarship associated with natural philosophy and transported into a context "where social, political, philosophical and technological

---

23 Jacques Derrida, *Spectres of Marx: The State of Debt, The Work of Mourning, and the New International,* trans. Peggy Kamuf (New York: Routledge, 1994), 18.

upheavals posed a challenge to established certainties."[24] The arrival of African bodies as vital cargo onto the shores of Virginia plays into the narrative of turbulence threatening to knock, as it were, reality off its hinges by creating times that grant America's first settlers' temporary license to "apply a multitude of perspectives to older histories, narratives, and themes."[25] In so doing, they generate corridors within reality for speculative figures to emerge that simultaneously carry forth the principles of the past alongside the prototypes of the future. At this rare moment of founding, there exists within America a condition of "'plurivocity' that very shortly thereafter will be forced to gather under one united roof" and effectively sing from the same hymn sheet.[26]

Elizabeth I's Christian settlers believed that it was only possible "to find a rule of cohabitation under such a roof" if it were "understood that this house will always be haunted rather than inhabited by the meaning of the original."[27] The design of the original "was to gather under a single roof the apparently disordered" animal-humans among them.[28] These disordered beings were to be, in effect, counted as things in order to authorize each one of these posthuman beings through their translation into English custom and registration under English law, thereby making their appearance "possible and intelligible without ever being reducible to them."[29] This enterprise "now extends in the 21st century to emergent and increasingly interactive screen-based mediums, such as augmented reality performances and social media."[30] These platforms concern themselves with the manipulation of time and space, and with the convergence and interconnection between worlds in a way that extends the commercial preoccupations of Elizabethan empire into post-digital

---

24 Julian Novitz, "'The Time Is Out of Joint': Interactivity and Player Agency in Videogame Adaptations of Hamlet," *Arts* 9, no. 4 (2020): 1.
25 Derrida, *Spectres of Marx*, 20.
26 Ibid., 25.
27 Ibid.
28 Ibid.
29 Ibid.
30 Novitz, "'The Time Is Out of Joint'," 2.

real-time; a situation where the boundaries between past, present, and future are blurred and intertwined and a stage where all subjects are governed by a logic of iterability.

Julian Novitz observes that "*Hamlet* is a play where the present is infected by the past in the form of the specter that puts the narrative in motion, where the indecisive protagonist suspends and repeats time to forestall an inevitably tragic future."[31] The play is entitled *Hamlet,* but he is not the only player nor the only element on the board within the space of Elsinore castle. Other characters are present that possess various "ethnic, gender and sexual identities," as well as their "own histories of familial abuse and neglect."[32] If these characters' agency is allowed to prevail, the consequence is one of the new possibilities being introduced into the outcome of the play, meaning that the hermeneutic world of *Hamlet* must end for these others to admitted into acts of interpretation alongside his. Laertes, Gertrude, Bernardo, and Ophelia, as equal players, pose different possible timelines and introduce difficulty into a world where simplicity had former reigned.

A course of action arises where "Hamlet is stalled as a character due to his uncertainty and trepidation," while others, like Ophelia, become "trapped by the ever-expanding multitude of interdependent choices available to them, where each playthrough is haunted by their awareness of alternative time-lines — the possible future outcomes that they have experienced in past play."[33] This is incredibly significant because it suggests that "Ophelia is the spectral figure amongst this doomed pairing and that she has been drowned in choice, while Hamlet prevaricates in the face of the awful future he is moving towards."[34] Hamlet's dread of the future mirrors our own insofar as it "anticipates that choices are increasingly dominated and constricted by [our] awareness of what is coming"; meaning a cli-

---

31 Ibid., 12.
32 Ibid.
33 Ibid., 13.
34 Ibid.

matic end of Anthropocenean time.[35] "Each repetition" of its inevitable narrative "further populates [our] collective timeline with future events that may occur because of [our] action or inaction."[36] We are "gradually made aware that there is no ideal or optimal outcome […] to discover" and that "every potential ending will result in death or unhappiness for at least some" of the world's population.[37] Those peoples, for dint of their race, gender, sexuality, or ability, will find themselves "trapped in a time-loop" dictated by a techno-scientific master narrative that is determined to ensure every potential outcome of the world that courses through this geological era "is tragic or bloody in some way."[38] Those same limited chorus of voices, similarly, are telling humanity, as a whole, that it is deprived of the ability "to definitively solve or 'win' in a scenario"; rather that it is up to only the world's most elite players "to decide what kind of tragedy they are willing to accept as a conclusion to the narrative."[39]

It is not the father, but the son, who gets to exercise first refusal here, a Hamlet-type, who insists upon some variation of an unhappy ending through the continued literalization of the peoples around him as helpless or passive and, thus, unable to meet the challenge of the future or to apply their own judgment in crafting at an ending. Going further still, they are deemed incapable of constructing a new beginning for themselves that would allow for growth and self-knowledge beyond the long projection of their simplicity as mere objects to manipulate and alter at will. When the time is "out of joint," the narrative against expectations shifts and subsumes Hamlet's world of privileged whiteness to the degree that it becomes possible for other worlds to branch off from this one, suggesting the possibility of the emergence of other problems that bring forward the spectral qualities of the past to bear on Blackness as the white world's source material. Without infinite access to that resource,

35   Ibid.
36   Ibid.
37   Ibid.
38   Ibid.
39   Ibid.

another posthuman version of the world is made plausible. Lee Edelman is right to cast *Hamlet* as a play about survival and about how the death drive is "recurrently projected onto those who occupy the position of the queer: those abjected as non-reproductive, anti-social, opposed to viability" that invite their categorical expulsion.[40] Edelman holds that queerness "occupies the place of the zero, the nothing, that invariably structures the logic of being that remains at once intolerable and inconceivable to it."[41]

In 2016, Simon Godwin staged a UK production of *Hamlet* for the Royal Shakespeare Company that queered the play by setting an imaginary postcolonial domain where "different kinds of diasporic blackness engaged with each other through the figure of Hamlet and his art."[42] Sujata Iyengar and Lesley Feracho assert that Godwin's deliberate choice to not specify the location of this presumably postcolonial African state afforded "the production mixed multiple referents of blackness (Eastern African, West African, Caribbean, South African, 1970s African American" to allow audiences to "examine how the concept of race changes with the transatlantic or transnational movement among spaces in this production."[43] The contemporary production featured an almost entirely Black and multicultural ethnic British migrant cast, whose presence was transported backward into a newly imagined, liberated postcolonial African republic of the 1970s. From the offing, it is clear that Hamlet, while Black, is still destined to be the educated product of the Atlantic world. As the play commences, we witness him receiving his degree at Wittenberg University, in Ohio. Unlike his peers, his progression is rudely interrupted by evidence that he himself is the pro-

40 Lee Edelman, "Against Survival: Queerness in a Time That's Out of Joint," *Shakespeare Quarterly* 62, no. 2, (2011): 148.
41 Ibid., 149.
42 Sujata Iyengar and Lesley Feracho, "Hamlet (RSC, 2016) and Representations of Diasporic Blackness," *Cahiers Élisabéthains* 99, no. 1 (July 2019): 147.
43 Ibid.

duction of a failed state where a coup has just occurred, forcing Hamlet backward toward his troubled country of origin.

Once there, it becomes apparent that all those surrounding him are the products of a similarly disordered relationship to space and time. Hamlet's father, dressed in traditional West African attire, is a relic of the evolutionary past with regard to his country's relationship to modern statehood. His contemporary, Laertes, aligns himself with an Afro-funk style suggestive of an aesthetic affinity with Afrofuturism. Senior figures like Claudius and his ghostly father Fortinbras, elaborate the need for a constant appearance of military might in order to prop up their postcolonial status. When these figures are grouped, the whole scene becomes carnivalesque. The rhythmic, ritualized, and mimed atmosphere of the play is danced into precarious existence upon the stage. All of this hectic movement is choreographed presumably for the pleasure of a predominantly white audience: the stage is littered with the rarefied graffiti art of the Black American artist Jean-Michel Basquiat. Even here, what passed as subversive iconography, is speculatively mediated by the specter of an elite whiteness. In this context, that figure is Andy Warhol.

The actors' accents function to translate what is taking place center stage through eccentric variation. Everything "from the Jamaican patois of the male gravedigger who deployed a human femur as a mock-microphone to the upper-middle-class British English of Hamlet's white college friends, Rosencrantz and (a female) Guildenstern, to the African-accented English of Marcellus in the opening scene and the London intonation of Gertrude and her son" is there to signpost an understanding of them being here because "we," meaning English white settlers, were there.[44] This includes the casting of Hiran Abeseyekere as Horatio to acknowledge "the large historic presence of Indian or South Asian immigrants in both the Caribbean and post/colonial African nations, turning Claudius into a potential ghost of Idi Amin," who, of course, declared himself an uncrowned King

---

44 Ibid., 148.

of Scotland as part of the deal of becoming Uganda's first post-colonial ruler.[45] Amin infamously expelled British Indian and South Asian immigrants in 1972, as part of his efforts to rid the country of any remnants of British colonial rule, by returning control of commerce and administration to native Ugandans.

What Amin was doing was effectively world making, not by altering or expanding the geography of Uganda, as any neo-imperial ruler might do, but by returning it to its solely African setting. He does this so that he might begin to create a new history to counter the one previously imposed on the country through Britain's colonial administration of it as a protectorate. Race is never taken out of the equation, per se, but it is accounted separately here for white history. Amin fashions his homeland out of a combination of will and imagination rather than defining it in formally geographic terms, as was the case with land demarcations of the colonial past.

Hamlet was never going to be this sort of revolutionary ruler. In the play, he is a product of the African aristocracy with the Black diasporic world at his command. He is an absent leader preferring to spend his youth abroad in the United States where his wealth, education, and foreignness effectively insulate him from the racism typically experienced by native African Americans. He is compelled to return home by his father's untimely death. Once there, he finds that his perspective on his Old World has been utterly transformed by his contact with the New World. Hamlet is fundamentally altered by that comparative perspective, which rattles his former certainties and alienates him from his origins.

There is no here or there within the world of neoliberal globalization for which Hamlet has now become the consummate product, a member of the cosmopolitan elite who believe themselves to citizens of the world that, as Theresa May famously remarked, means they are citizens of nowhere. Amin would admire May's observation that for the Hamlets of this world, citizenship means nothing and, therefore, they are hardly de-

---

45 Ibid.

serving of it. So, it goes that Hamlet keeps on struggling, on endless repeat, past the point of death to understand his place in a neoliberal world order where he faces a series of poor choices: to either submit to a state of perpetual exile or rage against the injustice of his impossible situation. Neither of these activities are world building, and for this reason, Hamlet becomes an unreasonable player, someone unable to grasp the moral dilemma of his times, to be or to not be human, that is the question. This question remains a superficial one because Hamlet's question is a derivative one. The neoliberal Black version of Hamlet is cast as a painter of huge graffiti-style canvases. These feature dinosaurs, lizards, and skulls, most wearing crowns, suggestive of some relational awareness of his place within a larger systematic classification of life and value. It is still a white audience that act as his external examiners, however invisible their presence may feel against the weight of his decision making within this projected setting. Race and the British state still have a very complicated relationship and require deeper redress.

As Ha-Joon Chang argues,

> since the late 1970s, there was a consensus that development is largely about the transformation of the productive structure (and the capabilities that support it) and the resulting transformation of social structure — urbanization, dissolution of the traditional family, changes in gender relationships, rise of the labor movement, the advent of the welfare state, and so on. This was mainly (although not exclusively) to be achieved through industrialization.[46]

After neoliberalism became the normative economic paradigm, the term "'development" came to mean poverty reduction, provision of basic needs, individual betterment, sustenance of ex-

---

46 Ha-Joon Chang, "Hamlet without the Prince of Denmark: How Development Has Disappeared from Today's Development," in *Global Governance at Risk*, eds. David Held and Charles Roger (Hoboken: John Wiley & Sons 2013), 130.

isting productive structure — that is, anything but 'development' in the traditional sense."[47] Curiously, Chang refers to this phenomenon as "*Hamlet* without the Prince of Denmark," implying an event that takes place despite the absence of its principle element.[48]

In 1607, a production of *Hamlet* was staged on board an English ship moored off the coast of West Africa.[49] Hamlet's course "had literally proceeded along the arteries of Empire," insofar as this performance took place in the same year that the colony of Virginia was founded in America.[50] England's formal involvement in bringing slaves from West Africa to Virginia would not formally commence for another seventy years. Accordingly, *Hamlet*'s arrival in West Africa was an exercise in "performing English in distant locations."[51] Peter Erikson argues that "Englishness," at this time in history, remained something strictly enacted at the level of performance rather than it being a code word for "political force."[52] In this context, what was being enacted was "whiteness" for its own sake. In the early modern world, whiteness, not Blackness, was the first racial category. During Shakespeare's lifetime, whiteness would remain very much a nascent form, "unstable and in flux."[53]

In Shakespeare's plays, whiteness was portrayed "as flawed" and "on the defensive."[54] It was whiteness that created "an active sense of insecurity and anxiety" within the action of what takes place on stage.[55] West Africa provides an exegetical space through which to elucidate whiteness as a negative characteristic. It was whiteness set against Blackness — figured here as the threat of disappearing into obscurity — that enliven the feelings

---

47 Ibid., 131.
48 Ibid.
49 Peter Erickson, "Can We Talk about Race in 'Hamlet'?," in *"Hamlet": Critical Essays,* ed. Arthur F. Kinney (London: Routledge, 2001), 210.
50 Ibid.
51 Ibid.
52 Ibid.
53 Ibid.
54 Ibid.
55 Ibid.

of vulnerability, weakness, and indecision that led to Hamlet's crisis upon returning to his country of origin, Denmark. All of these qualities acted as plagues upon Hamlet's lively body, preventing it from coming into full (well) being. Hamlet's salvation from his father's ill fate is premised on the preservation of this whiteness against the foreign blood that threatens to taint it. Danger is cast onto the scene through the arrival of the Norwegian Prince Fortinbras, exposing the degree to which the family line is corrupted by the expression of materiality and sexuality within the bodies of all three of these noblemen.

Fate in Hamlet is not determined by life, but death, and more specifically, murder, which cuts at the arteries of racial ascension. Emily J. Barlets observes that in this time in history, "race is a terrain for royal men only."[56] Hamlet is both the king and the prince, the father and the son, as is Fortinbras. What unites them are channels of attack, murder, and revenge all leading back to Elsinore castle and the scene of Prince Hamlet's failed ascension. *Hamlet* without the Prince of Denmark leaves behind not a ghost, but an endless looping of that same failed pathway. Hamlet's late declaration of himself as a Dane, rather than as a prince, comes at a time when he becomes aware that Norway will soon colonize Denmark through military conquest. The term "Dane" announces a particular geography to which Hamlet assigns himself only when facing almost certain ruin as the country's abortive leader. Denmark is a prison because it confines Hamlet to a particular fate and a particular set of duties that he is temperamentally unable to fulfill. He simply cannot bring himself to identify as someone who relishes conquering other lands.

His claim to Danishness only takes place once his on Danish soil, when abroad he assumed himself something else entirely, as a "Dansker," a sort of diasporic identification that allows him to travel from the soil of his birth without retaining an exact

---

56 Emily J. Bartels, "Identifying the 'Dane': Gender and Race in 'Hamlet,'" in *The Oxford Handbook of Shakespeare and Embodiment: Gender, Sexuality, and Race,* ed. Valerie Traub (Oxford: Oxford University Press, 2016), 203.

likeness to it. Unlike Fortinbras who has a strong relationship with Norway, Hamlet's national identity only truly flourishes in self-imposed exile. Hamlet is England's first iconic diasporic character. Ineffectuality and melancholia are the product of his rootlessness. They are evidence of the corruption of his loyalty to the state and preference for foreign allegiances. This is represented by his friendship with Horatio and, by extension, his loyalty to Wittenberg University. Through both relationships, Hamlet learns effectively how to become human. Upon his return to Elsinore Castle, Hamlet finds he has no remaining desire to become Prince of Demark; rather he aspires to become Horatio.

Therefore, Hamlet reenters Elsinore, not as its heir apparent, but "as a university student trained in devising arguments, forming syllogisms, and declaiming in disputations"; and as such, "he transfers all these skills learnt in Germany's seat of Protestantism to the bloody events at hand when he reenters" Catholic Denmark.[57] Andrew Hui makes of Horatio a prototype of the first humanist historian: "Horatio partakes in the structure of discourse in which storytelling, circulation of information, and memorializing are of the utmost importance and in which he will emerge as the final voice. What has happened in the play and how it will be represented must be reconciled with Horatio's memory."[58] Such is the magnitude of his authority, that, by the end of the play, it is he alone who is able "to turn Hamlet into *Hamlet*," to make a zero out of a one and to queer history to the effect that it has no power beyond that of continuous enunciation.[59] It is all there in the name, Horatio as, an "aspirated — a living, breathing oratio."[60] Horatio is a living thing that makes the life of other things possible — or in the instance life after death possible by giving over one's voice, one's breath to another as Hamlet does to Fortinbras to keep his story going.

---

57 Andrew Hui, "Horatio's Philosophy in 'Hamlet,'" *Renaissance Drama* 41, nos. 1–2 (2013): 153.
58 Ibid.
59 Ibid.
60 Ibid.

Something Rotten in the State of Being

If Horatio lends air to Hamlet's troubled thoughts, so too he does his unwholesome thinking that cannot readily separate mind from environment, inspiration from respiration, ventilation from expiration. As a consequence, Hamlet senses deadly enemies through their vital spirits with whom he is compelled to share the castle's rarefied air space. A king, a prince, a commoner share in common a susceptibility to plague. The something that is rotten in Denmark is the bodies that have fallen prey to it. Their lungs are but versions of the "fans, bellows, windmills, guns, mines, sails, powder houses" as technologies all vitally dependent on air.[61] The commercialization of air made of these men, including Hamlet, consumptive machines, "imagining themselves or others as mechanical objects or forms of artificial life."[62] Hamlet's is a toxic and infected world of state power and foreign capital, much like our own in this century. Air mattered as much as it does today, as a medium of information and communication, and equally of pestilence and corruption. Hamlet may indeed be the first Anthropocenean play insofar as it forces its audiences to confront "a decayed world," riven in "climatological corrupt," and to imagine ways to live on past such dire conditions by "consuming the dead" as means through which to interpret the greater contagion of this world.[63] Hamlet can no more outrun his suffocating fate at the beginning of this performance than we can at its end stage.

The word *ham* in Danish means skin. It is the very place that Hamlet cannot imagine himself feeling at home in. Perhaps this is the case because this, overall, is a time of transmutation where the center cannot hold and materiality melts, like Marx famously said, into air. Hamlet's unnatural birth comes from Horatio's mouth. It is therefore his skin, not Hamlet's, that voices the fi-

---

61 Carlo Mazzio, "The History of Air: 'Hamlet' and the Trouble with Instruments," *South Central Review* 26, nos. 1–2 (2009): 158.

62 Ibid.

63 Ibid., 177.

nal pronouncement of death and in that same breath transmits his legacy solely into performance. The project of colonization might be viewed as the product of just so many failed dramas, just so many rough attempts to instrumentalist life. All of which came at the cost of the body. By turning parts of it into a non-human entity, it allowed others to simply rush past into a ter-rifying unknown future, one that ultimately screamed out for understanding. The early modern world was just the start of these venting problems that would engulf and then drown all these many centuries to come. Imagine Hamlet standing there as one of its first raced body, smoldering amidst all its potential atmospheric casualties, pretending that it was he alone who had social agency as one of its rarefied specimens modelled to repeat a thousand times over its ambivalence; its whiteness a technol-ogy out of control.

Few recall that Hamlet was an adolescent form of whiteness and that it was predicated to harden and mature into what his father's apparition suggested. Fenn Elan Stewart suggests "that *Hamlet,* through the interventions of its main character, thwarts the assumption that the relationship between a nobleman and his land is natural, that the desire for possession and rule is inherent."[64] Hamlet's "wind and whirling words" make such a denunciation possible of conventional social relations in favor of "speaking to an unknowing world."[65] The complexity of that arrangement is evident in "Hamlet's relationship to Horatio as it crisscrosses the dense hierarchical network of vertical and horizontal relationships in the play — parent-child, sovereign-subject, divine-human, lord-servant, brother-sister, comrade-and-sentry, rival states" and, finally, finds itself in the territory of the posthuman because it is both associative and preoccupied with its inherent position of subordinancy.[66] This subordinan-cy transfers to his status as Hamlet's friend because he is there

---

64 Fenn Elan Stewart, "'The King Is a Thing': Hamlet and the Prostheses of Nobility," (PhD diss., University of British Columbia, 2008), 3.

65 Ibid.

66 Hui, "Horatio's Philosophy in 'Hamlet,'" 158.

merely as witness to Hamlet's radical transformation from an aristocratic member of "a martial class to a highly literate, governing" body.[67]

Elizabeth Hanson contends that "humanist ideas of friendship," were "called upon and even enacted in order to manage the contradiction between different protocols of ordering men."[68] In early modern England, the right and ability to learn afforded one the status of human. The nobility's incursion into the university could be construed as a prototype for settler colonialism insofar as their presence within this learning environment would have been wholly unknown prior to this era. Their conspicuous arrival coincided with the entrance of poor students, such as Horatio, onto the scene of higher education. This contact with other classes of men introduced a sort of heterogeneity into the aristocracy's conception of its own status. It permitted them to adopt an understanding of themselves as ranked beings within a new order of whiteness that was now formulating itself around this older institution. At the university, the well- and low-born convened, creating the specter of class contagion, conditions of which spoke to an unnatural intimacy breeding within this experimental environment.

Horatio's presence reminds him of his princely status and his protean identity as a king. This in itself is a racializing and engendering act. What is exceptional about his presence is his indistinction, as compared to Rosencrantz and Guildenstern, who are made obvious by virtue of their elevated social status. Horatio is no such thing. Neither is he gentle in his manner. It remains fundamentally unclear what value he presents within this environment. Does his standing within the castle diminish its fortification? Does it threaten to erase its traditional structure altogether? Where does that potential land with Hamlet? Horatio provides Hamlet with a template for the everyman of modern life who has the privilege to exist hypothetically, un-

---

67 Elizabeth Hanson, "Fellow Students: Hamlet, Horatio, and the Early Modern University," *Shakespeare Quarterly* 62, no. 2 (2011): 207.

68 Ibid., 208.

til he is actually recognized. Hamlet, by contrast, cannot enjoy such anonymity because he is always already actualized through his title, Prince. It is that title that literally places him within the walls of the castle, as opposed to the university. His title suggests an understanding that resides most greatly in particularly. What he longs for, instead, is the universalism associated with mankind, a belonging in every place within a category that values the attainable versus the transcendent, and, thus, appeals to Hamlet's dilemma to be or not to be *human.* It is this tangible status that is the very thing denied to Hamlet and his ghost of a father.

Horatio's clerical patrilineage carries no such burden, nor ambiguity, between materiality and immateriality, responsibility, and care. Horatio is strangely self-possessed, meaning that he is possessed by Hamlet's rank insofar as he can lay claim to selfhood, whereas Horatio remains at the level of pupil gazing up towards that level of actualization but never achieving it for himself — at least within the confines of the performative. Throughout the play, Horatio remains ancillary to the action, functioning as a basis for judgment, while not necessarily having discreet ownership of own his perspective. His position remains fundamentally baseless as is Hamlet's tarried remainder at university into his thirtieth year. This elongation of time seems to resemble some of the qualities that are missing in Horatio's character development, as a real late starter to the property of advancement and mature occupation. The king is a thing that Hamlet can't permit himself to become. The recording machine as a relation, instead, he must dearly possess. Hamlet's postmortem address forces us, as its meta-audience, to press play and bring into action once more his particular form of un/living that disjointed his present, making of him a problem of both time and materiality to be reckoned with at some future date; that time is now. Like Hamlet, we now dwell within "contagious ecologically precarious, at times unlivable spaces shaped by inequalities directly impacting the Earth system."[69] Even as Hamlet attempts to contemplate animals, plants, soils, and minerals, he does so as if those these lives are otherwise to his own. His perspective foreshadows a time when racialized and colonized bodies are seen as more belonging to the categories of life and are similarly judged from a contemplative distance as distinct things to the morose likes of him.

Proponents of posthumanism such as Bruno Latour savor the prospect of their own ecological and technological moment increasingly mirroring that of Shakespeare as he brings Hamlet into an artificial staging in the sixteenth century. In the twenty-first century, that action is brought on by a desire to lend animacy to the lives of all sorts of entities, without requiring the world in which they live to adjust in any way—other than backward to the pre-history of modernity. If such a thing were possible, a miserable Hamlet perhaps would be the last person to recommend it. Hamlet, as one of early modernity's great disabled figures, would fail when measured against today's new materialist bias "toward lively, responsive dynamism […] in the objects, animals, and humans it assembles."[70] On that score, Hamlet succeeds only in being devitalized by the others he encounters in proto-posthumanist ways that allow the viscera of matter to hold the world together, even when it shows itself to fall apart. Finally, Hamlet represents the dark side of early modern humanism, which displays itself in the belief that the inability to conduct oneself properly was a marker of a certain type of degeneration that would later emerge as categorical difference. The mass cultivation of the body would eventually synchronize with particular kinds of difference that, over time, would limit one into a lesser category of being. The posthuman takes off as a divergence from the path of upward mobility and off course from the route of expansive development. At the crossroad of these coordinates, it is possible to discern the relationship between

---

69 Steven Swarbrick and Karen Raber, "Introduction: Renaissance Posthumanism and Its Afterlives," *Criticism* 62, no. 3 (Summer 2020): 315.

70 Ibid., 319.

differentiation and discrimination as one of progressive alienation and constructive isolation.

Bibliography

Aho, Tanja, Liat Ben-Moshe, and Leon J. Hilton. "Mad Futures: Affect/Theory/Violence." *American Quarterly* 69, no. 2 (June 2017): 291–302. DOI: 10.1353/aq.2017.0023.

Ahuja, Neel. *Bioinsecurities: Disease Interventions, Empire, and the Government of Species.* Durham: Duke University Press, 2016.

———. "Post-mortem on Race and Control." In *Control Culture: Foucault and Deleuze after Discipline,* edited by Frida Beckman, 34–43. Edinburgh: Edinburgh University Press, 2018.

Aikens, Natalie, Amy Clukey, Amy K. King, and Isadora Wagner. "South to The Plantationocene." *ASAP Journal,* October 17, 2019. http://asapjournal.com/south-to-the-plantationocene-natalie-aikens-amy-clukey-amy-k-king-and-isadora-wagner/.

"AlterEgo: Frequently Asked Questions." *FAQs, MIT Media Lab.* https://www.media.mit.edu/projects/alterego/frequently-asked-questions/#faq-what-is-alterego.

"AlterEgo: Similarity Makes the Difference for People Suffering from Social Disorders." *European Commission,* September 10, 2015. https://ec.europa.eu/digital-single-market/en/news/alterego-similarity-makes-difference-people-suffering-social-disorders.

Ante-Contreras, Daniel Michael. "Autism as Metaphor: The Affective Regime of Neoliberal Masculinity." PhD diss., University of California Riverside, 2017. https://escholarship.org/uc/item/5jp50764.

Aronoff, Kate. "BP Is Not Woke. It's an Imperialist Success Story." *The New Republic,* June 6, 2020. https://newrepublic.com/article/158052/bp-not-woke-its-imperialist-success-story.

Auvinen, Tero. "Organizing for the End of the World with Nothing More to Say? The Implications of Non-consensually Implemented Singularity and 'Trans-/Posthuman' Technologies for Social Movements." *Journal of Alternative Perspectives in the Social Sciences* 8, no. 2 (2016): 209–47.

Babich, Babette. "Radio Ghosts: Phenomenology's Phantoms and Digital Autism." *Thesis Eleven* 153, no. 1 (August 2019): 57–74. DOI: 10.1177/0725513619863852.

Baldwin, Andrew, and Bruce Erickson. "Introduction: Whiteness, Coloniality, and the Anthropocene." *Environment and Planning D: Society and Space* 38, no. 1 (2020): 3–11. DOI: 10.1177/0263775820904485.

Baldwin, James. "Letter from a Region of My Mind." *The New Yorker,* November 9, 1962. https://www.newyorker.com/magazine/1962/11/17/letter-from-a-region-in-my-mind.

Baptiste, Edward E. "'Cuffy', 'Fancy Maids', and 'One-Eyed Men': Rape, Commodification, and the Domestic Slave Trade in the United States." In *The Chattel Principle: Internal Slave Trades in the Americas,* edited by Walter Johnson, 165–202. New Haven: Yale University Press, 2004.

Barkin, Kenneth. "W.E.B. Du Bois' Love Affair with Imperial Germany." *German Studies Review* 28, no. 2 (May 2005): 285–302. https://www.jstor.org/stable/30038150.

Bartels, Emily, J. "Identifying the 'Dane': Gender and Race in 'Hamlet'." In *The Oxford Handbook of Shakespeare and Embodiment: Gender, Sexuality, and Race,* edited by Valerie Traub, 197–210. Oxford: Oxford University Press, 2016.

Bascom, Ben. "Queer Anachronism: Jeffrey Brace and the Racialized Republic." *Arizona Quarterly: A Journal of American Literature, Culture, and Theory* 75, no. 1 (2019): 23–47. DOI: 10.1353/arq.2019.0001.

Bateman, Bradley W. "There Are Many Alternatives: Margaret Thatcher in the History of Economic Thought." *Journal of the History of Economic Thought* 24, no. 3 (2002): 307–11. DOI: 10.1080/104277102200004758.

Bates, David W. "Unity, Plasticity, Catastrophe: Order and Pathology in the Cybernetic Era." In *Catastrophes: A History and Theory of an Operative Concept,* edited by Nitzan Lebovic and Andreas Killen, 32–54. Berlin: De Gruyter, 2014.

Battle-Baptiste, Witney, and Britt Rusert, eds. *W.E.B. Du Bois's Data Portraits: Visualizing Black America.* Hudson: Princeton Architectural Press, 2018.

BCRW Videos. "Anna Lowenhaupt Tsing — A Feminist Approach to the Anthropocene: Earth Stalked by Man." *Vimeo,* December 18, 2015. https://vimeo.com/149475243.

Beck, Hamilton, "W.E.B. Du Bois as a Study Abroad Student in Germany, 1892–1894." *Frontiers* 2, no. 1 (1996):45–63. DOI: 10.36366/frontiers.v2i1.25.

Beller, Jonathan. *The Message Is Murder: Substrates of Computational Capital.* London: Pluto Press, 2017.
———. *The World Computer: Derivative Conditions of Racial Capitalism.* Durham: Duke University Press, 2021.

Berardi, Franco. *Futurability: The Age of Impotence and the Horizon of Possibility.* London: Verso Books, 2017.
———. *Heroes: Mass Murder and Suicide.* London: Verso Books, 2015.

Bhambra, Gurminder K. "Locating Brexit in the Pragmatics of Race, Citizenship and Empire." In *Brexit: Sociological Responses,* edited by William Outhwaite, 91–100. London: Anthem Press, 2017.

"Biography Arnav Kapur." *Stern Strategy Group,* n.d. https://sternspeakers.com/speakers/arnav-kapur/.

Black, Edwin. "In Germany's Extermination Program for Black Africans, a Template for the Holocaust." *The Times of Israel,* May 5, 2016. https://www.timesofisrael.com/in-germanys-extermination-program-for-black-africans-a-template-for-the-holocaust/.

Bledsoe, Adam, and Willie Jamaal Wright. "The Anti-Blackness of Global Capital." *Environment and Planning D: Society and Space* 37, no. 1 (2019): 8–26. DOI: 10.1177/0263775818805102.

Bould, Mark. "The Ships Landed Long Ago: Afrofuturism and Black SF." *Science Fiction Studies* 34, no. 2 (2007): 177–86. https://www.jstor.org/stable/4241520.

Bove, Alexander. "What Happens When Replicants become Extimate? On the Uncanny Cut of the Capitalocene in 'Blade Runner 2049.'" In *Lacanian Perspectives on "Blade Runner 2049,"* edited by Calum Neill, 139–66. Basingstoke: Palgrave Macmillan, 2020.

Brandt, Marisa. "Janelle Monae's Liberationist Posthuman Pop." *Difference Engines,* March 5, 2014. http://www.differenceengines.com/?p=1021.

Browne, Simone. *Dark Matters: On the Surveillance of Blackness.* Durham: Duke University Press, 2015.

Bryce, James. *The American Commonwealth, Vol. 3: The Party System and Public Opinion.* New York: Macmillan, 1888.

———. *The Relations of the Advanced and the Backward Races of Mankind.* Oxford: Clarendon Press, 1902.

Butler, Philip. "Making Enhancement Equitable: A Racial Analysis of the Term 'Human Animal' and the Inclusion of Black Bodies in Human Enhancement." *Journal of Posthuman Studies* 2, no. 1 (2018): 106–21. DOI: 10.5325/jposstud.2.1.0106.

Canny, Nicholas P. "The Ideology of English Colonization: From Ireland to America." *The William and Mary Quarterly* 30, no. 4 (1973): 575–98. https://www.jstor.org/stable/1918596.

Censer, Jack R., and Lynn Hunt. "'The Code Noir' (The Black Code)." In *Liberty Equality, and Fraternity: Exploring the*

French Revolution. University Park: Penn State University Press, 2001. CD-ROM.

Centre for Contemporary Cultural Studies. *Empire Strikes Back: Race and Racism in 70s Britain.* London: Taylor & Francis, 2004.

Chang, Edmond Y. "Technoqueer: Re/Con/Figuring Posthuman Narratives." PhD diss., University of Washington, 2012. https://digital.lib.washington.edu/researchworks/handle/1773/22631.

Chang, Ha-Joon. "Hamlet without the Prince of Denmark: How Development Has Disappeared from Today's Development." In *Global Governance at Risk,* edited by David Held and Charles Roger, 129–48. Hoboken: John Wiley & Sons, 2013.

Chen, Mel Y. *Animacies: Biopolitics, Racial Mattering, and Queer Affect.* Durham: Duke University Press, 2012.

"Christen Clifford Answers a Few of Our #Fiftyquestions." *Some Serious Business,* n.d. https://someseriousbusiness.org/ssbaway-artist-christen-clifford-answers-a-few-of-our-fifty-questions/.

Chude-Sokei, Louis. *The Sound of Culture: Diaspora and Black Technopoetics.* Middletown: Wesleyan University Press, 2015.

Ciment, James. *Another America: The Story of Liberia and the Former Slaves Who Ruled It.* New York: Farrar, Straus and Giroux, 2013.

Cohen, Julie E. "Emergent Limbic Media System." In *Life and the Law in the Era of Data-Driven Agency,* edited by Mireille Hildebrandt and Kieron O'Hara, 60–79. Cheltenham: Edward Elgar, 2019.

Colebrook, Claire. *Deleuze: A Guide for the Perplexed.* London: Continuum, 2006.

———. "Slavery and the Trumpocene: It's Not the End of the World." *Oxford Literary Review* 41, no. 1 (2019): 40–50. DOI: 10.3366/olr.2019.0264.

Coley, Rob. "'A World Where Nothing is Solved': Investigating the Anthropocene in 'True Detective.'" *The Journal of*

Popular Television 5, no. 2 (2017): 135–57. DOI: 10.1386/
jptv.5.2.135_1.

Consterdine, Erica. "Hostile Environment: The UK
Government's Draconian Immigration Policy Explained."
The Conversation, April 26, 2018. https://theconversation.
com/hostile-environment-the-uk-governments-draconian-
immigration-policy-explained-95460.

Cooper, Melinda. *Life as Surplus: Biotechnology and Capitalism
in the Neoliberal Era.* Seattle: University of Washington
Press, 2011.

Couldry, Nick, and Ulises A. Mejias. *The Costs of Connection:
How Data is Colonizing Human Life and Appropriating It for
Capitalism.* Stanford: Stanford University Press, 2019.

Cox, Geoff. *David Link: Das Herz der Maschine*. Stuttgart:
Hatje Cantz, 2011.

Crano, Ricky D'Andrea. "Posthuman Capital: Neoliberalism,
Telematics, and the Project of Self-Control." PhD diss., The
Ohio State University, 2014. http://rave.ohiolink.edu/etdc/
view?acc_num=osu1405531247.

Crawford, Ashley. "Kathy Acker & McKenzie Wark Review:
Their Emails Are Fascinating and Ghoulish." *The Sydney
Morning Herald,* March 21 2015. https://www.smh.com.au/
entertainment/books/kathy-acker--mckenzie-wark-review-
their-emails-are-fascinating-and-ghoulish-20150318-143gnn.
html.

Crawley, Ashon T. *Blackpentecostal Breath: The Aesthetics of
Possibility.* New York: Fordham University Press, 2016.

Crow, Matthew. "Jefferson's Whale: Race, Climate, and
Commerce in Early America." *Journal of the Early Republic*
40, no. 3 (Fall 2020): 435–63. DOI: 10.1353/jer.2020.0064.

da Silva, Denise Ferreira. *Toward a Global Idea of Race.*
Minneapolis: University of Minnesota Press, 2007.

Dalton, Nicholas S., Rebecca Moreau, and Ross K. Adams.
"Resistance Is Fertile: Design Fictions in Dystopian Worlds."
*Proceedings of the 2016 CHI Conference Extended Abstracts
on Human Factors in Computing Systems,* May 2016, San
Jose, CA, USA, 365–74. DOI: 10.1145/2851581.2892572.

Danewid, Ida. "The Fire This Time: Grenfell, Racial Capitalism and the Urbanisation of Empire." *European Journal of International Relations* 26, no. 1 (2020): 289–313. DOI: 10.1177/1354066119858388.

Davies, William. "Sabotaging Progress: The Cultural Economy of Resentment in Late Neoliberalism." *Society for the Advancement of Socio-economics.* http://sase.org/wp-content/uploads/2018/04/1-Davies-final.pdf.

Davis, Janae, Alex A. Moulton, Levi Van Sant, and Brian Williams. "Anthropocene, Capitalocene, Plantationocene? A Manifesto for Ecological Justice in an Age of Global Crises." *Geography Compass* 13, no.5 (2019): e12438. DOI: 10.1111/gec3.12438.

Davison, Sally, and George Shire. "Race, Migration and Neoliberalism." *Soundings* 59 (2015): 81–95. DOI: 10.3898/136266215814890558.

Deckard, Sharae. "'Open Veins': Latin America in the World-Ecology." In *Ecological Crisis and Cultural Representation in Latin America,* edited by Mark Anderson and Zélia M. Bora, 3–21. Lanham: Lexington Books, 2016.

Deleuze, Gilles. *Pure Immanence: Essays on a Life.* Translated by Anne Boyman. New York: Zone Books, 2001.

Deleuze, Gilles., and Félix Guattari. *A Thousand Plateaus: Capitalism and Schizophrenia.* Translated by Brian Massumi. London: Continuum, 2009.

Derrida, Jacques. *Spectres of Marx: The State of Debt, The Work of Mourning, and the New International.* Translated by Peggy Kamuf. New York: Routledge, 1994.

Dery, Mark. "Black to the Future: Interviews with Samuel R. Delany, Greg Tate, and Tricia Rose." In *Flame Wars: The Discourse of Cyberculture,* edited by Mark Dery, 179–222. Durham: Duke University Press, 1994.

Dhaliwal, Sukhwant, and Kirsten Forkert. "Deserving and Undeserving Migrants." *Soundings* 61 (2015): 49–61. DOI: 10.3898/136266215816772205.

Dickens, Charles. *Bleak House.* London: Bradbury and Evans, 1853.

———. *Our Mutual Friend.* London: Chapman and Hall, 1903.

Dikant, Thomas. "Settler Colonial Statistics: Jefferson, Biopolitics, and 'Notes on the State of Virginia.'" *Early American Literature* 54, no. 1 (2019): 69–96. DOI: 10.1353/eal.2019.0006.

Dillon, Elizabeth Maddock. "Zombie Biopolitics." *American Quarterly* 71, no. 3 (2019): 625–52. DOI: 10.1353/aq.2019.0047.

Doucet-Battle, James. "Bioethical Matriarchy: Race, Gender, and the Gift in Genomic Research." *Catalyst: Feminism, Theory, Technoscience* 2, no. 2 (2016): 1–28. DOI: 10.28968/cftt.v2i2.28802.

Douglass, Frederick. *Narrative of the Life of Frederick Douglass, An American Slave, Written by Himself.* New York: W.W. Norton, 1997.

Dowdall, Lisa. "Treasured Strangers: Race, Biopolitics, and the Human in Octavia E. Butler's 'Xenogenesis' Trilogy." *Science Fiction Studies* 44, no. 3 (2017): 506–25. https://www.jstor.org/stable/10.5621/sciefictstud.44.3.0506.

Driggers, Edward Allen. "The Chemistry of Blackness: Benjamin Rush, Thomas Jefferson, Everard Home, and the Project of Defining Blackness through Chemical Explanations." *Critical Philosophy of Race* 7, no. 2 (2019): 372–91. https://muse.jhu.edu/article/730004.

Du Bois, W.E.B. "Black Folk and Birth Control." *Birth Control Review* 16, no. 6 (1932): 166–67, https://documents.alexanderstreet.com/d/1000670632.

———. "Letter from W.E.B. Du Bois to James Bryce, June 22, 1909." W.E.B. Du Bois Papers (MS 312), Special Collections and University Archives, University of Massachusetts Amherst Libraries, http://credo.library.umass.edu/view/full/mums312-b001-i338.

———. "My Evolving Program for Negro Freedom." In *What the Negro Wants,* edited by Rayford W. Logan, 31–70. Chapel Hill: The University of North Carolina Press, 1944.

———. *On Sociology and the Black Community.* Edited by Edwin D. Drive and Dan. S. Green. Chicago: University of Chicago Press, 2013.

———. "Strivings of the Negro People." *The Atlantic,* August 1987. https://www.theatlantic.com/magazine/archive/1897/08/strivings-of-the-negro-people/305446/.

———. *The Philadelphia Negro: A Social Study.* Edited by Isabel Eaton. University Park: The Pennsylvania State University, 1899.

———. *The Souls of Black Folk.* Mineola: Dover Publications, 2012.

Dunbar-Ortiz, Roxanne. *An Indigenous Peoples' History of the United States.* Boston: Beacon Press, 2014.

Edelman, Lee. "Against Survival: Queerness in a Time That's out of Joint." *Shakespeare Quarterly* 62, no. 2 (2011): 148–69. https://www.jstor.org/stable/23025624.

Edwards, Erin E. *The Modernist Corpse: Posthumanism and the Posthumous.* Minneapolis: University of Minnesota Press, 2018.

Elizabeth I. "Charter to Sir Walter Raleigh: 1584." *Yale Law School Lillian Goldman Law Library, The Avalon Project.* https://avalon.law.yale.edu/16th_century/raleigh.asp.

Ellis, Cristin. *Antebellum Posthuman: Race and Materiality in the Mid-nineteenth Century.* New York: Fordham University Press, 2018.

English, Daylanne K. *Unnatural Selections: Eugenics in American Modernism and the Harlem Renaissance.* Chapel Hill: University of North Carolina Press, 2005.

Epstein, Catherine A. *Nazi Germany: Confronting the Myths.* Hoboken: John Wiley & Sons, 2014.

Erickson, Peter. "Can We Talk about Race in 'Hamlet'?" In *"Hamlet": Critical Essays,* edited by Arthur F. Kinney, 207–14. London: Routledge, 2001.

Erwin, Sean. "Microbiopolitics: 'Security Mechanisms', the 'Hela Cell', and The Human Strain." *Humanities and Technology Review* 3 (2014): 106–30. https://philarchive.org/archive/ERWMSM.

"Extract: Reverse Cowgirl." *Tank Magazine,* n.d. https://
    tankmagazine.com/tank/2020/02/reverse-cowgirl/.
Fancher, Patricia. "Embodying Turing's Machine:
    Queer, Embodied Rhetorics in the History of Digital
    Computation." *Rhetoric Review* 37, no. 1 (2018): 90–104. DOI:
    10.1080/07350198.2018.1395268.
Fanon, Frantz. *The Wretched of The Earth.* Translated by
    Richard Philcox. New York: Grove Weidenfeld, 1963.
Febler, Garrett. "Black Zionism, Reparations, and the 'Palestine
    Problem.'" *Black Perspectives,* August 28, 2016. https://www.
    aaihs.org/black-zionism-reparations-and-the-palestine-
    problem/.
Feiner, Sean Michael. "A Rave at the End of World: The Politics
    of Queer Hauntology and Psychedelic Chronomancy." PhD
    diss., State University of New York at Buffalo, 2019.
Fernando, Jude L. "The Virocene Epoch: The Vulnerability
    Nexus of Viruses, Capitalism and Racism." *Journal of
    Political Ecology* 27, no. 1 (2020): 635–84. DOI: 10.2458/
    v27i1.23748.
Flavell, Julie. *When London Was Capital of America.* New
    Haven: Yale University Press, 2010.
Flisfeder, Matthew. "Object Oriented Subjectivity: Capitalism
    and Desire in 'Blade Runner 2049.'" In *Lacanian Perspectives
    on "Blade Runner 2049,"* edited by Calum Neill, 121–38.
    Basingstoke: Palgrave Macmillan, 2020.
"Full Text: Theresa May's Conference Speech." *The Spectator,*
    October 5, 2016. https://blogs.spectator.co.uk/2016/10/full-
    text-theresa-mays-conference-speech/.
Gabbard, Krin. *Jazz Among the Discourses.* Durham: Duke
    University Press, 1995.
Gaboury, Jacob. "A Queer History of Computing: Part Three."
    Rhizome, April 9, 2013. https://rhizome.org/editorial/2013/
    apr/9/queer-history-computing-part-three/.
Gill-Peterson, Julian. "The Technical Capacities of the
    Body: Assembling Race, Technology, and Transgender."
    Transgender Studies Quarterly 1, no. 3 (2014): 402–18. DOI:
    10.1215/23289252-2685660.

Gillette, Aaron. "The Rise of Environmental Behaviorism." In *Eugenics and the Nature-Nurture Debate in the Twentieth Century,* 107–20, Basingstoke: Palgrave Macmillan, 2007.

Gilman, Ernest B. "The Subject of the Plague." *Journal for Early Modern Cultural Studies* 10, no. 2 (2010): 23–44. https://www.jstor.org/stable/23242139.

Gilroy, Paul. *Against Race: Imagining Political Culture beyond the Color Line.* Cambridge: Belknap Press of Harvard University Press, 2000.

Giuliani, Gaia. *Monsters, Catastrophes and the Anthropocene: A Postcolonial Critique.* London: Routledge, 2020.

Goss, Emma Audrey. "The Artificially Intelligent Woman: Talking Down to the Female Machine." PhD diss., Barnard College, 2015. DOI: 10.7916/D8Q23ZBF.

Graeber, David. "Turning Modes of Production Inside Out: Or, Why Capitalism Is a Transformation of Slavery." *Critique of Anthropology* 26, no. 1 (2006): 61–85, DOI: 10.1177/0308275X06061484.

Griffin, Peter. "A Device That Taps into Your Thoughts." *The Hindu,* April 15, 2018. https://www.thehindu.com/sci-tech/science/a-device-that-taps-into-your-thoughts/article23550039.ece.

Guinness, Katherine. "The Coloniser and *corpus nullius*." *Parallax* 26, no. 1 (2020): 76–88. DOI: 10.1080/13534645.2019.1685782.

Hack, Daniel. "Close Reading at a Distance: The African Americanization of 'Bleak House.'" *Critical Inquiry* 34, no. 4 (Summer 2008): 729–53. DOI: 10.1086/592542.

Hanson, Elizabeth. "Fellow Students: Hamlet, Horatio, and the Early Modern University." *Shakespeare Quarterly* 62, no. 2 (2011): 205–29. https://www.jstor.org/stable/23025628.

Haraway, Donna. "Anthropocene, Capitalocene, Plantationocene, Chthulucene: Making Kin." *Environmental Humanities* 6, no. 1 (2015): 159–65. DOI: 10.1215/22011919-3615934.

Harde, Roxanne. "'The Savage Inscription': Abolitionist Writers and the Reinscription of Slavery." *Mosaic: An*

Interdisciplinary Critical Journal 37, no. 2 (2004): 1–20.
https://www.jstor.org/stable/44030378.

Harvey, Sandra. "The HeLa Bomb and the Science of
Unveiling." *Catalyst: Feminism, Theory, Technoscience* 2, no.
2 (2016): 1–30. DOI: 10.28968/cftt.v2i2.28803.

Hayles, N. Katherine. *How We Became Posthuman: Virtual
Bodies in Cybernetics, Literature, and Informatics.* Chicago:
University of Chicago Press, 1999.

Helmreich, Stefan, and Nicole Labruto. "Species of Biocapital,
2008, and Speciating Biocapital, 2017." In *The Palgrave
Handbook of Biology and Society,* edited by Maurizio
Meloni, John Cromby, Des Fitzgerald, and Stephanie Lloyd,
851–76. Basingstoke: Palgrave Macmillan, 2017.

Herschthal, Eric. "Antislavery Science in the Early Republic:
The Case of Dr. Benjamin Rush." *Early American Studies:
An Interdisciplinary Journal* 15, no. 2 (Spring 2017): 274–307.
DOI: 10.1353/eam.2017.0014.

Hinkson, John. "Trump's Trajectory?" *Arena Magazine* 145
(December 2016). https://arena.org.au/trumps-trajectory-
by-john-hinkson/.

"Hormone Therapy Increases Breast Cancer Risk among
Transgender Women." *Healio Endocrinology,* May 16, 2019.
https://www.healio.com/news/endocrinology/20190516/
hormone-therapy-increases-breast-cancer-risk-among-
transgender-women.

Horne, Gerald. *The Apocalypse of Settler Colonialism: The Roots
of Slavery, White Supremacy, and Capitalism in Seventeenth-
Century North America and the Caribbean.* New York:
Monthly Review Press, 2018.

Houdek, Matthew, and Kendall R. Phillips. "Rhetoric and the
Temporal Turn: Race, Gender, Temporalities." *Women's
Studies in Communication* 43, no. 4 (2020): 369–83. DOI:
10.1080/07491409.2020.1824501.

Hui, Andrew. "Horatio's Philosophy in 'Hamlet.'" *Renaissance
Drama* 41, nos. 1–2 (2013): 151–71. DOI: 10.1086/673910.

"iCub Robot." *Istituto Italiano di Tecnologia.* https://icub.iit.it/
products/icub-robot.

"In the Mood: Glenn Miller." *Lyrics,* n.d. https://www.lyrics. com/track/1064441/Glenn+Miller/In+the+Mood.

Iyengar, Sujata, and Lesley Feracho. "'Hamlet' (RSC, 2016) and Representations of Diasporic Blackness." *Cahiers Élisabéthains* 99, no. 1 (July 2019): 147–60. DOI: 10.1177/0184767819837738.

Izhar, Siraj. "*Corpus nullius*: How Europe Recast the Migrant Body Narratives of the Migrant-Refugee Crisis Reflect the Interests They Serve." *Public Seminar,* May 23, 2019. https:// publicseminar.org/essays/corpus-nullius-how-europe-recast-the-migrant-body/.

Jackson, Zakiyyah Iman. *Becoming Human: Matter and Meaning in an Antiblack World.* New York: New York University Press, 2020.

Jacobs, Michael. "Margaret Thatcher and the Inner Cities." *Economic and Political Weekly* 23, no. 38 (1988): 1942–44. https://www.jstor.org/stable/4379049.

"Jefferson, Race, and Democracy." *American Academy of Arts & Sciences, Spring 2018 Bulletin.* https://www.amacad.org/ news/jefferson-race-and-democracy.

Jefferson, Thomas. *Notes on the State of Virginia.* Edited by William Peden. Chapel Hill: University of North Carolina Press, 1982. https://www.jstor.org/ stable/10.5149/9780807899809_jefferson.

———. "Thomas Jefferson to Alexander von Humboldt, 6 December 1813." *National Archives: Founders Online.* https:// founders.archives.gov/documents/Jefferson/03-07-02-0011.

Johnson, Jessica Marie. "Introduction: The Women in the Water." In *Wicked Flesh: Black Women, Intimacy, and Freedom in the Atlantic World,* 1–15. Philadelphia: University of Pennsylvania Press, 2020.

Johnson, Poe. "Racial Technologies in the Time of Black Cyborgnetic Consciousness." In *The Routledge Companion to Biology in Art and Architecture,* edited by Charissa N. Terranova and Meredith Tromble, 368–84. London: Routledge, 2017.

Kazanjian, David. "Racial Governmentality: Thomas Jefferson and African Colonisation in the United States before 1816." *Alternation* 5, no. 1 (1998): 39–84. https://hdl.handle.net/10520/AJA10231757_83.

Keay, Douglas. "Margaret Thatcher: Interview for 'Woman's Own' ('No Such Thing as Society')." *Margaret Thatcher Foundation,* September 23, 1987. https://www.margaretthatcher.org/document/106689.

Kelly, Martin, "History and Founding of Virginia Colony." *ThoughtCo,* December 4, 2020. https://www.thoughtco.com/virginia-colony-103882.

Kendi, Ibram X. "Post-Traumatic Slave Syndrome Is a Racist Idea." *Black Perspectives,* June 21, 2016. https://www.aaihs.org/post-traumatic-slave-syndrome-is-a-racist-idea/.

Kil, Sang Hea. "Reporting from the Whites of Their Eyes: How Whiteness as Neoliberalism Promotes Racism in the News Coverage of 'All Lives Matter." *Communication Theory* 30, no. 1 (2020): 21–40. DOI: 10.1093/ct/qtz019.

King, Tiffany Lethabo. "Humans Involved: Lurking in the Lines of Posthumanist Flight." *Critical Ethnic Studies* 3, no. 1 (Spring 2017): 162–85. https://www.jstor.org/stable/10.5749/jcritethnstud.3.1.0162.

———. *The Black Shoals: Offshore Formations of Black and Native Studies.* Durham: Duke University Press, 2019.

Klaver, Claudia C. *A/moral Economics: Classical Political Economy and Cultural Authority in Nineteenth-Century England.* Columbus: Ohio State University Press, 2003.

Koch, Alexander, Chris Brierley, Mark Maslin, and Simon Lewis. "European Colonization of the Americas Killed 10 Percent of World Population and Caused Global Cooling." *The Conversation,* January 31, 2019. https://theconversation.com/european-colonisation-of-the-americas-killed-10-of-world-population-and-caused-global-cooling-110549.

Koegler, Caroline. "Posthumanism and Colonial Discourse: Nineteenth-Century Literature and Twenty-First-Century Critique." *Open Library of Humanities* 6, no. 2 (2020): 1–37. DOI: 10.16995/olh.613.

Kouri, Scott. "Counselling in an Age of Empire." PhD diss., University of Victoria, 2019. http://hdl.handle.net/1828/10954.

Kraus, Chris. "'Cancer Became My Whole Brain': Kathy Acker's Final Year." *The New Yorker,* August 11, 2017. https://www.newyorker.com/books/page-turner/cancer-became-my-whole-brain-kathy-acker-final-year.

Kroker, Arthur. *Exits to the Posthuman Future.* Hoboken: John Wiley & Sons, 2014.

Lamb, Jonathan. *The Things Things Say.* Princeton: Princeton University Press, 2021.

Lammy, David. "Don't Let Rudd's Departure Distract from a Toxic Policy that Needs to Die." *The Guardian,* April 30, 2018. https://www.theguardian.com/commentisfree/2018/apr/30/amber-rudd-departure-toxic-policy-windrush-generation-home-secretary-david-lammy.

Lawrence, Errol. "Just Plain Common Sense: The 'Roots' of Racism." In *The Empire Strikes Back: Race and Racism in 70s Britain,* 47–94. London: Hutchinson, 1982.

Lemire, Elise. *"Miscegenation": Making Race in America.* Philadelphia: University of Pennsylvania Press, 2010.

Lemke, Sieglinde. "Berlin and Boundaries: *sollen* versus *geschehen.*" *boundary 2* 27, no. 3 (2000): 45–78. DOI: 10.1215/01903659-27-3-45.

Lewis, David Levering. *W.E.B. Du Bois: The Fight for Equality and the American Century, 1919–1963.* New York: Henry Holt and Company, 2001.

Linne, Karsten. "The 'New Labour Policy' in Nazi Colonial Planning for Africa." *International Review of Social History* 49, no. 2 (2004): 197–224. https://www.jstor.org/stable/44583437.

Little, Peter C. "On the Micropolitics and Edges of Survival in a Technocapital Sacrifice Zone." *Capitalism Nature Socialism* 28, no. 4 (2017): 62–77. DOI: 10.1080/10455752.2016.1257037.

Lockridge, Kenneth A. *On the Sources of Patriarchal Rage: The Commonplace Books of William Byrd and Thomas Jefferson and the Gendering of Power in the Eighteenth Century.* New York: New York University Press, 1992.

Lowe, Lisa. *The Intimacies of Four Continents.* Durham: Duke University Press, 2015.

Loza, Susana. "Sampling (Hetero)Sexuality: Diva-Ness and Discipline in Electronic Dance Music." *Popular Music* 20, no. 3 (2001): 349–57. DOI: 10.1017/S0261143001001544.

Mackey, Nathaniel. "Other: From Noun to Verb." In *Jazz Among the Discourses,* edited by Krin Gabbard, 76–99. Durham: Duke University Press, 1995.

Madley, Benjamin. "From Africa to Auschwitz: How German South West Africa Incubated Ideas and Methods Adopted and Developed by the Nazis in Eastern Europe." *European History Quarterly* 35, no. 3 (2005): 429–64. DOI: 10.1177/0265691405054218.

Malaklou, M. Shadee. "'Dilemmas' of Coalition and the Chronopolitics of Man: Towards an Insurgent Black Feminine Otherwise." *Theory & Event* 21, no. 1 (2018): 215–58. https://muse.jhu.edu/article/685977.

Malik, Nasrine. "'I Felt a Nausea of Fury' — How I Faced the Cruelty of Britain's Immigration System." *The Guardian,* March 5, 2018. https://www.theguardian.com/uk-news/2018/mar/05/immigrant-home-office-british-citizenship-hostile-environment.

Manning, Erin. *The Minor Gesture.* Durham: Duke University Press, 2016.

Mansfield, Becky, and Julie Guthman. "Epigenetic Life: Biological Plasticity, Abnormality, and New Configurations of Race and Reproduction." *Cultural Geographies* 22, no. 1 (2015): 3–20. DOI: 10.1177/1474474014555659.

Maurer, Bill. "Re-risking in Realtime: On Possible Futures for Finance after the Blockchain." *Behemoth: A Journal on Civilisation* 9, no. 2 (2016): 82–96. DOI: 10.6094/behemoth.2016.9.2.917.

Mazzio, Carlo. "The History of Air: 'Hamlet' and the Trouble with Instruments." *South Central Review* 26, nos. 1–2 (2009): 153–96. https://www.jstor.org/stable/40211295.

Mbembe, Achille. "The Universal Right to Breathe." Translated by Carolyn Shread. *Critical Inquiry* 47, no. S2 (Winter 2021): S58–S62. DOI: 10.1086/711437.

McBrien, Justin. "Accumulating Extinction: Planetary Catastrophism in the Necrocene." In *Anthropocene or Capitalocene: Nature, History, and the Crisis of Capitalism,* edited by Jason W. Moore, 116–37, Oakland: PM Press, 2016.

@mckenziewark. *Twitter,* May 4, 2019. https://twitter.com/mckenziewark/status/1124721644759547905?lang=en.

———. *Twitter,* July 23, 2019. https://twitter.com/mckenziewark/status/1153725614584684550?lang=en.

McKittrick, Katherine. *Dear Science and Other Stories.* Durham: Duke University Press, 2020.

———. "Plantation Futures." *Small Axe* 17, no. 3 (November 2013): 1–15. https://muse.jhu.edu/article/532740.

———. *Sylvia Wynter: On Being Human as Praxis.* Durham: Duke University Press, 2015.

McReynolds, Phillip. "Zombie Cinema and the Anthropocene: Posthuman Agency and Embodiment at the End of the World." *Cinema: Journal of Philosophy and the Moving Image* 7 (2015): 149–68. DOI: 10.34619/n8w6-r547.

Means, Alexander J., and Graham B. Slater. "The Dark Mirror of Capital: On Post-neoliberal Formations and the Future of Education." *Discourse: Studies in the Cultural Politics of Education* 40, no. 2 (2019): 162–75. DOI: 10.1080/01596306.2019.1569876.

Michaels, Walter Benn. "Plots against America: Neoliberalism and Antiracism." *American Literary History* 18, no. 2 (2006): 288–302. DOI: 10.1093/alh/ajj017.

Miller, F. Bart. *Rethinking Négritude through Léon-Gontran Damas.* Amsterdam: Editions Rodopi, 2014.

Miller, Monica L. "W.E.B. Du Bois and the Dandy as Diasporic Race Man." *Callaloo* 26, no. 3 (2003): 738–65. https://www.jstor.org/stable/3300725.

Mirzoeff, Nicholas. "Artificial Vision, White Space and Racial Surveillance Capitalism." *AI & Society* 36 (2021): 1295–305. DOI: 10.1007/s00146-020-01095-8.

Moore, Jason W. "The Capitalocene Part II: Accumulation by Appropriation and the Centrality of Unpaid Work/Energy." *The Journal of Peasant Studies* 45, no. 2 (2018): 237–79. DOI: 10.1080/03066150.2016.1272587.

Moore, Marlon Rachquel. "Opposed to the Being of Henrietta: Bioslavery, Pop Culture and the Third Life of HeLa Cells." *Medical Humanities* 43, no. 1 (2017): 55–61. DOI: 10.1136/medhum-2016-011072.

Moore, Sophie Sapp, Monique Allewaert, Pablo F. Gómez, and Gregg Mitman. "Plantation Legacies." *Edge Effects,* October 12, 2019. https://edgeeffects.net/plantation-legacies-plantationocene/.

Moran, Michael G. *Inventing Virginia: Sir Walter Raleigh and the Rhetoric of Colonization, 1584–1590.* Bern: Peter Lang, 2007.

Moten, Fred. *In the Break: The Aesthetics of the Black Radical Tradition.* Minneapolis: University of Minnesota Press, 2003.

Muñoz, José Esteban, Jinthana Haritaworn, Myra Hird, Zakiyyah Iman Jackson, Jasbir K. Puar, Eileen Joy, Uri McMillan, Susan Stryker, Kim TallBear, Jami Weinstein, and Jack Halberstam. "Theorizing Queer Inhumanisms." GLQ 21, nos. 2–3 (2015): 209–48. DOI: 10.1215/10642684-2843323.

Nascimento, Giovana Xavier da Conceição. "Os perigos dos Negros Brancos: cultura mulata, classe e beleza eugênica no pós-emancipação (EUA, 1900–1920)." *Revista Brasileira de História* 35, no. 69 (2015): 1–22. DOI: 10.1590/1806-93472015v35n69008.

NCC Staff. "How Things Have Changed in Philadelphia since the 1787 Convention." *The Constitution Daily,* May 25, 2016. https://constitutioncenter.org/blog/how-things-have-changed-since-1787.

Nemenyi, Daniel. "What Is an Internet? Norbert Wiener and the Society of Control." PhD diss., Kingston University, 2019. https://ethos.bl.uk/OrderDetails.do?uin=uk.bl.ethos.789360.

Nhemachena, Artwell, Nokuthula Hlabangane, and Maria B. Kaundjua. "Relationality or Hospitality in Twenty-First Century Research? Big Data, Internet of Things, and the Resilience of Coloniality on Africa." *Modern Africa: Politics, History and Society* 8, no. 1 (2020): 105–39. DOI: 10.26806/modafr.v8i1.278.

Nishime, LeiLani. "The Mulatto Cyborg: Imagining a Multiracial Future." *Cinema Journal* 44, no. 2 (Winter 2005): 34–49. http://www.jstor.org/stable/3661093.

Nixon, Rob. *Slow Violence and the Environmentalism of the Poor.* Cambridge: Harvard University Press, 2011.

Nordlund, Christer. "Endocrinology and Expectations in 1930s America: Louis Berman's Ideas on New Creations in Human Beings." *The British Journal for the History of Science* 40, no. 1 (2007): 83–104. https://www.jstor.org/stable/4500683.

Novitz, Julian. "'The Time Is Out of Joint': Interactivity and Player Agency in Videogame Adaptations of 'Hamlet.'" *Arts* 9, no. 4 (2020): 1–17. DOI: 10.3390/arts9040122.

Nyong'o, Tavia. *Afro-Fabulations: The Queer Drama of Black Life.* New York: New York University Press, 2019.

Ogle, Vanessa. "The End of Empire and the Rise of Tax Havens." *The New Statesman,* December 18, 2020. https://www.newstatesman.com/international/2020/12/end-empire-and-rise-tax-havens.

Palmer, Tyrone S. "'What Feels More Than Feeling?': Theorizing the Unthinkability of Black Affect." *Critical Ethnic Studies* 3, no. 2 (Fall 2017): 31–56. DOI: 10.5749/jcritethnstud.3.2.0031.

Parent, Anthony S. *Foul Means: The Formation of a Slave Society in Virginia, 1660–1740.* Chapel Hill: University of North Carolina Press, 2003.

Parker, Geoffrey. *Global Crisis: War, Climate Change and Catastrophe in the Seventeenth Century.* New Haven: Yale University Press, 2013.

Peterson, Charles F. *DuBois, Fanon, Cabral: The Margins of Elite Anti-colonial Leadership.* Lanham: Lexington Books, 2007.

Pettit, Michael. "Becoming Glandular: Endocrinology, Mass Culture, and Experimental Lives in the Interwar Age." *The American Historical Review* 118, no. 4 (2013): 1052–76. https://www.jstor.org/stable/23785434.

"Philippa Schuyler — Genius or Genetic Experiment?" *On An Overgrown Path,* August 1, 2011. https://www.overgrownpath.com/2011/08/philippa-schuyler-genius-or-genetic.html.

Pikó, Lauren. "'We're Full': Capacity, Finitude, and British Landscapes, 1945–1979." *Australian Journal of Politics & History* 64, no. 3 (2018): 450–63. DOI: 10.1111/ajph.12485.

Preciado, Paul B. *Testo Junkie: Sex, Drugs, and Biopolitics in the Pharmacopornographic Era.* New York: Feminist Press at the City University of New York, 2013.

Price, Neil. "'James His Towne' and Village Nations: Cognitive Urbanism in Early Colonial America." In *The Urban Mind: Cultural and Environmental Dynamics,* edited by Paul J. Sinclair, Gullög Nordquist, Frands Herschend, Christian Isendahl, and Laura Wrang, 471–97. Uppsala: Uppsala University Press, 2010.

Puar, Jasbir K. "Bodies with New Organs Becoming Trans, Becoming Disabled." *Social Text* 33, no. 3 (2015): 45–73. DOI: 10.1215/01642472-3125698.

———. *The Right to Maim: Debility, Capacity, Disability.* Durham: Duke University Press, 2017.

Quashie, Kevin. *The Sovereignty of Quiet: Beyond Resistance in Black Culture.* New Brunswick: Rutgers University Press, 2012.

Rai, Milan. "Columbus in Ireland." *Race & Class* 34, no. 4 (1993): 25–34. DOI: 10.1177/030639689303400403.

Raiford, Wanda. "Race, Robots, and the Law." In *New Boundaries in Political Science Fiction,* edited by Donald M. Hassler and Clyde Wilcox, 93–114. Columbia: University of South Carolina Press, 2008.

Reed, Adolf. "Antiracism: A Neoliberal Alternative to a Left." *Dialect Anthropology* 42 (2018): 105–15. DOI: 10.1007/s10624-017-9476-3.

Rees, Tobias. "From the Anthropocene to the Microbiocene." *Noema,* June 10, 2020. https://www.noemamag.com/from-the-anthropocene-to-the-microbiocene/.

Reséndez, Andrés. *The Other Slavery: The Uncovered Story of Indian Enslavement in America.* Boston: Mariner Books, 2016.

Richard, François G. "Traveling Theory: Mark Leone, Slavery, and Archaeology's Critical Imagination." *Journal of African Diaspora Archaeology and Heritage* 3, no. 1 (2014): 81–102. DOI: 10.1179/2161944114Z.00000000014.

Ridpath, John Clark. *Ridpath's History of the World: Being an Account of the Ethnic Origin, Primitive Estate, Early Migrations, Social Conditions, and Present Promise of the Principal Families of Men, Vol 2: The West Aryans.* Boonsboro: Jones Bros. Publishing Company, 1893.

Robinson, Cedric J. "W.E.B. Du Bois and Black Sovereignty." In *Imagining Home: Class, Culture, and Nationalism in the African Diaspora,* edited by Sidney Lemelle and Robin. D.G. Kelley, 145–57. London: Verso, 1994.

Rollo, Toby. "The Color of Childhood: The Role of the Child/Human Binary in the Production of Anti-Black Racism." *Journal of Black Studies* 49, no. 4 (May 2018): 307–29. DOI: 10.1177/0021934718760769.

Rolston, Bill, "The Training Ground: Ireland, Conquest and Decolonisation." *Race and Class* 34, no. 4 (1993): 13–24. https://ssrn.com/abstract=2690793.

Rosenberg, Jordana. "The Molecularization of Sexuality: On Some Primitives of the Present." *Theory and Event* 17, no. 2 (2014): n.p.

Rosenthal, Caitlin. *Accounting for Slavery: Masters and Management.* Cambridge: Harvard University Press, 2019.

Rush, Benjamin. "Observations Intended to Favour a Supposition That the Black Color (As It Is Called) of the Negroes Is Derived from the Leprosy." *Transactions of the American Philosophical Society* 4 (1799): 289–97. https://www.jstor.org/stable/1005108.

Samuelson, Meg. "Thinking with Sharks: Racial Terror, Species Extinction, and Other Anthropocene Fault Lines." *Australian Humanities Review* 63 (2018): 37–50. http://australianhumanitiesreview.org/2018/12/02/issue-63-november-2018/.

Scher, Avichai. "Study Finds Health Risks for Transgender Women on Hormone Therapy." *NBC News,* July 19, 2018. https://www.nbcnews.com/health/health-news/study-finds-health-risks-transgender-women-hormone-therapy-n890031.

Schuller, Kyla. "Biopower below and before the Individual." *GLQ: A Journal of Lesbian and Gay Studies* 22, no. 4 (2016): 629–36. DOI 10.1215/10642684-3603126.

Schultz, Tristan. "Mapping Indigenous Futures: Decolonising Techno-colonising Designs." *Strategic Design Research Journal* 11, no. 2 (2018): 79–91. DOI: 10.4013/sdrj.2018.112.04.

Seigworth, Gregory J. "Wearing the World like a Debt Garment: Interface, Affect, and Gesture." *ephemera: theory & politics in organization* 16, no. 4 (2016): 15–31. http://ephemerajournal.org/contribution/wearing-world-debt-garment-interface-affect-and-gesture.

Senna, Danzy. "George Schuyler: An Afrofuturist before His Time." *The New York Review,* January 19, 2018. https://www.nybooks.com/daily/2018/01/19/george-schuyler-an-afrofuturist-before-his-time/.

Serlin, David. *Replaceable You: Engineering the Body in Postwar America.* Chicago: University of Chicago Press, 2004.

Shahabuddin, Mohammad. "The Colonial 'Other' in the Nineteenth Century German Colonisation of Africa, and International Law." *African Yearbook of International Law Online / Annuaire Africain de droit international online* 18, no. 1 (2010): 15–39. DOI: 10.1163/22116176-01801004.

Sharpe, Christina. *In the Wake: On Blackness and Being.* Durham: Duke University Press, 2016.

Sharath, Arjun Narayan. "The Anglo-American Relationship: Race and the Perception of German Threat 1890–1910." PhD diss., University of Cambridge. DOI: 10.17863/CAM.48589.

Shaxson, Nicholas. *Treasure Islands: Uncovering the Damage of Offshore Banking and Tax Havens.* New York: St. Martin's Press, 2011.

Sherman, Shantella Y. "In Search of Purity: Popular Eugenics and Racial Uplift among New Negroes 1915–1935." PhD diss., University of Nebraska 2014.

Shilliam, Robbie. "Enoch Powell: Britain's First Neoliberal Politician." *New Political Economy* 26, no. 2 (2020): 239–49. DOI: 10.1080/13563467.2020.1841140.

Shiva, Vandana. *Monocultures of the Mind: Perspectives on Biodiversity and Biotechnology.* London: Zed Books, 1993.

Siemens, Daniel. "'Sword and plough': Settling Nazi Stormtroopers in Eastern Europe, 1936–43." *Journal of Genocide Research* 19, no. 2 (2017): 191–213. DOI: 10.1080/14623528.2017.1313518.

Silberman, Steve. "The Geek Syndrome." *Wired,* December 1, 2001. https://www.wired.com/2001/12/aspergers/.

Simmons, Kristen. "Settler Atmospherics." *Fieldsights,* November 20, 2017. https://culanth.org/fieldsights/settler-atmospherics.

Singh, Nikhil Pal. *Black Is a Country: Race and the Unfinished Struggle for Democracy.* Cambridge: Harvard University Press, 2009.

———. *Race and America's Long War.* Berkeley: University of California Press, 2017.

———. "The Pervasive Power of the Settler Mindset." *Boston Review,* November 26, 2019, http://bostonreview.net/war-security-race/nikhil-pal-singh-pervasive-power-settler-mindset.

Smith, Shawn Michelle. *Photography on the Color Line: W.E.B. Du Bois, Race, and Visual Culture.* Durham: Duke University Press, 2004.

Snaza, Nathan. *Animate Literacies: Literature, Affect, and the Politics of Humanism.* Durham: Duke University Press, 2019.

Soncul, Yiğit, and Grant Bollmer. "Networked Liminality." *Parallax* 26, no. 1 (2020): 1–8. DOI: 10.1080/13534645.2019.1685775.

Spillers, Hortense J. "Mama's Baby, Papa's Maybe: An American Grammar Book." *Diacritics* 17, no. 2 (1987): 65–81. DOI: 10.2307/464747.

"ss Experimental Agricultural Facility 'Herb Garden'/ 'Plantation.'" *KZ-Gedenkstaette-Dachau,* n.d., https://www. kz-gedenkstaette-dachau.de/en/historical-site/virtual-tour-of-the-surrounding-area/ss-experimental-facility-herb-garden-plantation/.

St. John de Crevecoeur, J. Hector. "Letters from an American Farmer." *Yale Law School Lillian Goldman Law Library, The Avalon Project.* https://avalon.law.yale.edu/subject_menus/ letters.asp.

Staudenmaier, Peter. "Organic Farming in Nazi Germany: The Politics of Biodynamic Agriculture, 1933–1945." *Environmental History* 18, no. 2 (2013): 383–411. DOI: 10.1093/ envhis/ems154.

Stewart, Fenn Elan. "'The King Is a Thing': Hamlet and the Prostheses of Nobility." PhD diss., University of British Columbia, 2008. http://hdl.handle.net/2429/2645.

Steyerl, Hito. "Dramaturgies of Resistance: Collectivity, Performance, Dialectics." Performance at the Jackman Humanities Institute's Program for the Arts at the University of Toronto, Ontario, Canada, January 22, 2021.

———. "Hito Steyerl: I Will Survive Sep 26, 2020 — Jan 10, 2021." *Kunstsammlung Nordrhein Westfalen,* n.d. https:// www.kunstsammlung.de/en/steyerl/.

Sumner, Judith. *Plants Go to War: A Botanical History of World War II.* Jefferson: McFarland and Co., 2019.

Swarbrick, Steven, and Karen Raber. "Introduction: Renaissance Posthumanism and Its Afterlives." *Criticism* 62, no. 3 (Summer 2020): 313–28. DOI: 10.13110/ criticism.62.3.0313.

Taylor, Jesse Oak. "The Novel as Climate Model: Realism and the Greenhouse Effect in 'Bleak House.'" *Novel: A Forum on Fiction* 46, no. 1 (2013): 1–25. https://www.jstor.org/ stable/43829934.

Taylor, Julie. "Mechanical Reproduction: The Photograph and the Child in 'The Crisis' and the 'Brownies' Book.'" *Journal of American Studies* (2019): 737–774. DOI: 10.1017/S0021875819000045.

TED. "How AI Could Become an Extension of Your Mind | Anvar Kapur." *YouTube,* June 6, 2019. https://www.youtube.com/watch?v=TrofjEAetVs.

Teixeira Pinto, Ana. "Capitalism with a Transhuman Face: The Afterlife of Fascism and the Digital Frontier." *Third Text* 33, no. 3 (2019): 315–36. DOI: 10.1080/09528822.2019.1625638.

The Free Enterprise Group. *A Time for Choosing: Free Enterprise in Twenty-First Century Britain.* Edited by Kwasi Kwarteng, Ryan Bourne, and Jonathan Dupont. Basingstoke: Palgrave Macmillan, 2016.

Tiara, Roxanne. "Digital Territory, Digital Flesh: Decoding the Indigenous Body." *A Peer-Reviewed Journal about Machine Feeling* 8, no. 1 (2019): 70–80. DOI: 10.7146/aprja.v8i1.115416.

Tinline, Phil. "Back to the Future: What the Turmoil of the 1970s Can Teach Us Today." *The New Statesman,* May 8, 2019. https://www.newstatesman.com/politics/uk/2019/05/back-future-what-turmoil-1970s-can-teach-us-today.

Tomlinson, Sally. "Enoch Powell, Empires, Immigrants and Education." *Race Ethnicity and Education* 21, no. 1 (2017): 1–14. DOI: 10.1080/13613324.2017.1365055.

Turing, Alan M. "Computing Machinery and Intelligence." *Mind* 59, no. 236 (1950): 433–60. DOI: 10.1093/mind/LIX.236.433.

Tyler, Imogen. "Designed to Fail: A Biopolitics of British Citizenship." *Citizenship Studies* 14, no. 1 (2010): 61–74. DOI: 10.1080/13621020903466357.

van Veen, tobias c., and Reynaldo Anderson. "Future Movements: Black Lives, Black Politics, Black Futures – An Introduction." *Topia: Canadian Journal of Cultural Studies* 39 (2018): 5–21. DOI: 10.3138/topia.39.00.

van Vuuren, Hennie. *Apartheid Guns and Money: A Tale of Profit.* London: Hurst and Co., 2019.

Vogel, Shane. *The Scene of Harlem Cabaret: Race, Sexuality, Performance.* Chicago: University of Chicago Press, 2009.

Walsh, Joanna. "Kathy Acker's Pioneering Adventures in the Internet's Erogenous Zone." *The Guardian,* April 21, 2015. https://www.theguardian.com/books/booksblog/2015/apr/21/kathy-acker-internet-erogenous-zone-emails-book.

Wark, McKenzie. "Unfellows: Transsexual Aesthetics after the Human." *Proceedings of the Australian Academy of the Humanities 50th Academic Symposium, Humanizing the Future.* Brisbane, Australia, November 13–15, 2019.

———. Interviewed by Michelle Esther O'Brien. NYC *Trans Oral History Project,* June 20, 2019.

Washington, H.A. *The Writings of Thomas Jefferson, Vol. VI.* Frankfurt: Outlook Verlag, 2018.

Wells, Justine. "W.E.B. Du Bois and the Conservation of Races: A Piece of Ecological Ancestry." *Rhetoric Society Quarterly* 49, no. 4 (2019): 342–64. DOI: 10.1080/02773945.2019.1634830.

Wilderson III, Frank B. *Afropessimism.* New York: Liveright, 2020.

Wilentz, Amy. "A Zombie Is a Slave Forever." *The New York Times,* October 30, 2012. https://www.nytimes.com/2012/10/31/opinion/a-zombie-is-a-slave-forever.html.

Wilkin, Sam. "Iran Considered Nuclear Weapons during 1980s Iraq War, Ex-president Says." *Reuters,* October 29, 2015. https://www.reuters.com/article/us-iran-nuclear-rafsanjani/iran-considered-nuclear-weapons-during-1980s-iraq-war-ex-president-says-idUSKCN0SN0E720151029.

Williams, Lyneise E. *Latin Blackness in Parisian Visual Culture, 1852–1932.* London: Bloomsbury, 2019.

Wilson, Elizabeth A. *Affect and Artificial Intelligence.* Seattle: University of Washington Press, 2010.

———. "'Would I Had Him with Me Always': Affects of Longing in Early Artificial Intelligence." *Isis* 100, no. 4 (2009): 839–47. DOI: 10.1086/652023.

Wolfe, George C., dir. *The Immortal Life of Henrietta Lacks.* Chicago: Harpo Films. Aired April 22, 2017, on HBO.

Wolfe, Patrick. "*Corpus nullius*: The Exception of Indians and Other Aliens in US Constitutional Discourse." *Postcolonial Studies* 10, no. 2 (2007): 127–51. DOI: 10.1080/13688790701348540.

Wright, Michelle M. *Becoming Black: Creating Identity in the African Diaspora.* Durham: Duke University Press, 2004.

Wright, Willie Jamaal. "The Morphology of Marronage." *Annals of the American Association of Geographers* 110, no. 4 (2020): 1134–49. DOI: 10.1080/24694452.2019.1664890.

Wyly, Elvin. "Cities, Consent, and Coercion." *Society + Space,* May 1, 2019. https://www.societyandspace.org/articles/cities-consent-and-coercion.

Wynter, Sylvia. "Unsettling the Coloniality of Being/Power/Truth/Freedom: Towards the Human, after Man, its Overrepresentation — An Argument." *CR: The New Centennial Review* 3, no. 3 (2003): 257–337. DOI: 10.1353/ncr.2004.0015.

Yakushko, Oksana. *Scientific Pollyannaism: From Inquisition to Positive Psychology.* New York: Springer, 2019.

Yusoff, Kathryn. *A Billion Black Anthropocenes or None.* Minneapolis: University of Minnesota Press, 2018.

Zimmerman, Andrew. *Alabama in Africa: Booker T. Washington, the German Empire, and the Globalization of the New South.* Princeton: Princeton University Press, 2010.

———. "German Sociology and Empire: From Internal Colonization to Overseas Colonization and Back Again." In *Sociology and Empire: The Imperial Entanglements of a Discipline,* edited by George Steinmetz, 166–87. Durham: Duke University Press, 2013.

Zinkievich, Craig. "Star Trek Online: Second Interview." By Gareth Von Kallenbach. *Skewed & Reviewed,* Apr 27, 2009. https://www.sknr.net/2009/04/25/star-trek-online-second-interview/.

Zuboff, Soshana. *The Age of Surveillance Capitalism: The Fight for a Human Future at the New Frontier of Power.* New York: Public Affairs, 2019.